# Advances in Petroleum Technology

# Advances in Petroleum Technology

edited by

**Subrata Borgohain Gogoi**

JENNY STANFORD
PUBLISHING

*Published by*

Jenny Stanford Publishing Pte. Ltd.
Level 34, Centennial Tower
3 Temasek Avenue
Singapore 039190

Email: editorial@jennystanford.com
Web: www.jennystanford.com

**British Library Cataloguing-in-Publication Data**
A catalogue record for this book is available from the British Library.

**Advances in Petroleum Technology**



ISBN 978-981-4877-19-0 (Hardcover)
ISBN 978-1-003-04993-7 (eBook)

*To my parents . . .*

*My Maa, Dibya Borgohain,*

*and Deauta, Late Narayan Chandra Borgohain*

# Contents

*Preface* xxiii

**1 A Preliminary Study on Grain Size Analyses of Reservoir Rocks of the Upper Assam Basin** **1**
*Pranab Boral, Subrata Borgohain Gogoi, Tapan Jyoti Gogoi and Arunabh Borpatra Gohain*
1.1 Introduction 2
1.2 Study Area 3
1.3 Methodology 4
1.4 Results and Discussion 9
1.5 Conclusion 28

**2 Depositional Environment and Hydrocarbon Source Potential of the Barail and Disang Groups of Rocks along the Tuli–Mokokchung Road Section, Nagaland** **29**
*Manash Pratim Gogoi, Yadav Krishna Gogoi, Devojit Bezbaruah and Pradip Borgohain*
2.1 Introduction 30
2.2 Geology of the Area 33
2.3 Geological Mapping 35
2.4 Depositional Environment 38
2.4.1 Granulometric Analysis 38
2.4.1.1 Cumulative curves 38
2.4.2 Relationship between Textural Parameters 39
2.4.2.1 Kurtosis ($K_G$) vs. skewness ($S_K$) 39
2.5 Evaluation of the Hydrocarbon Source Potential 42
2.5.1 Elemental Analysis 42
2.5.2 Rock-Eval Analysis 44
2.6 Conclusion 47

3 **Characterisation of Barail Sandstones of Mizoram: A Sedimentological Approach** **49**
*Parakh Protim Phukan, Pradip Borgohain, Bubul Bharali and Devojit Bezbaruah*
3.1 Introduction 50
3.2 Field Characteristics 53
3.3 Methodology 55
3.3.1 Depositional Environment 57
3.3.1.1 Lithofacies analysis 57
3.3.1.2 Granulometric analysis 59
3.3.2 Porosity Determination 62
3.3.3 Petrography 63
3.3.3.1 Observations 63
3.3.3.2 Porosity 68
3.4 Conclusion 75

4 **Chemical Enhanced Oil Recovery with Special Reference to the Upper Assam Basin** **77**
*Kalpajit Hazarika and Subrata Borgohain Gogoi*
4.1 Introduction 78
4.2 Experimental 79
4.2.1 Material 79
4.2.2 Methodology 80
4.2.2.1 Core flooding 80
4.2.2.2 Selection of a CEOR slug 81
4.2.2.3 Recovery $\eta$ 81
4.2.2.4 Areal displacement efficiency ($E_d$) 81
4.2.2.5 Areal sweep efficiency ($E_a$) 82
4.2.2.6 Overall efficiency ($E_o$) 82
4.2.2.7 Thermogravimetric analysis 82
4.3 Results and Discussion 83
4.3.1 Core Flooding 83
4.3.2 Mobility Ratio 86
4.3.3 Effect of Polymer on the CEOR Slug 88
4.3.4 Effect of Alkali on the CEOR Slug 88
4.3.5 Stability of ASP Slugs 90
4.4 Conclusion 90

**5 Formulation of a Stable Microemulsion Slug for Enhanced Oil Recovery in the Upper Assam Basin** **93**
*Shilpi Sarmah, Subrata Borgohain Gogoi and Fan Xianfeng*
5.1 Introduction 94
5.2 Materials and Methods 96
5.2.1 Materials 96
5.2.2 Methods 96
5.2.2.1 Experiment on IFT measurements 96
5.2.2.2 Experiment on phase behaviour 97
5.2.2.3 Characterisation of microemulsions 97
5.2.2.4 Core flooding 97
5.3 Results and Discussion 97
5.3.1 IFT Measurements 97
5.3.2 Phase Behaviour 99
5.3.3 Characterisation of Microemulsions 99
5.3.4 Core Flooding 101
5.4 Conclusion 101

**6 Systematic Study of Low-Cost, High-Performance Chemical Flooding as EOR for Depleted Oilfields of the Upper Assam Basin** **103**
*Miranda Kakoty and Subrata Borgohain Gogoi*
6.1 Introduction 104
6.2 Experimental 106
6.2.1 Apparatus 106
6.2.2 Materials 106
6.2.2.1 Surfactants 106
6.2.2.2 Alkalis 106
6.2.2.3 Crude oil, formation water and synthetic soft brine 107
6.2.3 Procedure 107
6.2.3.1 IFT determination 107
6.2.3.2 Aqueous stability 108
6.2.3.3 Foam stability 108
6.2.3.4 Thermal stability 109
6.2.3.5 Core flooding 109
6.3 Results and Discussion 110
6.3.1 Chemical Formula Design 110

6.3.2 Aqueous Stability 114
6.3.3 Foam Stability 115
6.3.4 Thermogravimetric Analysis 116
6.3.5 Core Flooding 117
6.4 Conclusion 118

**7 Characterisation of Reservoir Rock and Fluids for $CO_2$ Foam Enhanced Oil Recovery Application 121**
*Ranjan Phukan, Subrata Borgohain Gogoi and Pankaj Tiwari*
7.1 Introduction 122
7.2 Experimental 126
7.2.1 Materials 126
7.2.2 Core Analysis 128
7.2.3 Crude Oil Analysis 129
7.2.4 Formation Water Analysis 129
7.3 Results and Discussion 130
7.3.1 Core Analysis 130
7.3.2 Determination of Porosity 130
7.3.2.1 Determination of permeability ($K$) 132
7.3.3 Reservoir Rocks 136
7.3.4 Crude Oil Analyses 137
7.3.4.1 API gravity 138
7.3.4.2 Crude oil viscosity 139
7.3.4.3 Pour point 140
7.3.4.4 Resin, asphaltene and wax content 140
7.3.4.5 Acid number 142
7.3.4.6 Formation water analyses 143
7.4 Conclusion 144

**8 Formulation of an Optimum Surfactant Combination for EOR Applications 147**
*Sekhar Gogoi, Shilpi Sarmah and Subrata Borgohain Gogoi*
8.1 Introduction 148
8.2 Experimental 150
8.2.1 Material 150
8.2.2 Methodology 151
8.2.2.1 Measuring the surface tension 151

8.2.2.2 Measuring the interfacial tension 151
8.2.2.3 Measuring the ST and IFT 151
8.3 Results and Discussion 151
8.4 Conclusion 154

**9 Microbial Enhanced Oil Recovery: A Technological Perspective 157**
*Saurav Haloi and Tapas Medhi*
9.1 Introduction 158
9.2 Crude Oil and Its Importance in India 160
9.3 Conventional Oil Recovery Methods 161
9.4 EOR 162
9.4.1 Thermal EOR 162
9.4.2 CEOR 163
9.4.3 Miscible Flooding 164
9.5 MEOR 164
9.5.1 History of MEOR 164
9.5.2 Types of MEOR 166
9.5.3 Microbes and Their Metabolites in MEOR 167
9.6 Biosurfactants 169
9.6.1 Glycolipids 170
9.6.1.1 Rhamnolipids 170
9.6.1.2 Trehalolipids 170
9.6.1.3 Sophorolipids 172
9.6.2 Lipopeptides 172
9.6.3 Phospholipids, Fatty Acids and Neutral Lipids 173
9.7 Biosurfactant-Mediated MEOR Study at the Laboratory Level 174
9.8 Biosurfactant-Mediated MEOR in Field Studies 176
9.9 Why Biosurfactant over Chemical Surfactant in MEOR 177
9.10 Production of Biosurfactants 178
9.10.1 Usage of Waste or Cheap Substrates 178
9.10.2 Optimisation 178
9.10.3 Bioprocess Engineering 179
9.10.4 Genetic Engineering 181
9.11 Conclusion 182

**10 A Preliminary Theoretical Study on Microbial Enhanced Oil Recovery** **185**
*Saurav Bhattacharjee and Borkha Mech Das*
10.1 Introduction 186
10.2 Process 187
10.3 Mechanisms 188
10.4 Factors under Consideration 188
10.4.1 Selecting the Reservoir 188
10.4.2 Selecting Microbes 189
10.4.3 Selecting Nutrients 189
10.4.4 Preparation of a Microbial Slug 189
10.5 Advantages 190
10.6 Problems 190
10.7 Conclusion 191

**11 Revisiting Evaluation of Crude Oil in a Cost-Effective Way** **193**
*Pankaj Boruah, Miranda Kakoty, Walid M. Alakhdar and Subrata Borgohain Gogoi*
11.1 Introduction 194
11.2 Experimental 195
11.2.1 Materials and Instruments 195
11.2.2 Methodology 195
11.2.2.1 ASTM distillation 195
11.3 Results 200
11.4 Conclusion 202

**12 A Study on the Effects of Iron Nanoparticles in Water-Based Drilling Mud** **205**
*Debashree Dutta and Borkha Mech Das*
12.1 Introduction 206
12.2 Experimental 208
12.2.1 Materials 208
12.2.2 Methods 208
12.2.2.1 Formulation of bentonite-based drilling fluid 208
12.2.2.2 Synthesis of iron NPs 208
12.2.2.3 Characterisation of iron NPs 209
12.2.2.4 Formulation of nanobased mud 209

12.3 Results and Discussion 209
12.3.1 Characterisation of Iron NPs 209
12.3.1.1 Particle size determination 209
12.3.2 Rheological Properties 210
12.3.2.1 Rheological properties of base mud without NPs 210
12.3.2.2 Effect of iron NPs on rheological properties 212
12.3.2.3 Comparative analyses of rheological properties of 0.1 M and 0.2 M iron NPs in drilling fluid 213
12.3.3 Filtration Properties 214
12.3.3.1 Filtration properties of base mud without NPs 214
12.3.3.2 Effect of iron NPs on filtration properties 214
12.3.3.3 Comparative analyses of filtration properties of 0.1 M and 0.2 M iron NPs in drilling fluid 215
12.3.4 Cationic Exchange Capacity of NBM 217
12.4 Conclusion 219

**13 Use of Environment-Friendly Additives for Improving the Properties of Water-Based Mud for Drilling Reactive Formations 221**
*Bichakshan Borah and Borkha Mech Das*
13.1 Introduction 222
13.2 Experimental 224
13.2.1 Materials 224
13.2.2 Instruments 224
13.2.3 Methodology 224
13.2.3.1 Preparation of base mud 224
13.2.3.2 Preparation of CMC mud and PEG4000 mud 225
13.2.3.3 Preparation of CMC-PEG4000 mud 225

13.2.4 Property Analyses of Prepared Mud Samples 225
13.2.4.1 Density 225
13.2.4.2 Funnel viscosity 225
13.2.4.3 Rheological properties 226
13.2.4.4 Filtrate loss and mud cake thickness 226
13.2.4.5 Shale stability 227
13.3 Results and Discussion 227
13.3.1 Density 227
13.3.2 Funnel Viscosity 228
13.3.3 Rheological Properties 229
13.3.4 Filtration Properties 232
13.3.5 Shale Stability 235
13.4 Conclusion 235

**14 Effect of Different Plant Extracts on Enhancing Rheological Properties of Water-Based Drilling Fluids 239**
*Deepjyoti Mech, Borkha Mech Das, Rahul Kumar M., Meet Priyavadan Patel, Sandra Jayan and Rishabh Pardeshi*
14.1 Introduction 240
14.2 Experimental 242
14.2.1 Materials 242
14.2.2 Experimental Setup 242
14.2.3 Experimental Procedure 244
14.3 Results and Discussion 246
14.3.1 Formulation of Base Mud 246
14.3.2 Plant Extract 249
14.3.3 Plastic Viscosity 254
14.3.4 Yield Point 254
14.4 Conclusion 256

**15 Effects of Temperature on the Transportation of Assam Crude Oil through Pipelines 259**
*Bondita Robidas and Subrata Borgohain Gogoi*
15.1 Introduction 260
15.2 Experimental 262
15.2.1 Materials and Instruments 262

15.2.2 Methodology 263
15.2.2.1 Rheological properties 263
15.2.2.2 Shear stress, shear rate and viscosity 263
15.3 Results and Discussion 263
15.3.1 Pour Point 263
15.3.2 Effect of Shear Rate on Viscosity 263
15.3.3 Effect of Temperature on Shear Stress 264
15.3.4 Effect of Temperature on Viscosity 269
15.4 Conclusion 272

**16 Average Angle Method with Microsoft Excel for Directional Well Path Planning and Risk Monitoring 275**
*Devtusti Rath and Prasun Banik*
16.1 Introduction 276
16.1.1 Reasons for Directional Drilling 277
16.1.2 Fundamental Concepts of Directional Drilling 278
16.1.2.1 Directional surveying 278
16.1.2.2 Types of survey instruments 278
16.1.2.3 Methods of calculating wellbore trajectories 279
16.2 Average Angle Method for Directional Drilling Calculations 279
16.2.1 Calculations Involved 279
16.2.1.1 Wellbore geometry using the average angle method 279
16.3 Microsoft Excel Programme Development 280
16.3.1 Design 280
16.3.2 Introduction to Coding in Visual Basic 281
16.3.2.1 Field data interface 281
16.3.3 Survey Input Data and Coordinate Generation Interface 282
16.4 Results and Discussion 284
16.4.1 Input Data 284
16.4.2 Programme-Generated Results 284
16.5 Conclusion 285

**17 Using Digital Capabilities in Geoscience Software for Estimating the Resource Volume of a Reservoir Using the Map-Based Method: A Quick Evaluation Procedure 287**
*Nitin Lahkar and Shamina Sheena*
17.1 Introduction 288
17.2 Concept and Need of Screening Volumetrics in Geoscience Software for Quick Resource Evaluation by the Map-Based Method 289
17.3 Methodology and Work Done 290
17.3.1 Map Loading 291
17.3.2 Digitising the Map Using Geoscience Software 292
17.3.3 Making the Reservoir Top Surface 293
17.3.4 Prospect Differentiation 295
17.3.5 Volume Calculation 296
17.3.6 Uncertainty Analysis 297
17.4 Conclusion 298

**18 Chemical Consolidation as a Sand Control Method in Oilfields of the Upper Assam Basin 299**
*Aditi Gogoi and Pradip Borgohain*
18.1 Introduction 300
18.2 Experiments 301
18.2.1 Materials 301
18.2.2 Experimental Setup 301
18.2.3 Methodology 301
18.3 Results and Discussion 303
18.3.1 Sieve Analysis 303
18.3.2 Rock Thin Slide Analysis 304
18.4 Conclusion 307

**19 A Review of Microbial Degradation of Petroleum Hydrocarbons 309**
*Amarjit Rajbongshi and Subrata Borgohain Gogoi*
19.1 Introduction 310
19.2 Identification of Microorganisms 311
19.3 Microbial Degradation of Hydrocarbons 312

19.4 Biosurfactants 312
19.5 Factors Influencing the Degradation of Hydrocarbons 317
19.5.1 Temperature 317
19.5.2 pH 317
19.5.3 Activity of Water 317
19.5.4 Salinity 317
19.5.5 Oxygen Availability 317
19.5.6 Nutrient Availability 318
19.6 Conclusion 318

**20 Evaluation of the Corrosion and Scaling Potential of Oilfield Produced Water of the Upper Assam Basin 319**
*Debasish Konwar, Subrata Borgohain Gogoil and Tapan Jyoti Gogoi*
20.1 Introduction 320
20.2 Materials and Methods 323
20.2.1 Materials and Equipment 323
20.2.2 Physical and Chemical Characterisation 323
20.2.2.1 Water analyser 323
20.2.2.2 Titration 324
20.2.2.3 Alkalinity 324
20.2.2.4 Total hardness 325
20.2.2.5 O&G 325
20.2.2.6 Flame photometer 326
20.2.2.7 Atomic absorption spectrophotometer 326
20.2.3 Correlation Index 326
20.2.4 Water Stability Indices 328
20.2.4.1 LSI 328
20.2.4.2 RSI 328
20.2.4.3 PSI 329
20.2.4.4 AI 329
20.3 Results and Discussion 330
20.3.1 Physical and Chemical Characterisation 330
20.3.2 Spearman Rank Order Correlation Coefficient 333

20.3.3 Water Stability Indices 335
20.3.3.1 LSI 335
20.3.3.2 RSI 335
20.4 Conclusion 339

**21 A Review on Treatment and Management of Oilfield Produced Water** **341**
*Tapan Jyoti Gogoi, Subrata Borgohain Gogoi and Pranab Boral*
21.1 Introduction 342
21.2 Characteristics 343
21.2.1 Constituents 343
21.2.1.1 Dispersed oil 343
21.2.1.2 Dissolved organic compounds 344
21.2.1.3 Treatment chemicals 344
21.2.1.4 Produced solids 344
21.2.1.5 Metals 345
21.2.2 Theory of Oil–Water Emulsion 345
21.2.3 Particle Size Analyses 345
21.3 Treatment Methods 346
21.3.1 Physical Treatment 346
21.3.1.1 Sand filter 346
21.3.1.2 Hydrocyclones 347
21.3.1.3 Evaporators 347
21.3.1.4 Dissolved air precipitation 347
21.3.1.5 Freeze-thaw process 347
21.3.1.6 Physical adsorption 348
21.3.2 Chemical Treatment 348
21.3.2.1 De-emulsifiers 348
21.3.2.2 Precipitation 349
21.3.2.3 Oxidation 349
21.3.2.4 Electrochemical process 349
21.3.2.5 $O_3$ 350
21.3.3 Biological Process 350
21.3.4 Membrane Separation 352
21.4 Management Practices 354
21.4.1 Reuse 355
21.4.2 Injection into Disposal Wells 355

21.4.3 Direct Discharge 356
21.4.4 Consumption for Beneficial Use 356
21.5 Opportunities and Challenges 356
21.6 Conclusion 358

**22 Experimental Work to Evaluate the Effects of Disposed Produced Water on the Surrounding Environment: A Case Study in Libya 361**
*Jeevarupini Balakrishnan, Juhairi Aris Muhamad Shuhili, Tarek Arbi Omar Ganat, Subrata Borgohain Gogoil, Mohamad Nasir Mohamad Ibrahim and Raoof Gholami*
22.1 Introduction 362
22.2 Experimental 363
22.2.1 Sample Collection 363
22.2.2 Methodology 363
22.2.2.1 Spectrophotometry 363
22.2.2.2 Mohr's method 367
22.2.2.3 Gravimetry and titration 367
22.3 Results and Discussion 367
22.3.1 Sulphate Concentration 367
22.3.2 Bicarbonate Concentration 367
22.3.3 Calcium and Sodium Concentration 368
22.4 Conclusion 369

**23 Role of Sand Particle Size in the Retention of Total Petroleum Hydrocarbons and Heavy Metals 371**
*Hassan ElGharbi, Fatma Aloulou, Wajdi Haj Ali, Subrata Borgohain Gogoi and Monem Kallel*
23.1 Introduction 372
23.2 Materials and Methods 373
23.2.1 Materials 373
23.2.2 Methodology 373
23.2.2.1 Experimental 373
23.3 Results and Discussion 374
23.3.1 Characterisation of PW 374
23.3.2 Filtration 375
23.4 Conclusion 382

**24 A Review on Characterisation and Treatment of Oilfield Produced Water** **385**
*Pratiksha Rajkumari and Rishav Chetia*
24.1 Introduction 386
24.1.1 Production of OFPW 386
24.2 Composition of Produced Water 387
24.2.1 Dissolved and Dispersed Oil Components 387
24.2.2 Dissolved Minerals 388
24.2.3 Production Chemicals 389
24.2.4 Produced Solids 389
24.2.5 Dissolved Gases 389
24.3 Technologies 389
24.3.1 Bioremediation 389
24.3.2 Membranes 390
24.3.2.1 RO and NF 391
24.3.3 Adsorption 392
24.3.4 Electrochemistry 392
24.4 Conclusion 395

**25 Application of Membrane Bioreactors for the Modification of Microfiltration Membrane Surface for Enhanced Antibiofouling Capability in Wastewater Treatment** **397**
*Akshayjit Podder*
25.1 Introduction 398
25.1.1 Goal of the Study 401
25.1.2 The MBR Process 401
25.1.3 Characteristics of Industrial Wastewaters 402
25.2 Experimental 403
25.2.1 Materials for Membrane Modification 403
25.2.2 Methodology 403
25.2.2.1 Spectroscopic studies 404
25.2.2.2 Membrane antibiofouling potential 406
25.2.2.3 Stability of VA-modified membranes 406
25.2.2.4 CSLM 407

25.3 Results and Discussion 407
25.3.1 Membrane Modification 407
25.3.2 Modified Membrane Stability 410
25.3.3 Membrane Biofilm Production 410
25.4 Conclusion 412
25.5 Future Prospects 412

*Bibliography* 415

*Index* 465

# Preface

The idea of writing and editing this book came to me immediately after the International Seminar-cum-Workshop on Oil & Gas under the aegis of the UGC–UK India Education and Research Initiative Programme and the Indo–U.S. 21st Century Knowledge Initiative Programme, which was organised jointly by Dibrugarh University, Assam, India; the University of Edinburgh, U.K.; and the University of Louisiana at Lafayette, LA, USA, from 7 to 8 April 2019 at Dibrugarh University, which received great feedback from the participants. So I decided to compile some of the peer-reviewed papers presented in the seminar in the form of a book. It is the prompt and positive response from Jenny Stanford Publishing (formerly Pan Stanford Publishing), Singapore, that made my wish and endeavour of bringing out a book on the comprehensive research on the background and recent advances in petroleum technology a reality. The book deserves to be an excellent reference book on the various latest developments and technologies in the oil and gas sector, and I believe it will live up to the expectations of researchers actively involved in petroleum R&D, both in academics and in industry. The book comprises 25 chapters from different research groups, and each chapter represents the significance, objectives, novelty and technology in petroleum research.

The book has been checked thoroughly, but mistakes might still be there. I shall greatly appreciate if the readers kindly bring those to my notice. Suggestions for improvement are always welcome. Although the book is primarily intended for researchers in the field of petroleum technology, it will also be useful for postgraduate students and practising engineers in the oil and gas sector.

It is a great pleasure to put on record my indebtedness to my collaborators and students for the stimulus I received over the years.

I am thankful to Prof. Xianfeng Fan, School of Engineering, University of Edinburgh, U.K.; Prof. Ajanta B. Rajkonwar, Department of Commerce, Dibrugarh University; Prof. Fathi Boukadi, Department of Petroleum Engineering, University of Louisiana at Lafayette; Prof. Monem Kallel, Department of Geo-Resources and Environment, National Engineering School of Sfax, Sfax University, Tunisia; Dr. Kalpajit Hazarika, Petroleum Engineering Department, School of Engineering, Presidency University, Bangalore, India; and my PhD research scholars, Tapan Jyoti Gogoi, Debasish Konwar, Pranab Boral, Amarjit Rajbongshi, Shilpi Sarmah, Sekhar Gogoi, Miranda Kakoty, Bondita Robidas and Manash Pratim Gogoi, who helped me in many ways during the preparation of the book. Last but not least, I would like to extend my sincere gratitude to Karunakar Borah, Technical Officer, Dibrugarh University, and my children, Abhigyan Kumar Gogoi and Anannya Gogoi, who taught me LaTeX during the drafting of the book. Finally, the encouragement I received from my mother, Dibya Borgohain, and my husband, Kalyan Kumar Gogoi, proved invaluable in all stages of this work.

**Subrata Borgohain Gogoi**
2021

# Chapter 1

# A Preliminary Study on Grain Size Analyses of Reservoir Rocks of the Upper Assam Basin

**Pranab Boral,[a] Subrata Borgohain Gogoi,[a] Tapan Jyoti Gogoi[a] and Arunabh Borpatra Gohain[b]**

[a] *Department of Petroleum Technology, Dibrugarh University, Assam, India*
[b] *Department of Applied Geology, Dibrugarh University, Assam, India*
pranabboral8812@gmail.com

The grain size analyses for 18 conventional core samples obtained from various depths of the oil-producing horizons of the Upper Assam Basin were carried out. The mechanical sieving method using a Ro-tap shaker was chosen for the analysis. The cumulative frequency curve was obtained by plotting the grain size in the phi ($\phi$) scale versus the cumulative percentage using the Probability paper. The $\phi$ values of the percentiles 5, 16, 25, 50, 75, 84 and 95 were read off from the ogive curve and were plotted. These were used to calculate the statistical parameters of standard deviation, skewness, kurtosis and graphic mean. The mean value for the grain size distribution within the analysed sediments was 1.8. The graphic mean distribution for these sediments ranges from 2 $\phi$ to 3 $\phi$, which is indicative of the presence of mainly fine-grained sand. The analysed core samples show a standard deviation (sorting)

*Advances in Petroleum Technology*
Edited by Subrata Borgohain Gogoi

ISBN 978-981-4877-19-0 (Hardcover), 978-1-003-04993-7 (eBook)
www.jennystanford.com

over a spread of 0.35 to 1.4 $\phi$, indicating very fine skewed to near symmetrical. The frequent distributions of grain size from fine to medium to coarse grain are an excellent source of oil-bearing formations. The change in sediment grain also infers richness in interconnected porosity and variable permeability. The amounts of clay and silt-to-sand percentage in varying depth with a random distribution pattern laterally, obtained from the results, suggest that enhanced oil recovery methods are feasible in the basin.

## 1.1 Introduction

Consolidated conventional core samples were collected from a depth of 2299.6 to 3820.3 m from the producing horizons of the Upper Assam Basin. The studied samples are from 18 oil wells of different oilfields under Oil India Limited (OIL) and the Oil and Natural Gas Corporation Limited (ONGCL) operational areas of the Upper Assam Basin. The basin is a polyhistoric basin from where hydrocarbons (HCs) are being produced for more than a century. The first commercial oil discovery of this part of the world came from this basin in the form of the Digboi oilfield in 1889. Thereafter, the basin witnessed significant oil and gas discoveries. Exploration work, aimed at different plays, is still continuing in this basin.

A recent basin modelling study (carried out by OIL) integrating drilling data and other geo-scientific data to update the regional geologic model in context of basin evolution and its depositional history propounds that the basin was dipping towards the south and southeast after collision of the Indian plate (the Assam Shelf) with the Burmese plate and the sediments came primarily from the north and northeast. In the late Miocene and Paleocene periods, the Indian Plate (northern side of the Assam Shelf) docked with the Asian plate, and gradually the basin got tilted towards the north and northeast; and in the process the basin was locked from three sides, that is, in the north, east and south directions.

In recent times the basin has been receiving sediments almost equally from all sides, followed by subsequent subduction, and this situation is still being continued. In this chapter, dynamic evolution of this basin and its basinal depositional history is explained by regional seismic sections recorded at different parts of the basin. The environmental interpretation of grain size distributions found in sedimentary deposits has been, and still is, a fundamental goal of sedimentology [1].

## 1.2 Study Area

The Upper Assam Basin is one of the oldest petroleum-producing basins of the world, which is bounded in the north by the eastern Himalayas, in the east by the Mishmi Massif, in the south by Naga–Patkai Hills and in the west by the Mikir Hills and the Shillong Plateau. In the present-day context, tectonically, the basin can be subdivided into two parts, viz. the Assam Shelf fore-land and the Thrust fold/Schuppen Belt (Himalayan orogenic belt). The geologic formations in the basin primarily comprise sand and shale alterations of the sediment from the Paleocene/Eocene to the Recent age [2]. A thick pile of sediments ranging in age from the Eocene age

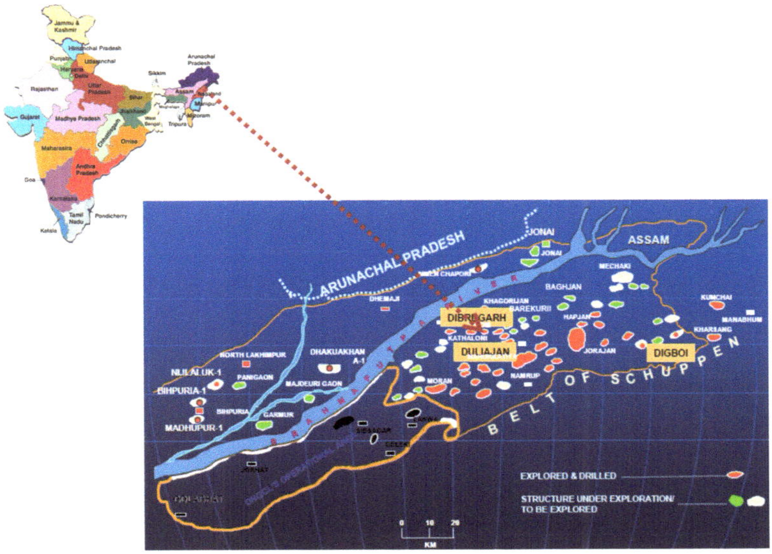


**Figure 1.1** Study area, sample collection and distribution of oilfields in the Upper Assam Basin.

has been deposited in the basin. Commercial HC production from the Upper Paleocene to the Mio-Pliocene has been established in this basin. In the subsurface, a ridge of Precambrian rocks known as the Brahmaputra arch roughly bisects the Assam Basin, and it is an important feature from the HC entrapment point of view [3]. The samples were collected from the oil-producing horizon at a depth range from 2299.6 to 3820.3 m. About 200 g of the samples were collected from the conventional core from Makum in the east to Lakwa in the west, as shown in Fig. 1.1. Preliminary research was done on Naharkatiya (NH) and Jorajan (JN) by Gogoi et al. (2017) [4].

## 1.3 Methodology

Grain size analysis was carried out on 18 core samples, as shown in Table 1.1, collected from the oil-producing horizon of the selected oilfields of the Upper Assam Basin. The analysis was aimed at measuring the individual grain sizes of the core samples. The samples were dried, and the mechanical sieving method using a Ro-tap shaker was chosen for the analysis. About 100 g of each sample was disaggregated using a porcelain mortar and pestle. The disaggregated samples were thoroughly mixed and split into quarters. This was weighed in an electronic balance with a precision 0.011 g. Each sample was then mechanically sorted into a set of US mesh sieves of 2.00, 1.18, 0.85, 0.60, 0.425, 0.30, 0.25, 0.15, 0.075 and 0.063 mm and a receiving pan using a Ro-tap shaker. The fraction retained in each sieve and the pan was weighed in a balance and its weight recorded and tabulated. The percentage aggregate was visually estimated and expressed in percentage points. The individual as well as the cumulative weight percentage was calculated and recorded. The grain size (in $\phi$) was then plotted on the abscissa, while the individual sample weight (in%) was plotted on the ordinate, thus forming a chart showing the grain size distribution (histogram) and frequency curve for each sample. The cumulative frequency curve (ogive) was plotted on a probability graph paper by plotting the grain size (in $\phi$ scale). The $\phi$ values of the percentiles 5%, 16%, 25%, 50%, 75%, 84% and 95% were

**Table 1.1** Results of 18 core samples

| S. no. | Sample | Depth (m) | $\sigma$ | $S_{Ki}$ | $M_Z$ | $K_G$ | Textural type |
|---|---|---|---|---|---|---|---|
| 1 | JN (Barail) | 2299.6 −2300.3 | 1.398485 (poorly sorted) | −0.81578 (very coarse skewed) | 2.136667 (fine grain $\geq \phi = 2.14 \leq$ coarse grain) | 0.554402 (very platykurtic) | Very fine sand |
| 2 | BP (Barail) | 3837.55 −3838.35 | 1.328788 (poorly sorted) | −0.01806 (near symmetrical) | 1.55 (fine grain $\geq \sigma = 1.55 \leq$ coarse grain) | 0.549324 (very platykurtic) | Coarse sand |
| 3 | NH (Tipam) | 2502.65 −2503.25 | 1.072803 (poorly sorted) | −0.06959 (near symmetrical) | 1.726667 (fine grain $\geq \sigma = 1.73 \leq$ coarse grain) | 0.869349 (platykurtic) | Fine sand |
| 4 | NH (Barail) | 3164.94 −3165.60 | 1.067803 (poorly sorted) | −0.12425 (very coarse skewed) | 1.883333 (fine grain $\geq \sigma = 1.88 <$ coarse grain) | 0.896516 (platykurtic) | Very fine sand |
| 5 | DK (Eocene) | 3297.82 −3298.48 | 1.060152 (poorly sorted) | −0.04951 (very coarse skewed) | 1.806667 (fine grain $\geq \sigma = 1.81 \leq$ coarse grain) | 0.854873 (platykurtic) | Very fine sand |
| 6 | KN (Eocene) | 3304.7 −3305.3 | 0.90310 (moderately sorted) | −0.28566 (very coarse skewed) | 2.226667 (fine grain $\geq \sigma = 2.23 \leq$ coarse grain) | 0.965815 (mesokurtic) Very | fine sand |
| 7 | MN (Barail) 1 | 2322.31 −2322.97 | 1.074545 (poorly sorted) | −0.41603 (very fine skewed) | 1.616667 (fine grain $\geq \sigma = 1.62 \leq$ coarse grain) | 0.771048 (platykurtic) | Medium sand |
| 8 | RS (Barail Main Sand) 1 | 3015.74 −3016.34 | 0.995455 (moderately sorted) | −0.1672 (fine skewed) | 1.783333 (fine grain $\geq \sigma = 1.78 \leq$ coarse grain) | 0.978222 (mesokurtic) | Fine sand |
| 9 | RS (Barail Main Sand) 2 | 2983.7 −2984.3 | 0.965909 (moderately sorted) | −0.15286 (fine skewed) | 1.783333 (fine grain $\geq \sigma = 1.78 \leq$ coarse grain) | 1.029356 (mesokurtic) | Fine sand |

*(Contd.)*

**Table 1.1** *(Continued)*

| S. no. | Sample | Depth (m) | $\sigma$ | $S_{Ki}$ | $M_Z$ | $K_G$ | Textural type |
|---|---|---|---|---|---|---|---|
| 10 | LP (Barail) 1 | 3518.7 −3519.3 m | 1.077576 (poorly sorted) | −0.1823 (fine skewed) | 1.913333 (fine grain $\geq \sigma = 1.91 \leq$ coarse grain) | 0.771321 (platykurtic) | Medium sand |
| 11 | LP (Barail) 2 | 3506.7 −3507.3 | 1.044394 (poorly sorted) | −0.12286 (fine skewed) | 1.866667 (fine grain $\geq \sigma = 1.87 \leq$ coarse grain) | 0.756440 (platykurtic) | Medium sand |
| 12 | GK (Tipam) 1 | 3075.65 −3076.30 | 1.2225 (poorly sorted) | −0.5058 (coarse skewed) | 1.736667 (fine grain $\geq \sigma = 1.74 \leq$ coarse grain) | 0.676230 (platykurtic) | Very fine sand |
| 13 | GK (Tipam) 2 | 2617.65 −2618.30 | 1.055152 (poorly sorted) | −0.25679 (coarse skewed) | 1.716667 (fine grain $\geq \sigma = 1.72 \leq$ coarse grain) | 0.622870 (very platykurtic) | Very fine sand |
| 14 | CMG (Tipam) 1 | 2983.70 −2984.35 | 1.012045 (poorly sorted) | 0.410703 (coarse skewed) | 1.383333 (fine grain $\geq \sigma = 1.72 \leq$ coarse grain) | 0.793231 (very platykurtic) | Medium sand |
| 15 | CMG (Tipam) 2 | 2986.70 −2987.30 | 0.994924 (poorly sorted) | −0.50126 (very fine skewed) | 1.32 (fine grain $\geq \sigma = 1.27 \leq$ coarse grain) | 0.860951 (platykurtic) | Medium sand |
| 16 | LK (Tipam) 1 | 3517.65 −3518 | 1.1125 (poorly sorted) | −0.26611 (coarse skewed) | 1.963333 (fine grain $\geq \sigma = 1.96 \leq$ coarse grain) | 0.715587 (platykurtic) | Very fine sand |
| 17 | LK (Tipam) 2 | 3824.70 −3825.30 | 1.031591 (poorly sorted) | −0.21415 (coarse skewed) | 1.966667 (fine grain $\geq \sigma = 1.97 \leq$ coarse grain) | 0.734795 (platykurtic) | Very fine sand |
| 18 | BJ (L+T) 1 | 3819.70 −3820.30 | 1.056894 (poorly sorted) | −0.32011 (very coarse skewed) | 2.06 (fine grain $\geq \sigma = 2.06 \leq$ coarse grain) | 0.888437 (platykurtic) | Very fine sand |

JN: Jorajan; BP: Bhogpara; NH: Naharkatiya; DK: Dikom; KN: Kothaloni; MN: Moran;
RS: Rudrasagar; LP: Laipingoan; GK: Galeki; CMG: Changmaigoan; LK: Lakwa; BJ: Baghjan

read off from the ogive curve. These were used to calculate the statistical parameters of standard deviation, skewness, kurtosis, mean and median, as shown in Eqs. 1.1–1.4, used to derive the various grain size parameters [5]. The graphic mean ($M_Z$) as in Eq. 1.1 is the mean size and is a function of the size range of available materials and the amount of energy impacted to the grains, which depends on the current velocity or turbulence of the transporting medium. The mean value for the grain size distribution within the analysed sediments was 1.802. The graphic mean distribution for these sediments ranged from 1.09 $\phi$ to 2.90 $\phi$, which is indicative of the presence of fine-grained sand to coarse-grained sand.

We have,

$$M_Z = \frac{\phi_{16} + \phi_{50} + \phi_{84}}{3}, \tag{1.1}$$

where

| | |
|---|---|
| $-12$ to $-8\ \phi$ | boulder |
| $-8$ to $-6\ \phi$ | cobble |
| $-6$ to $-2\ \phi$ | pebble |
| $-2$ to $-1\ \phi$ | granular |
| $-1$ to $-0.0\ \phi$ | very coarse-grained |
| 0.0 to 1.0 $\phi$ | coarse-grained |
| 1.0 to 2.0 $\phi$ | medium-grained |
| 2.0 to 3.0 $\phi$ | fine-grained |
| 3.0 to 4.0 $\phi$ | very fine-grained |
| 4.0 to 5.0 $\phi$ | coarse silt |
| 5.0 to 6.0 $\phi$ | medium silt |
| 6.0 to 7.0 $\phi$ | fine silt |
| 7.0 to 8.0 $\phi$ | very fine silt |
| 8.0 $\phi$ and smaller | clay |

The inclusive graphic standard deviation (sorting) as in Eq. 1.2 is a measure of the standard deviation, which is the spread of the grain size distribution with respect to the mean. Sorting is the most useful grain size data since it gives an indication of the effectiveness of the depositional medium in separating grains of different classes. The analysed core samples show a standard deviation. (Sorting shows a spread of 0.9 to 1.4 $\phi$ and a graphic mean value of 1.8.)

We have,

$$\sigma = \frac{\phi_{84} - \phi_{16}}{4} - \frac{\phi_{95} - \phi_{5}}{66}, \tag{1.2}$$

where

| | |
|---|---|
| $\leq 0.35\ \phi$ | very well sorted |
| 0.35 to 0.5 $\phi$ | well sorted |
| 0.50 to 0.71 $\phi$ | moderately well sorted |
| 0.71 to 1.0 $\phi$ | moderately sorted |
| 1.0 to 2.0 $\phi$ | poorly sorted |
| 2.0 to 4.0 $\phi$ | very poorly sorted |
| $\geq 4.0\ \phi$ | extremely poorly sorted |

Skewness ($S_{\mathrm{Ki}}$) as in Eq. 1.3 is a reflection of the depositional process. It is simply a measure of the symmetry of the distribution. Skewness is useful in environmental diagnosis because it is directly related to the fine and coarse tails of the size distribution and hence suggestive of energy of deposition. The skewness values of the samples ranged from $-0.0181$ to $-0.816$, thus indicating the presence of a near symmetrical and very coarse fraction in the particle population. The negative values indicate skewness towards the coarser grain sizes.

$$S_{\mathrm{Ki}} = \frac{\phi_{16} + \phi_{84} - 2\phi_{50}}{2(\phi_{84} - \phi_{16})} + \frac{\phi_{5} + \phi_{95} - 2\phi_{50}}{2(\phi_{95} - \phi_{5})}, \tag{1.3}$$

where

| | |
|---|---|
| 1 to 0.3 | very fine skewed |
| 0.3 to 0.10 | fine skewed |
| 0.10 to $-0.10$ | near symmetrical |
| $-0.10$ to $-0.30$ | coarse skewed |
| $-0.30$ to $-1$ | very coarse skewed |

Graphic kurtosis ($K_{\mathrm{G}}$) as in Eq. 1.4 is a measure of the peakedness of the curves towards the coarser grain sizes. The samples were predominantly platykurtic, the samples were better sorted in the tails than in the central portion and the curve was flat peaked or platykurtic. In numerical terms, the range of kurtosis was between 0.549 and 0.978, typifying that 61% of the samples are platykurtic, 22% are very platykurtic and 16.67% are mesokurtic.

$$K_G = \frac{\phi_{95} - \phi_5}{2.44(\phi_{75} - \phi_{25})}, \tag{1.4}$$

where

| | |
|---|---|
| $\leq 0.67$ | very platykurtic |
| $0.67 - 0.90$ | platykurtic |
| $0.90 - 1.11$ | mesokurtic |
| $1.11 - 1.50$ | leptokurtic |
| $-1.50 - 3.00$ | Very leptokurtic |
| $\geq 3.00$ | extremely leptokurtic |

## 1.4 Results and Discussion

Studies have been used to understand the provenance of the conventional core samples in Figs. 1.2–1.19 and in Table 1.1. A nearly unimodal frequency distribution indicates a single provenance for the sediments. Textural studies indicate that the sediments belong to mostly to very fine, fine, and medium-grained sand fractions, suggesting that the sediments were deposited under low-energy conditions, with the sediments being mostly very poorly sorted to moderately sorted, indicating texturally immature to submatured sediments of a fluvial environment [6]. The unimodality of the grain size distribution chart with the exception of some nearly bimodal chart reveals that the sands were not derived from more than a single source area. In general, and depending on average values, the sediments are fine-grained, poorly sorted sands, associated with moderately high current flows in a fluvial environment.

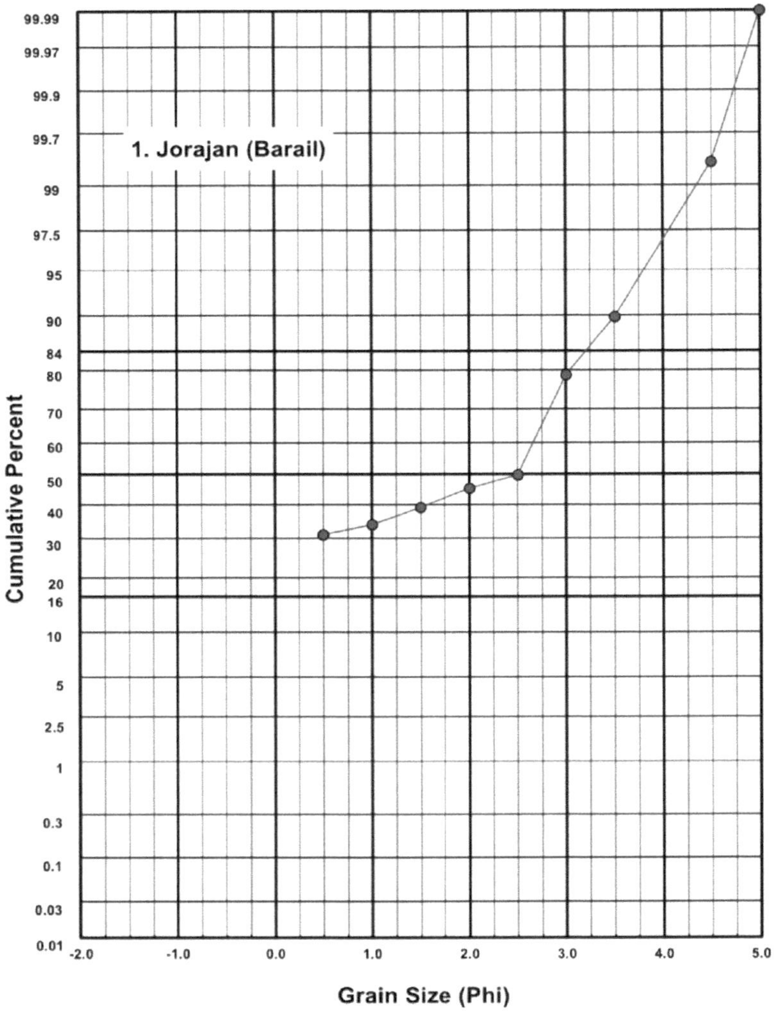


**Figure 1.2** Graphical representation of grain size analyses of Jorajan core samples.

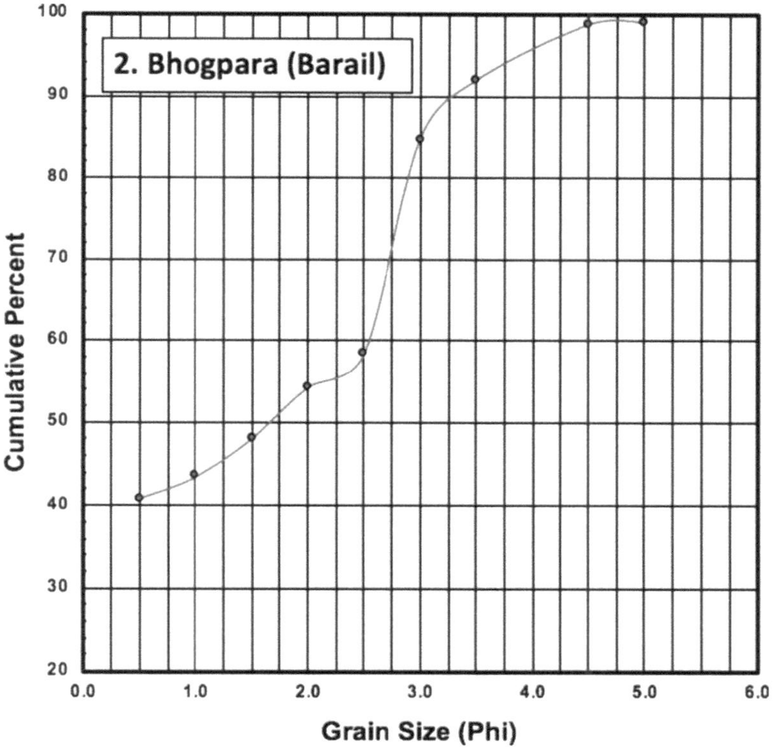


**Figure 1.3** Graphical representation of grain size analyses of Bhogpara (Barail) core samples.

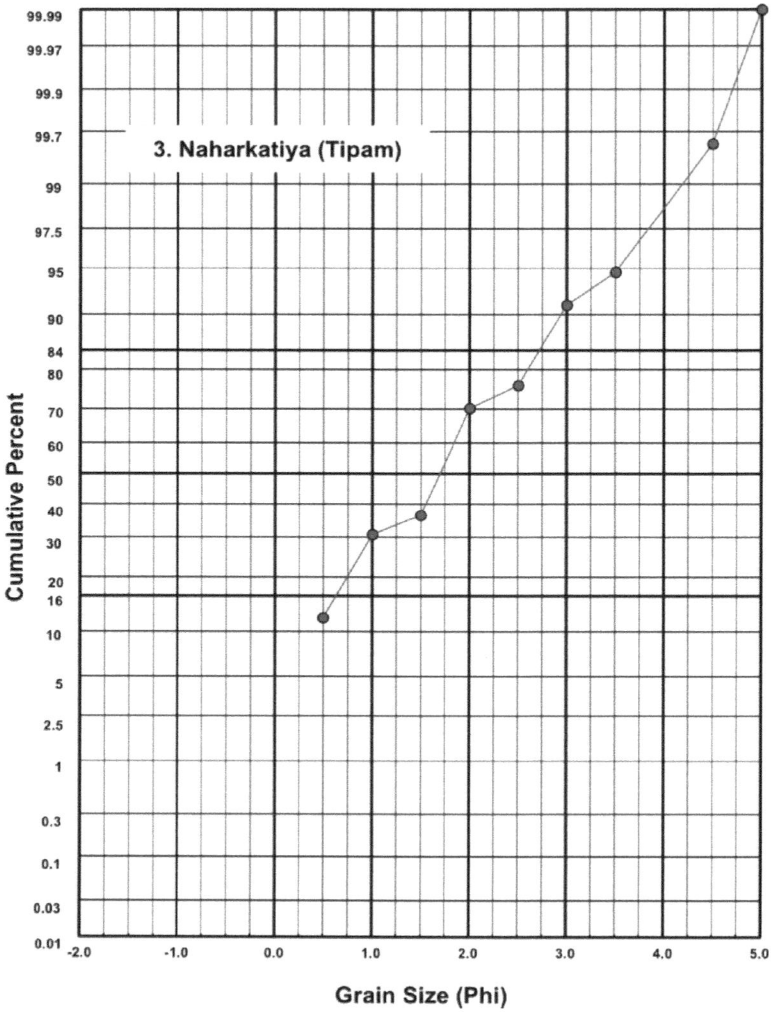


**Figure 1.4** Graphical representation of grain size analyses of Naharkatiya (Tipam) core samples.

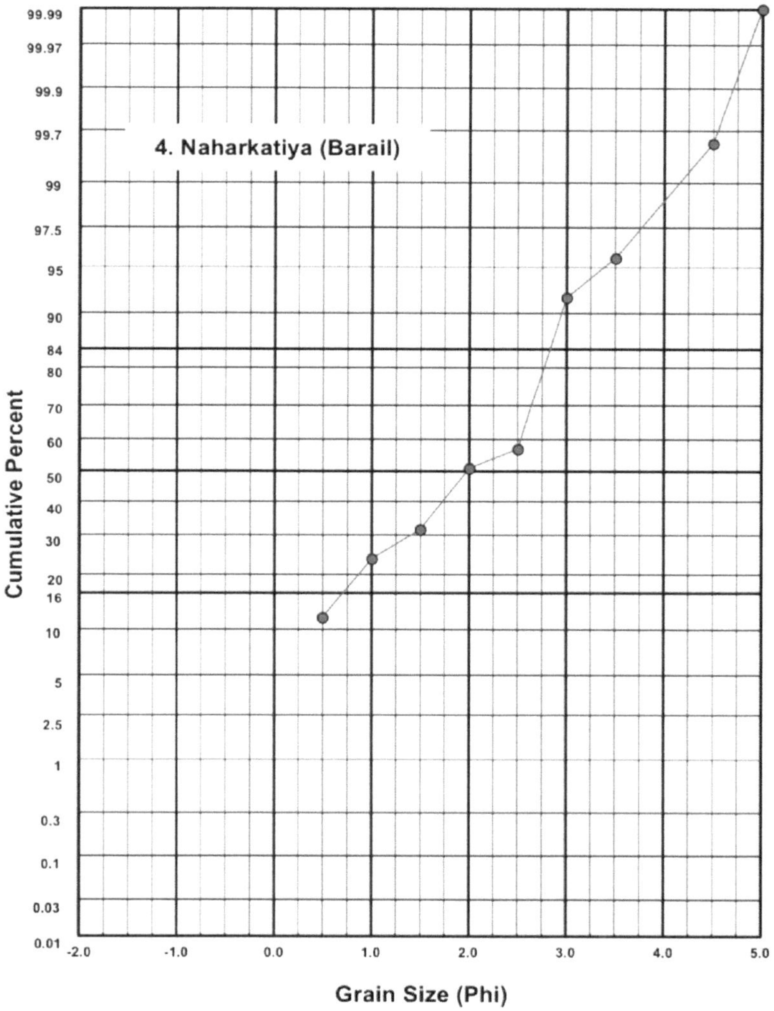


**Figure 1.5** Graphical representation of grain size analyses of Naharkatiya (Barail) core samples.

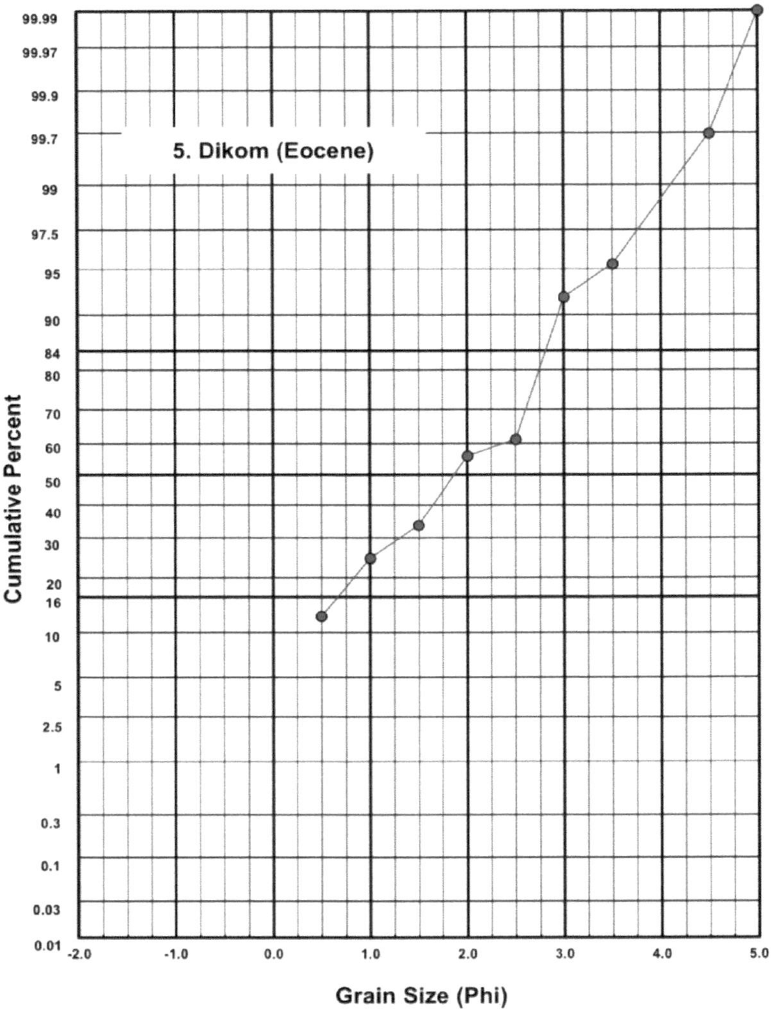


**Figure 1.6** Graphical representation of grain size analyses of Dikom (Eocene) core samples.

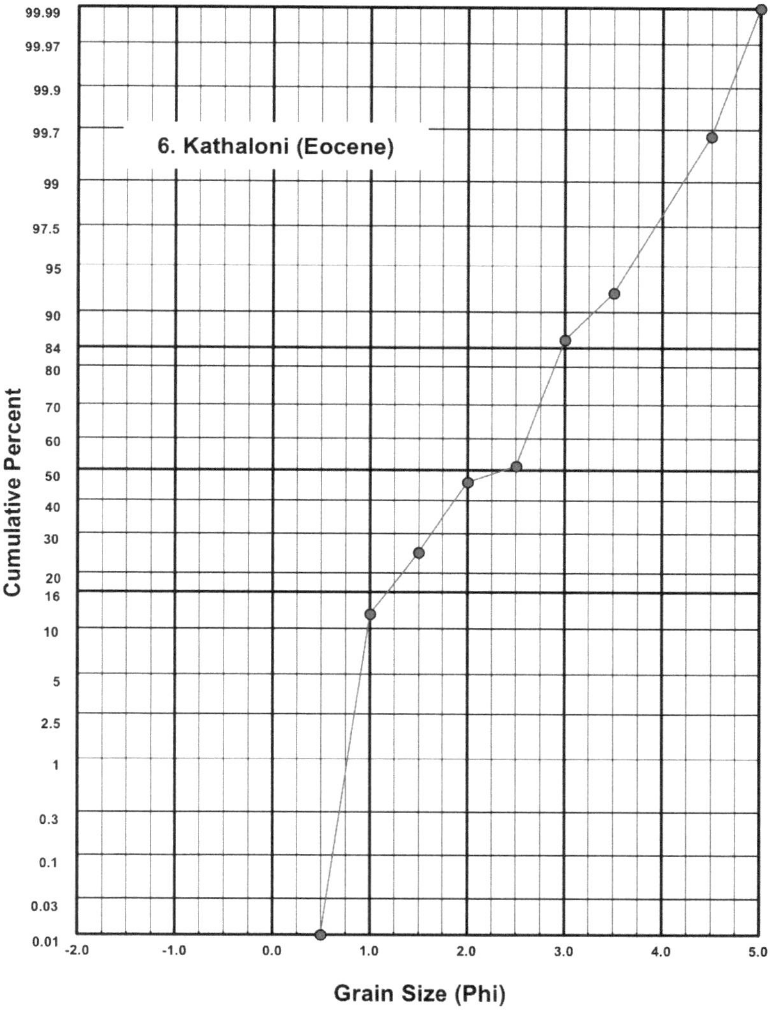


**Figure 1.7** Graphical representation of grain size analyses of Kathaloni (Eocene) core samples.

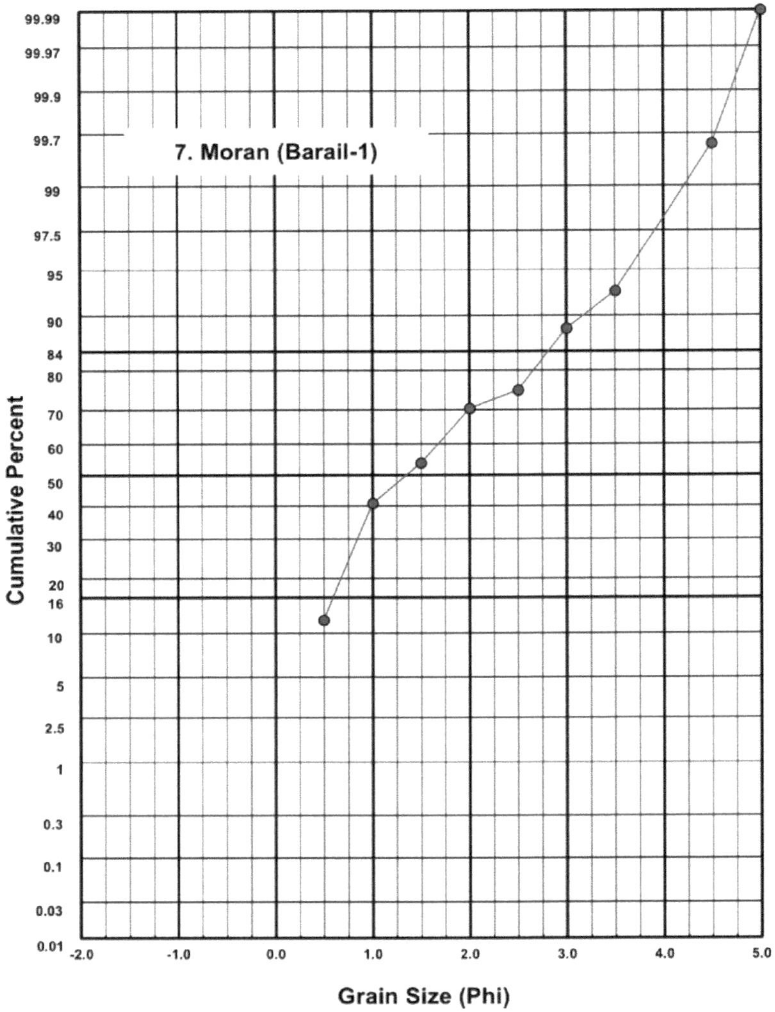


**Figure 1.8** Graphical representation of grain size analyses of Moran (Barail-1) core samples.

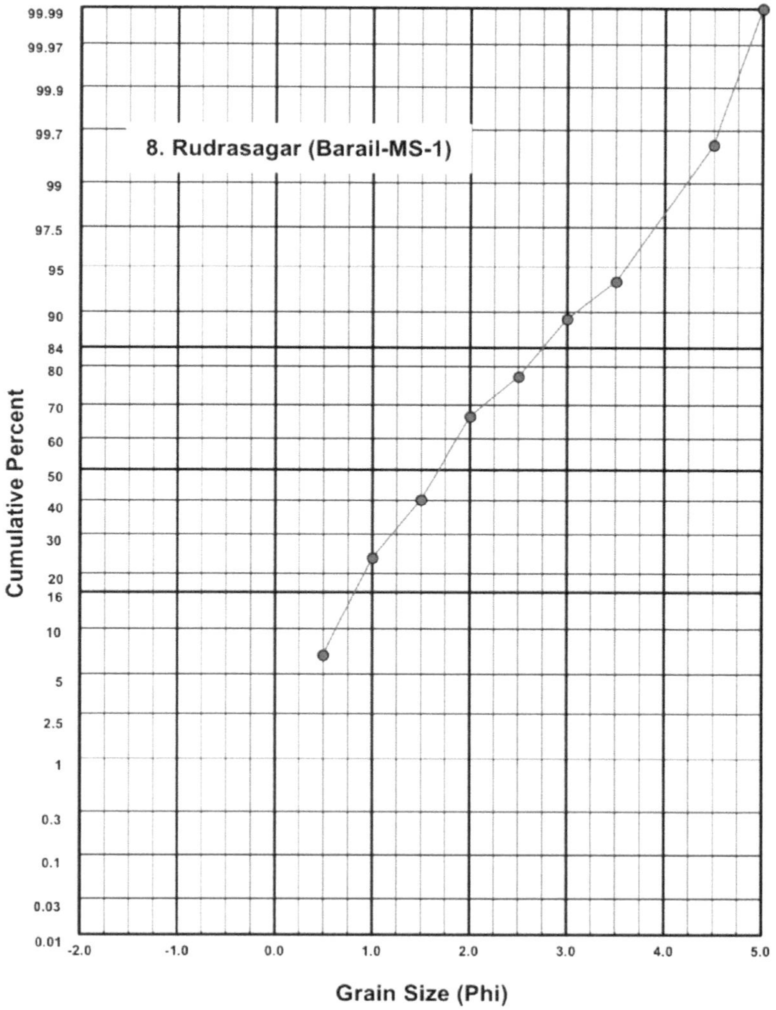


**Figure 1.9** Graphical representation of grain size analyses of Rudrasagar (Barail-MS-1) core samples.

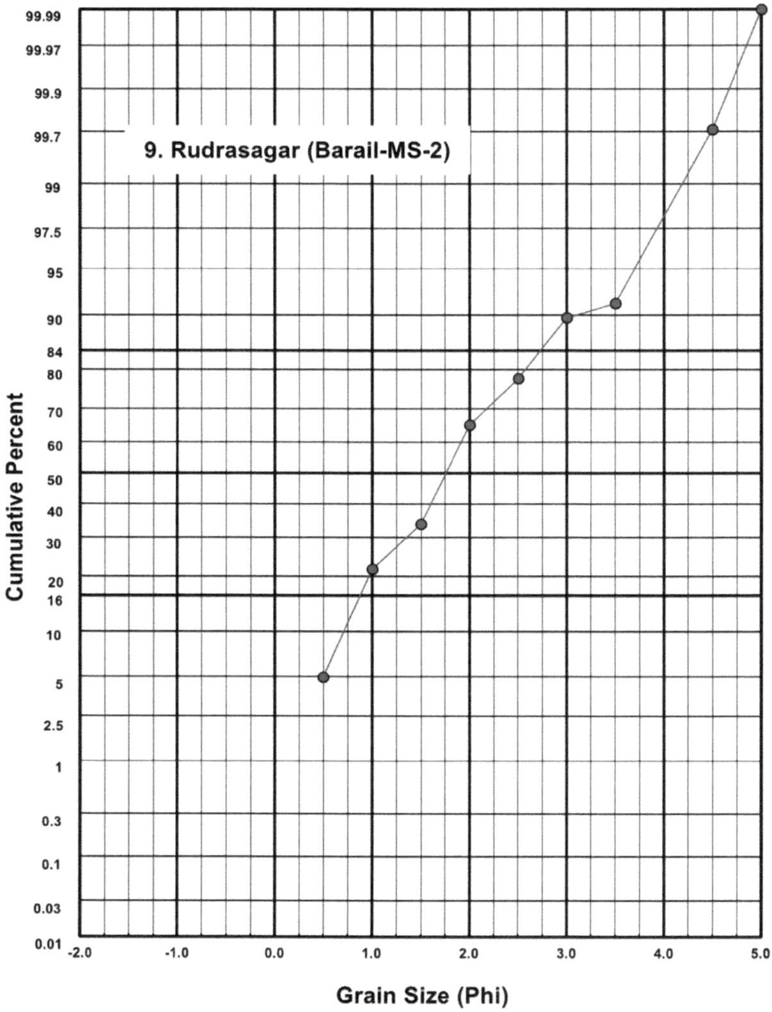


**Figure 1.10** Graphical representation of grain size analyses of Rudrasagar (Barail-MS-2) core samples.

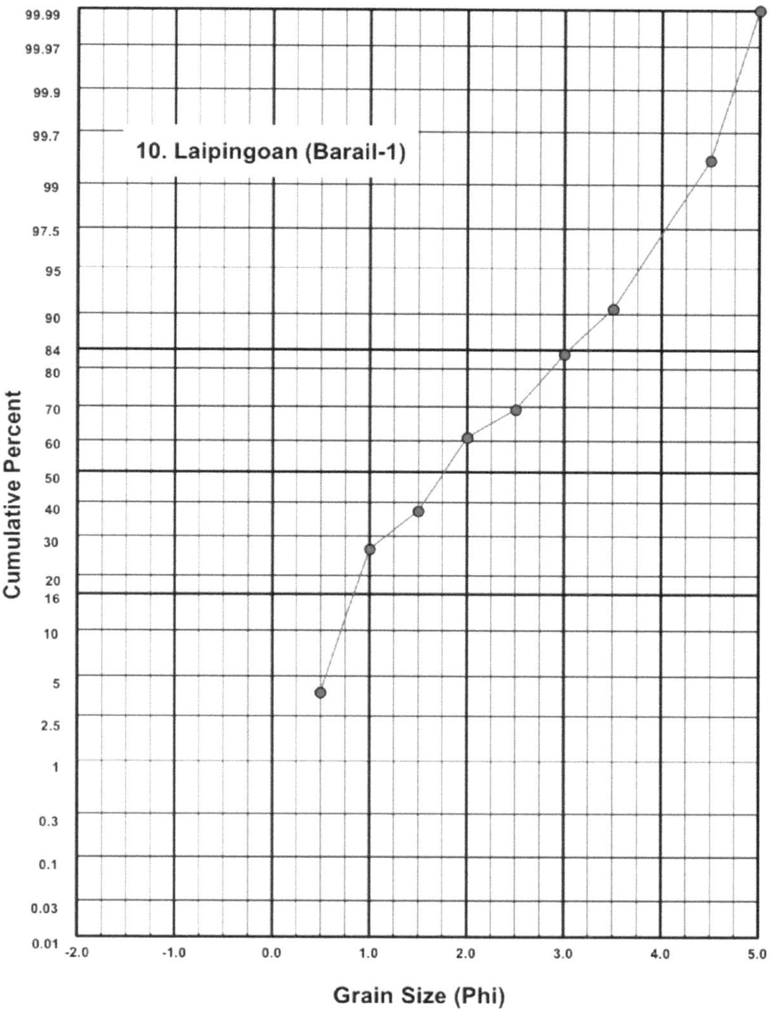


**Figure 1.11** Graphical representation of grain size analyses of Laipingoan (Barail-1) core samples.

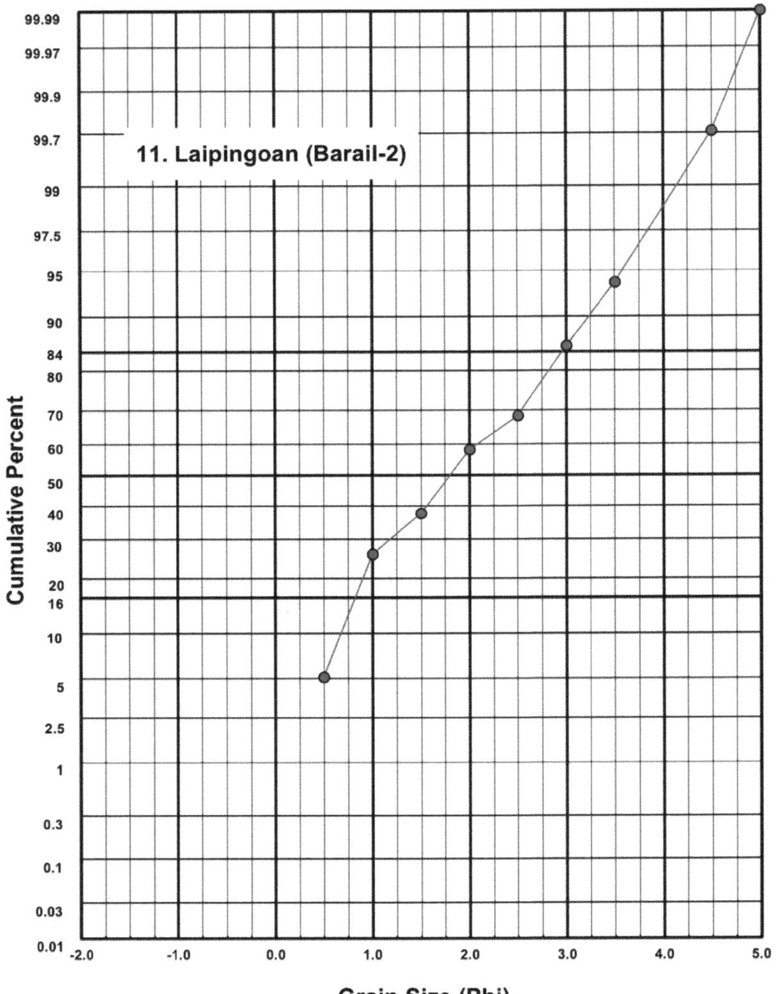


**Figure 1.12** Graphical representation of grain size analyses of Laipingoan (Barail-2) core samples.

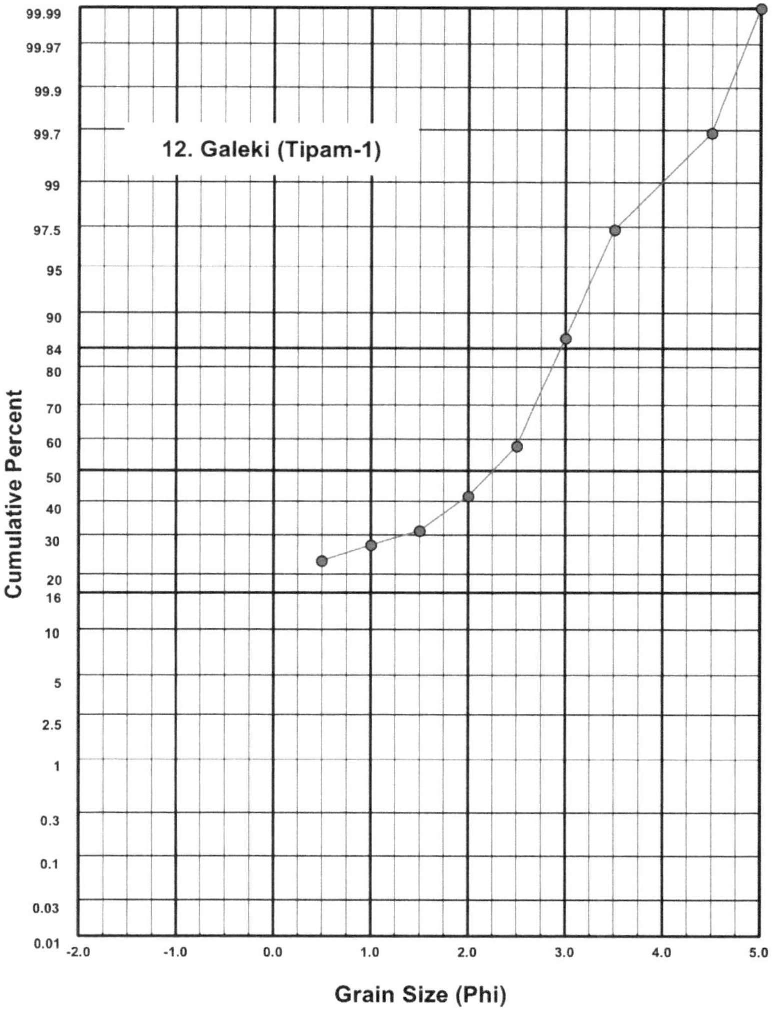


**Figure 1.13** Graphical representation of grain size analyses of Galeki (Tipam-1) core samples.

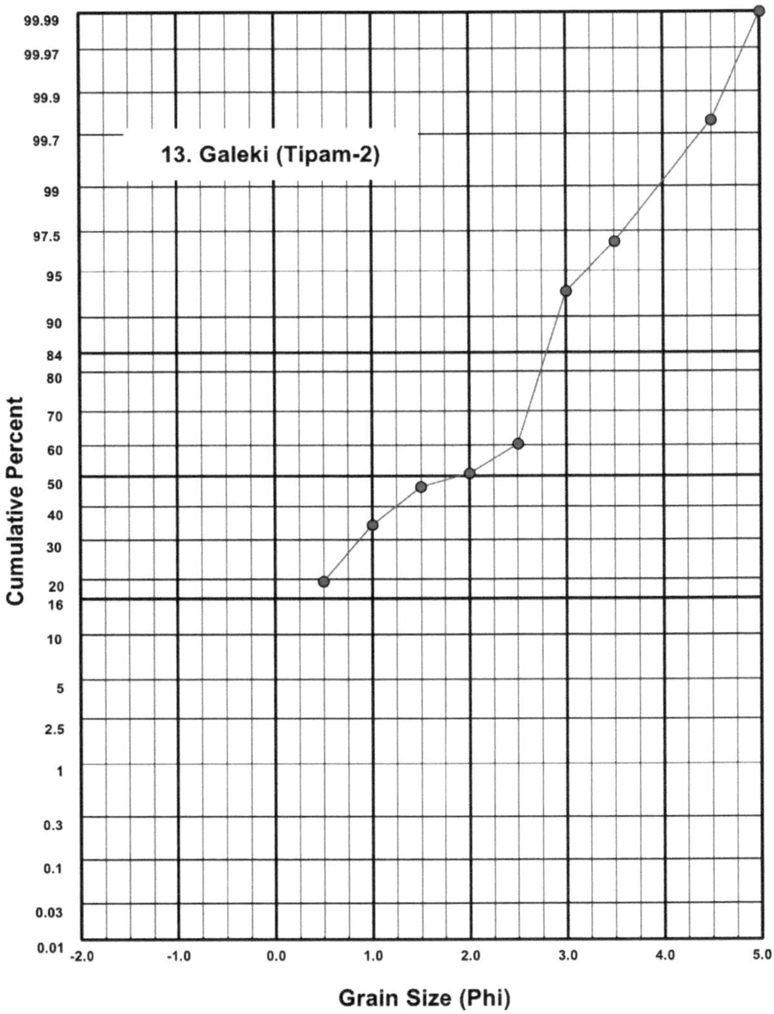


**Figure 1.14** Graphical representation of grain size analyses of Galeki (Tipam-2) core samples.

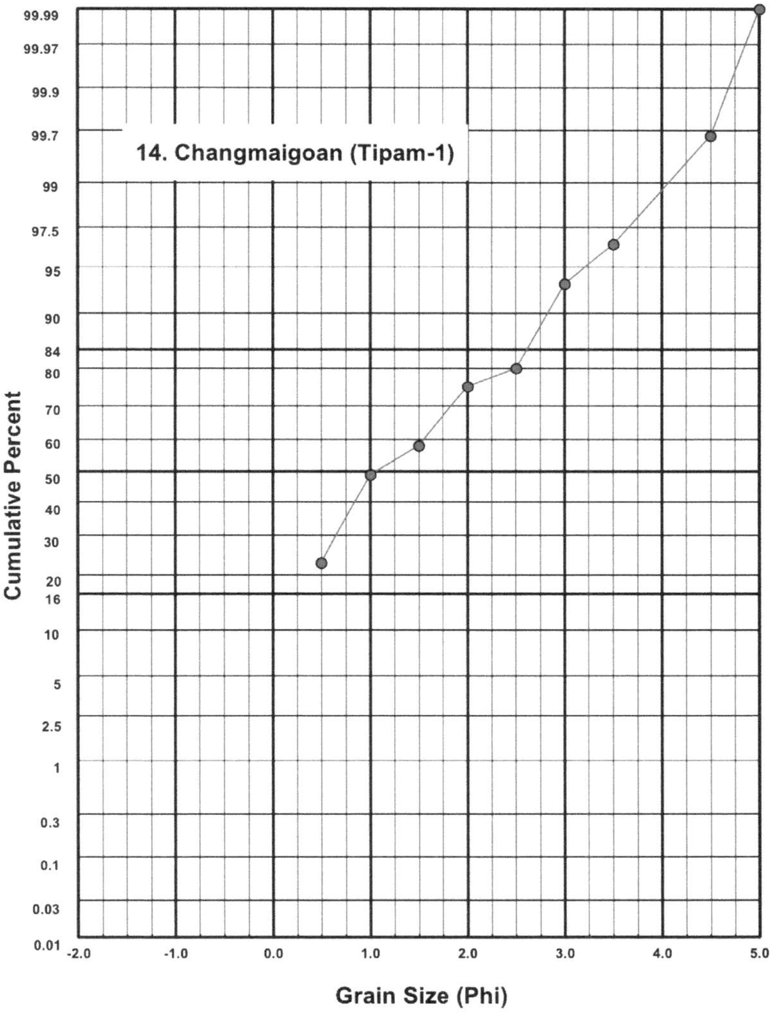


**Figure 1.15** Graphical representation of grain size analyses of Changmaigoan (Tipam-1) core samples.

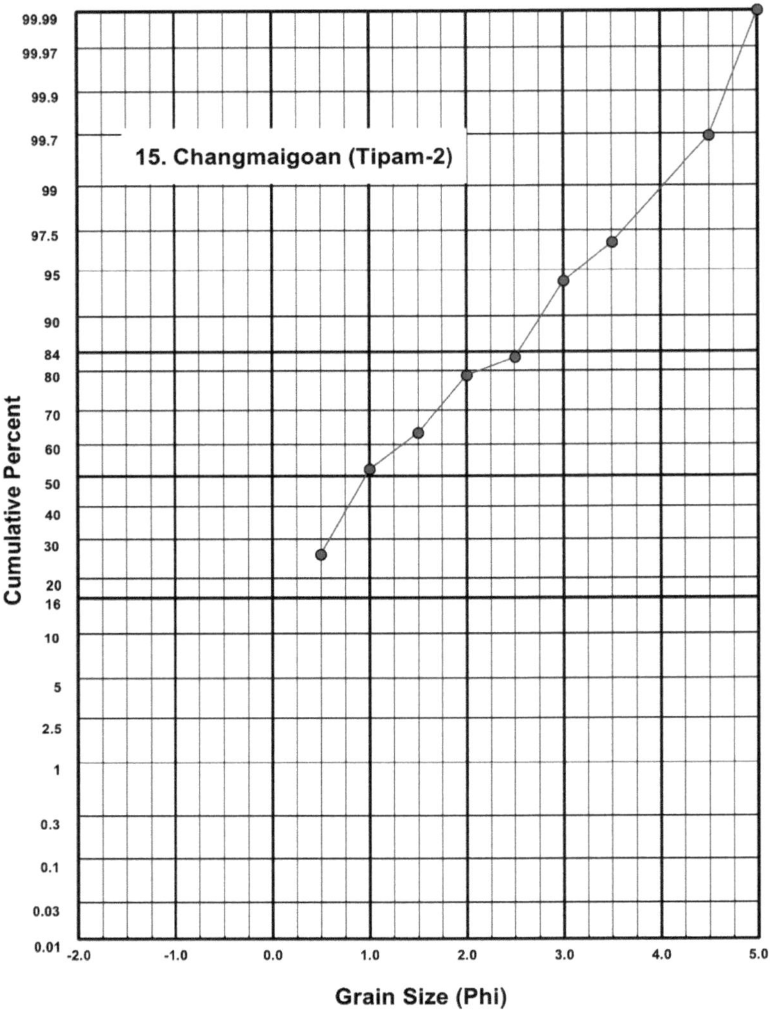


**Figure 1.16** Graphical representation of grain size analyses of Changmaigoan (Tipam-2) core samples.

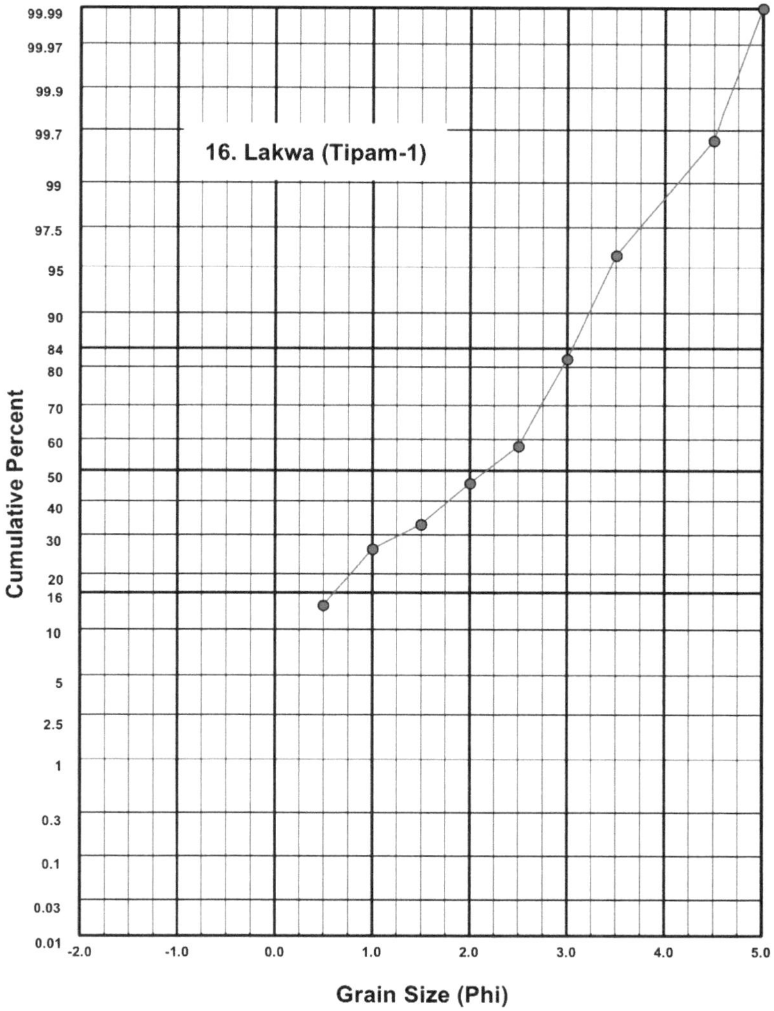


**Figure 1.17** Graphical representation of grain size analyses of Lakwa (Tipam-1) core samples.

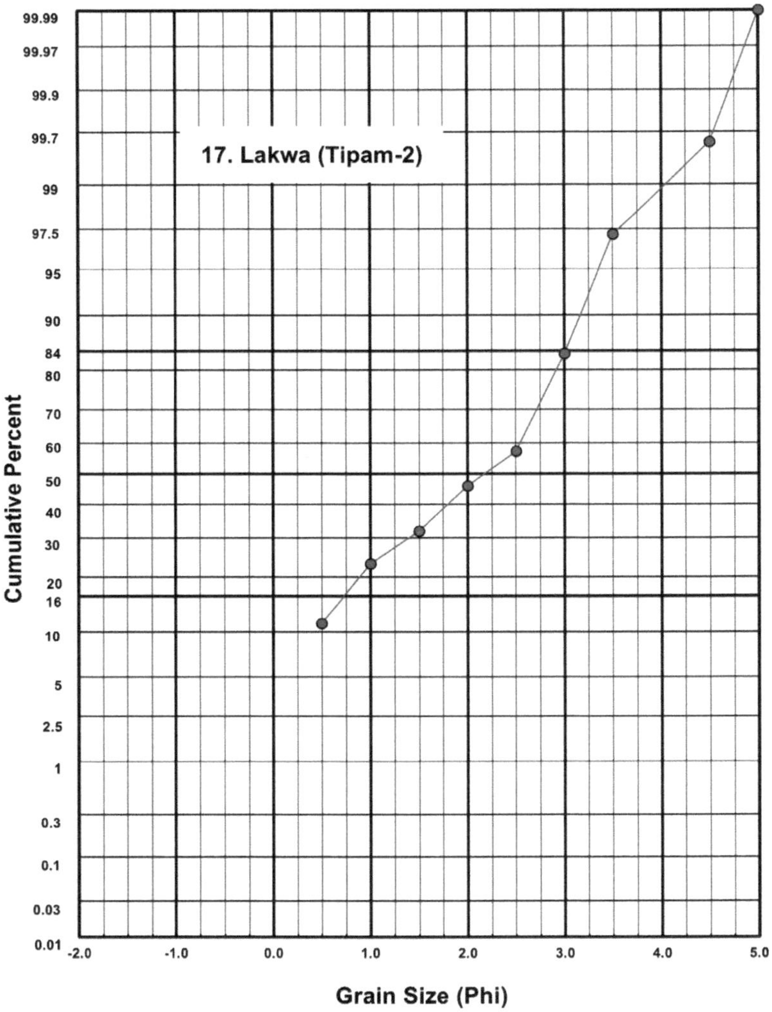


**Figure 1.18** Graphical representation of grain size analyses of Lakwa (Tipam-2) core samples.

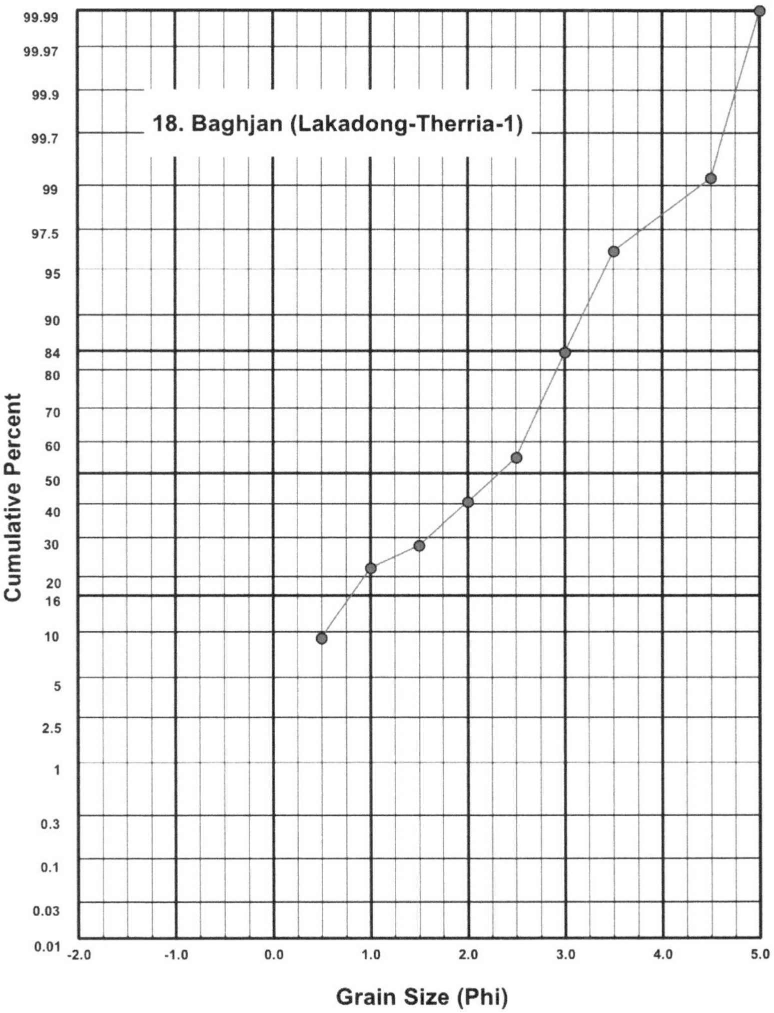


**Figure 1.19** Graphical representation of grain size analyses of Baghjan (Lakadong Therria-1) core samples.

## 1.5 Conclusion

The results obtained from grain size analyses depict heterogeneous oilfield reservoirs with a range of sedimentary structures. Here we use the 'heterogeneous' term in association with different grain sizes found in different oilfields, all of which fall under a similar depositional environment. For enhanced oil recovery, the clay, matrix and cement percentage is used to describe the reservoir rock condition for the performance factor. A change in the grain size with varying depth also indicates random distribution of porosity.

## Acknowledgements

We would like to acknowledge the funding received from the UK–India Educational Research Initiative (UKIERI) (No. F.No. 184-1/2018[IC]) entitled 'Novel EOR Technique for the Depleted Oil Reservoirs of Upper Assam Basin with Simultaneous $CO_2$ Capture and Sequestration', and Dibrugarh University for the laboratory facilities.

## Chapter 2

# Depositional Environment and Hydrocarbon Source Potential of the Barail and Disang Groups of Rocks along the Tuli–Mokokchung Road Section, Nagaland

**Manash Pratim Gogoi,[a] Yadav Krishna Gogoi,[a] Devojit Bezbaruah[a, b] and Pradip Borgohain[a]**

[a] *Department of Petroleum Technology, Dibrugarh University, India*
[b] *Department of Applied Geology, Dibrugarh University, India*
ta_manashgogoi@dibru.ac.in

The Belt of Schuppen of Nagaland has been identified by different workers as the most prospective area for hydrocarbon (HC) deposits. However, due to different logistics and the complex geological setup of the region, detailed exploration activities have not been done. The main objective of this present chapter is to assess the HC generation potential and reconstruction of the depositional environment of the Disang Group, which is considered as the main source rock of the Upper Assam Basin, and the Barail Group, which is considered as one of the important reservoir rocks, exposed along the Tuli–Mokokchung road section of Nagaland. The chapter

*Advances in Petroleum Technology*
Edited by Subrata Borgohain Gogoi

ISBN 978-981-4877-19-0 (Hardcover), 978-1-003-04993-7 (eBook)
www.jennystanford.com

includes detailed mapping integrated with granulometry, elemental and rock-eval analysis of the Barail and Disang Groups of rocks. The lithology of the study area comprises primarily alternate beds of shale and sandstone with some minor coal seams at places. However, both the groups of rocks are found to be sandstone dominated. The grain size analysis of the Barail and Disang Groups of rocks indicates that they are poorly to moderately sorted, strongly fine-skewed to coarse-skewed and platykurtic to leptokurtic. The statistical size parameters of the grains were further used to determine the depositional environment along with the energy conditions and the processes associated with it. The various bivariate diagrams indicate that the sediments were deposited under a shallow marine to deltaic environment with the influence of river as well as wave processes. Elemental analysis of the carbonaceous shale of the Barail and Disang Groups records the presence of $N_2$, carbon, $H_2$ and sulphur in the range of 0%–0.9%, 0%–6.87%, 0.23%–1.56% and 0.01%–2.8%, respectively. The bivariate plot of elemental analysis indicates that the Disang sediments were deposited in a marine environment, whereas the Barail Group of rocks was deposited in a restricted environment. The presence of coal seams within the Barail Group also supports this observation. Under rock-eval analysis, the S2 value for both the groups ranged between 0.01–0.3 mg/g, which indicated mostly poor source rock. However, the type II variety of organic matter and the total organic carbon (TOC) (0.05%–10.71%) content indicated certain shales are poor, while others are good source rock in the study area.

## 2.1 Introduction

Nagaland, also known as Switzerland of the East, lies between 25°6′N–27°4′N latitude and 93° 20′E–95°15′E longitude. Nagaland forms a part of the northern extension of the Arakan–Yoma Range, representing some of the Cretaceous and Tertiary orogenic up-heavals, forming a fairly young and mobile belt of Earth (Table 2.1) [7]. Nagaland may be divided into three major structural units which include the (i) Belt of Schuppen, (ii) the Inner Fold Belt and (iii) the Ophiolite Complex; all these units have northeast–

**Table 2.1** Stratigraphic succession of Nagaland (modified after Ref. [10])

| Age | Group | Litho-formations | | |
|---|---|---|---|---|
| | | Outer and Intermediate Hills | | Eastern High Hills |
| Recent-Pleistocene | | Alluvium and high level terraces | | |
| | Dihing | Boulder beds | | |
| | | ----------Unconformity---------- | | |
| Mio-Pliocene | Dupi Tila | Namsang Beds | | |
| | | ----------Unconformity---------- | | |
| Miocene | Tipam | Girujan Clay<br>Tipam Sandstone | | |
| | Surma | Upper Bhuban<br>Lower Bhuban | | |
| | | ----------Unconformity---------- | | |
| Oligocene | Barail | Renji | **Tikak Parbat** (Coal found) | <u>Jopi/Phokphur Formation</u><br>Tuffaceous shale, sandstone, greywacke, grit and conglomerate. Minor limestone and carbonaceous matter |
| | | Jenam | **Baragoloi** (Coal found) | |
| | | Laisong | Naogaon | |
| Upper Cretaceous - Eocene | Disang | Upper | | |
| | | Lower | | Shale/slate/phyllite with calcareous lenses in basal sections and invertebrate and plant fossils in upper sections with brine springs |
| | | -------Base not seen------- | | -------Fault/Thrust------- |
| Upper Jurassic | Ophiolite Complex | | | <u>Zepuhu Formation</u><br>Marine sediments (shale, greywacke, iron-rich sediments, chert and limestone with radiolaria and coccoliths), volcanics (basalt, spilite, volcaniclastics), metabasics greenchist, glaucophane, schist/glaucophane-bearing metachert, eclogite), layered cumulate sequence (peridotite, pyroxenite, gabbroids, plagiogranite, anorthosite), and peridotite tectonite and serpentinite associated with deposits of podiform chromite with deposits of podiform chromite and nickeliferous magnetite, minor Cu-Mo sulphides associated with late felsic intrusions and some dolerite dykes |
| | | | | -------Fault/Thrust------- |
| Pre-Mesozoic | Naga Metamorphic Complex | | | <u>Nimi Formation</u><br>Weakly metamorphosed limestone, phyllite, quartzite and quartz-sericite schist<br><u>Naga Metamorphics</u><br>Mica schist, granitoid gneiss and feldspathic metagreywacke with tectonic slices of ophiolite in variable dimensions |

southwest trends [8]. It is found that the Belt of Schuppen has got immense interest for geoscientists for its complex structure and tectonics and also the availability of crude oil and coal deposits. The burial of sediments beneath the thrust belt might have provided the requisite thermokinematic conditions for maturation of the source rock facies and accumulation of hydrocarbons (HCs) within the pore spaces of the Barail and Tipam Groups of rocks, which are prolific producers of oil and gas in the neighbouring Upper Assam Basin [9]. Indications of oil and gas throughout the Schuppen Belt warrant intensive research spread in the Schuppen Belt. HC exploration and production activities have been going on extensively in the adjacent Upper Assam Basin since the discovery of the Digboi oilfield in 1889, located immediately south of the Naga Thrust within the Belt of Schuppen (DGH); the stratigraphic succession of Nagaland, modified after Ref. [10], is as follows.

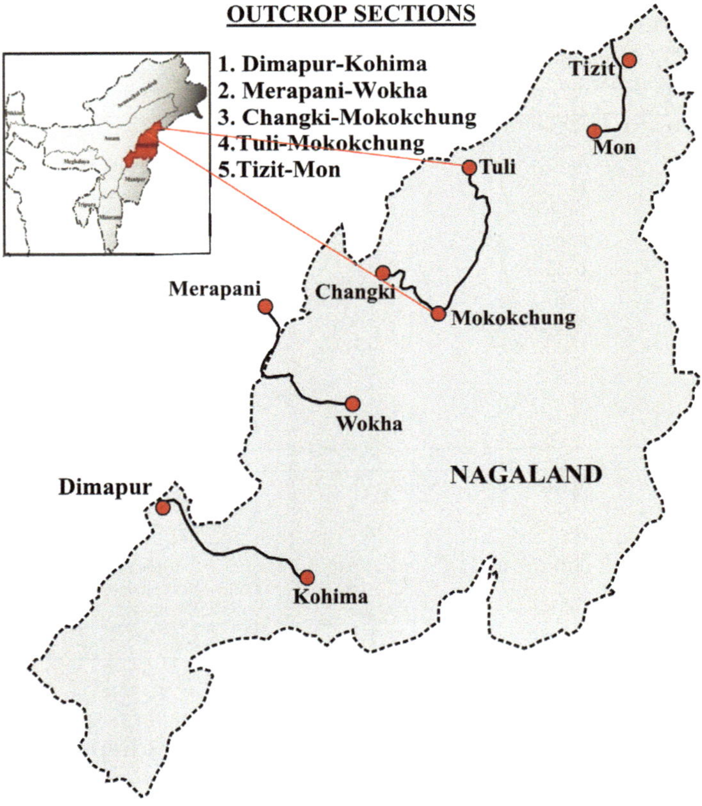


**Figure 2.1** Location map of the Tuli–Mokokchung study area.

Rock-eval pyrolysis is the most widely used method which provides information on the quantity, type and thermal maturity of the associated organic matter as well as its HC potential. The hydrogen index (HI) and $T_{max}$ are used to assess the oil and gas potential and maturity of the source rock, respectively [11].

## 2.2 Geology of the Area

Geological mapping and a detailed study of the exposed litho units were carried out along the Tuli–Mokokchung road section. Near the Assam–Nagaland border, the gentle undulatory topography changes abruptly into low hills with exposure of the Tipam Group of rocks, overlying unconformably by river terraces. This juxtaposition of older alluvium with rocks of the Miocene age and the change in topography indicate the presence of the Naga Thrust in the area [12]. The Dihing boulder beds lie unconformably over the Tipam sandstone beds (Fig. 2.2A). The sandstone is fine- to medium-grained. Beds are dipping at an angle of 25° towards the northeast direction (080°). Contact between Tipam sandstone formation and Girujan clay formation was observed (Fig. 2.2B). A change in the dip of the beds was observed in the Nankamtsu Nala. The beds are dipping at an angle of 40° in the southeast direction. This sudden change can be as a result of a probable thrust. The Barail Group of rocks is thrust over the Dihing Group of rocks. The lithology of the beds changes from sandstone bands within Dihing beds to alternate sandstone and shale (Fig. 2.2C) in the Barail Group of rocks.

A depression is found, which is due to the presence of a strike slip fault. Sedimentary structures like current beddings (Fig. 2.2D) are observed with gritty lenses. The sandstone is fine-grained to coarse-grained and gritty sandstone. The pebble beds are observed within sandstone beds and along with sandstone lenses and carbonised wood (Fig. 2.2E). Carbonaceous shales are present along with sandstone (Fig. 2.2F). The lithology of the boulders is different from one another. The beds dip at an angle of 30° towards the northeast direction (75°). Towards Mokokchung, sandstones are dominating with minor bands of shale. The thickness of sandstone beds ranges from 1 to 2 ft. A probable transitional zone between the Barail

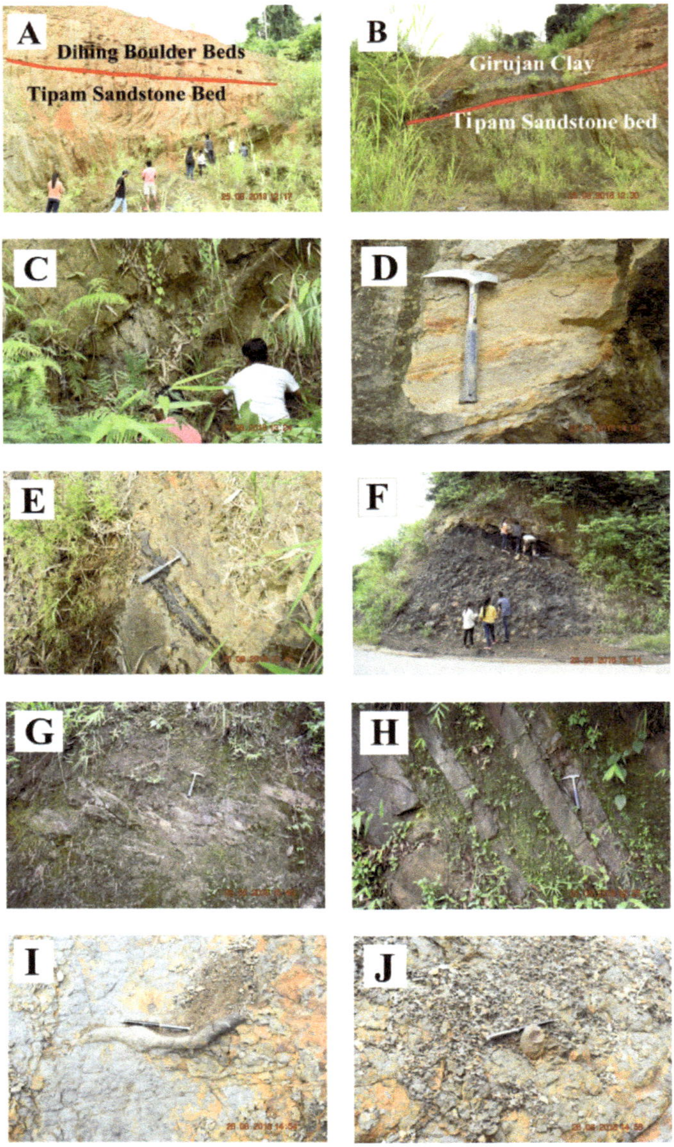


**Figure 2.2** (A) Dihing boulder bed overlying the Tipam sandstone bed, (B) Girujan clay overlying the Tipam sandstone bed, (C) alternate sandstone-shale, (D) current bedding, (E) carbonised wood, (F) carbonaceous shale, (G) recumbent fold, (H) alternate shale-sandstone with massive sandstone and (I and J) burrow.

and Disang thrust is observed at N26°37.462′, E94°39.253′. The sandstone beds show a coarsening upward sequence. The beds are dipping at an angle of 37° towards the southeast direction (143°).

Alternate sandstone and shale beds are encountered, which suffer from multiple phase folding, forming a recumbent fold (Fig. 2.2G) whose limbs are dipping at an angle of 25° towards the east (090°) and 35° towards the west (270°). A small-scale fault at the crest of a kink bed is seen. Also, the presence of vertical and parallel burrows to the bedding plane indicates a shallow marine environment (Figs. 2.2I and 2.2J).

## 2.3 Geological Mapping

The toposheets of the study area ($83\frac{J}{10}$ and $83\frac{J}{11}$) were geo-referenced and a mosaic was made with Global Mapper software. The cross-section (Fig. 2.3) and geological map (Fig. 2.5) of the area were also prepared using Global Mapper software.

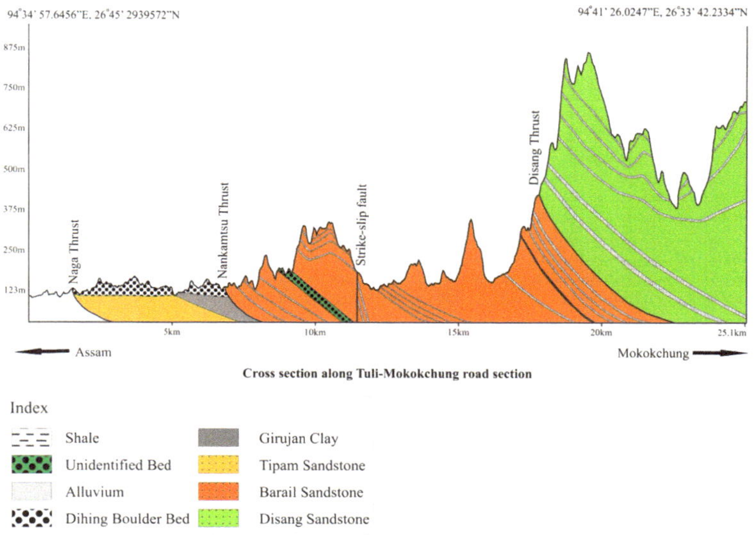


**Figure 2.3** Cross-section of the study area.

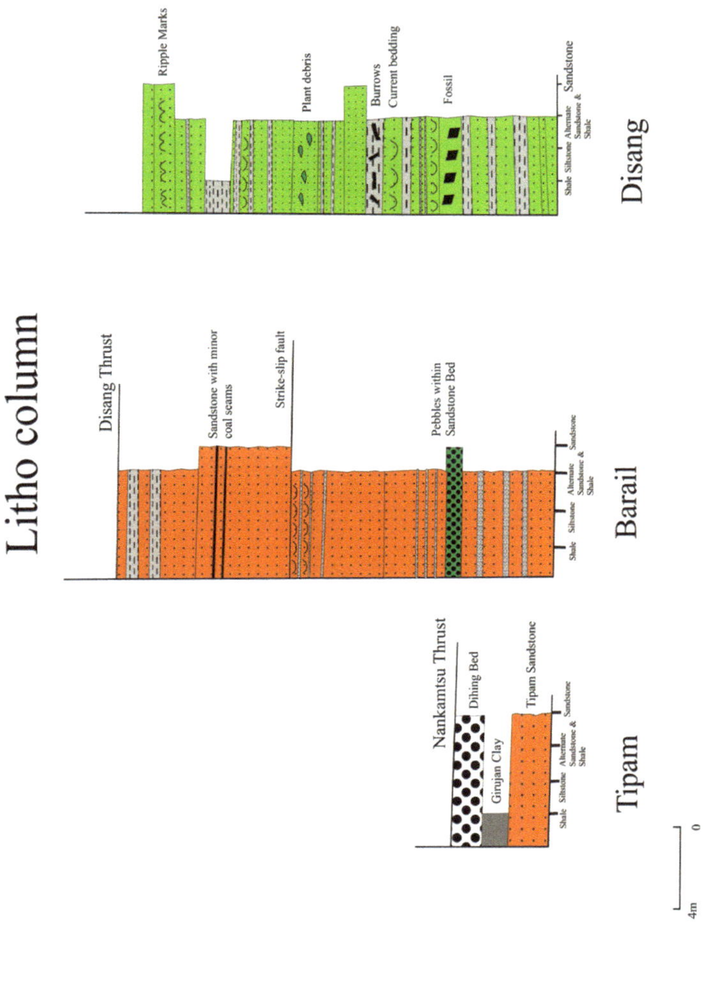


**Figure 2.4** Lithocolumn of the section.

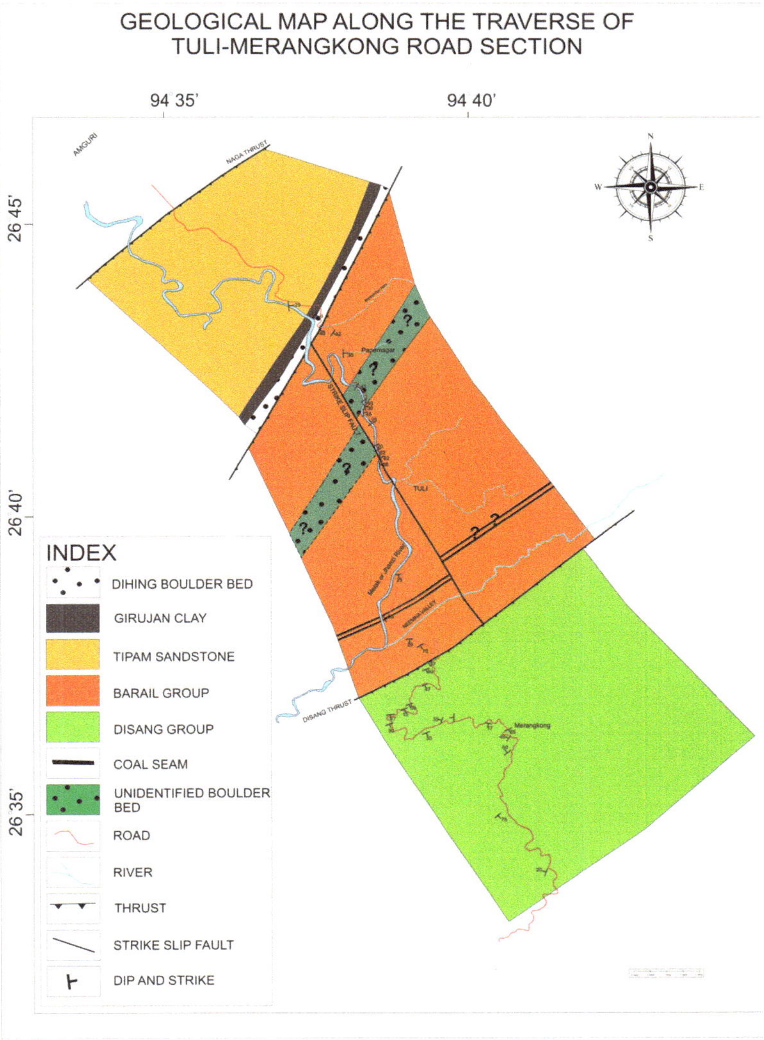


**Figure 2.5** Geological map of the study section.

## 2.4 Depositional Environment

### 2.4.1 Granulometric Analysis

In the present study 14 sandstone samples, Barail (8) and Disang (6) Groups, were sieved using a Ro-Top sieve shaker and standard sieve sets (ASTM: 35, 45, 60, 80, 120, 170, 230, 325 and pan), which measure the weight percentage of different sieve sets. Various grain size parameters are given in Table 2.2.

#### 2.4.1.1 Cumulative curves

Cumulative curves were plotted on the arithmetic scale with the combination of cumulative weight percentage (wt%) on the $Y$ axis and $\phi$ values ($\phi$) on the $X$ axis. These curves provide grain size information values of $\phi$, which is used in the mathematical calculation of statistical parameters (graphic mean, $M_Z$; graphic standard deviation, $\sigma_1$; graphic skewness, $S_K$; and graphic kurtosis, $K_G$) proposed by Folk and Ward (1957) [13]. When plotted on the arithmetic scale, cumulative curves typically show the S-shaped trend. The slope of the central portion of the curve indicates sorting of the sample; if there is a very gentle slope, it indicates poor sorting, and a very steep slope means good sorting.

**Table 2.2** Textural parameters of the samples

| Litho unit | Sample | $M_Z$ | $\sigma_1$ | Sorted | $S_{Ki}$ | $K_G$ | $C^*$ | $C^{**}$ |
|---|---|---|---|---|---|---|---|---|
| | 9 | 2.87 | 0.82 | poorly | −0.1 | 1.31 | 1.21 | 0.62 |
| | 10 | 1.35 | 1.03 | poorly | −0.05 | 1.24 | 0.97 | 0.98 |
| | 12 | 2.32 | 0.94 | moderately | −0.02 | 1.09 | 1.05 | 2.27 |
| Barail gp. | 13 | 2.41 | 0.87 | moderately | 0.35 | 1.12 | 1.15 | 0.35 |
| | 14 | 2.57 | 0.94 | moderately | −0.02 | 1.1 | 1.06 | 0.78 |
| | 16 | 2.13 | 0.84 | moderately | −0.05 | 1.07 | 1.19 | 0.35 |
| | 17 | 2.10 | 0.74 | moderately | 0.10 | 1.13 | 1.35 | 7.87 |
| | 22 | 2.22 | 1.15 | poorly | 0.12 | 0.65 | 0.87 | 0.23 |
| | 25 | 1.85 | 1.2 | poorly | −0.26 | 0.93 | 0.83 | 0.72 |
| | 27 | 0.93 | 1.9 | poorly | −0.05 | 0.95 | 0.52 | 3.67 |
| Disang gp. | 28 | 2.8 | 1.33 | poorly | −0.54 | 0.651 | 0.75 | 8.88 |
| | 31 | 2.18 | 1.87 | poorly | −0.21 | 0.67 | 1.15 | 2.85 |
| | 37 | 1.97 | 1.25 | poorly | −0.23 | 0.75 | 0.8 | 0.59 |
| | 39 | 1.75 | 1.07 | poorly | −0.09 | 1.17 | 0.93 | 0.76 |

$C^*$: $\sqrt{\sigma}^{-2}$; $C^{**}$: $(K_G\sigma^2)/M_z$; p: poorly; m: moderately.

### 2.4.2 Relationship between Textural Parameters

#### 2.4.2.1 Kurtosis ($K_G$) vs. skewness ($S_K$)

The $K_G$ Versus $S_K$ plot after [14] is considered as a strong tool for interpreting the depositional condition of sediments. The plot indicates that the Disang sandstones are beach-deposited (Fig. 2.6), while the Barail sandstones are river-deposited (Fig. 2.7).

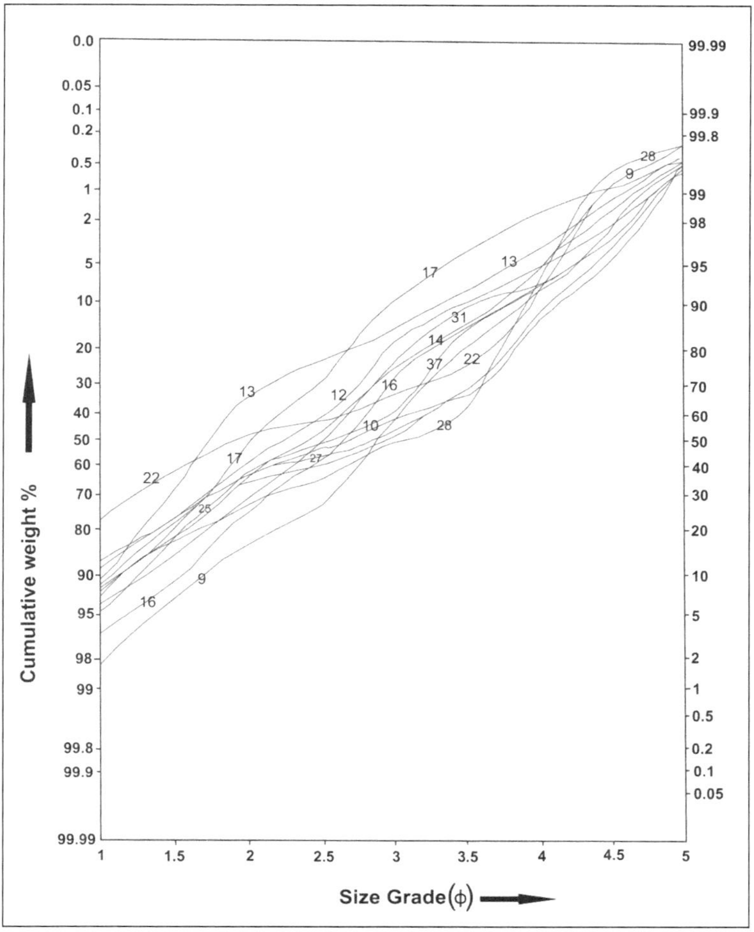


**Figure 2.6** Cumulative curves.

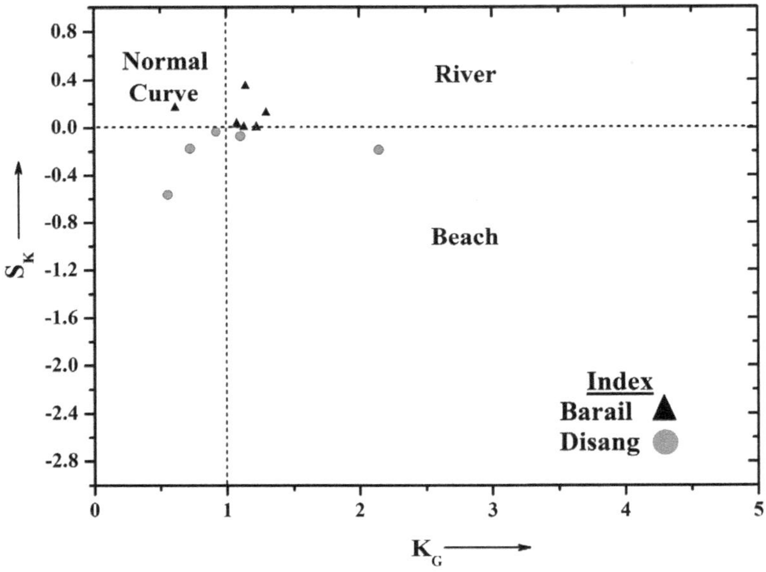


**Figure 2.7** Kurtosis vs. skewness binary plot [6].

$S_K$ versus $K_G$

Another bivariate plot of $S_K$ and the $K_G$ plot by Thompson et al. (1972) demarcates the energy condition prevailing at the time of deposition along with the particle size. The analysed samples range from a mixture of fine to medium range. The energy conditions as depicted by the plot show the condition to be high, hereby indicating that the sediments were deposited in a marine basin, as suggested by high-energy conditions (Fig. 2.8).

Sahu (1964) figured out a plot for interpreting the depositional environment on the basis of grain size values by using the bivariate plot of $\sqrt{\sigma}^{-2}$ versus $(K_G)/M_Z$ [15].

Barail sandstones while plotting in a log–log plot (Fig. 2.9). It was observed that most of the samples were deposited in deltaic conditions, followed by Disang sandstones deposited in shallow marine conditions.

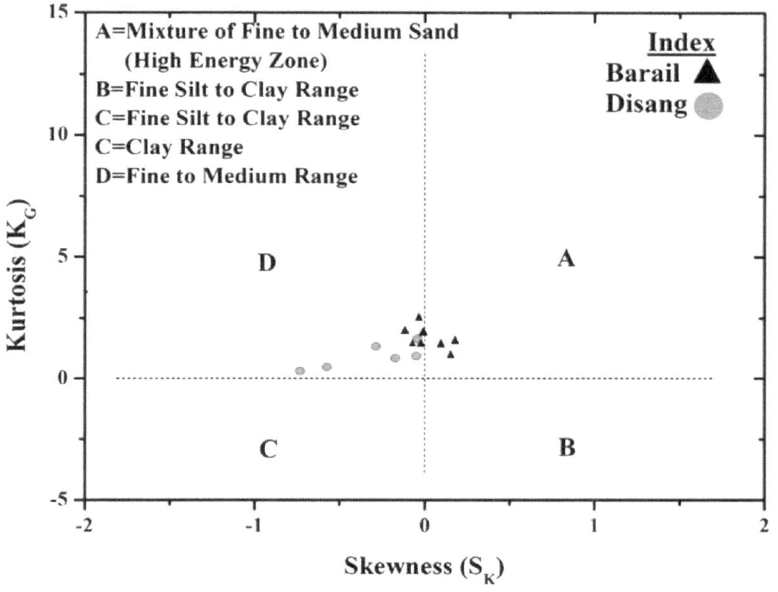


**Figure 2.8** Skewness vs. kurtosis binary plot [20].

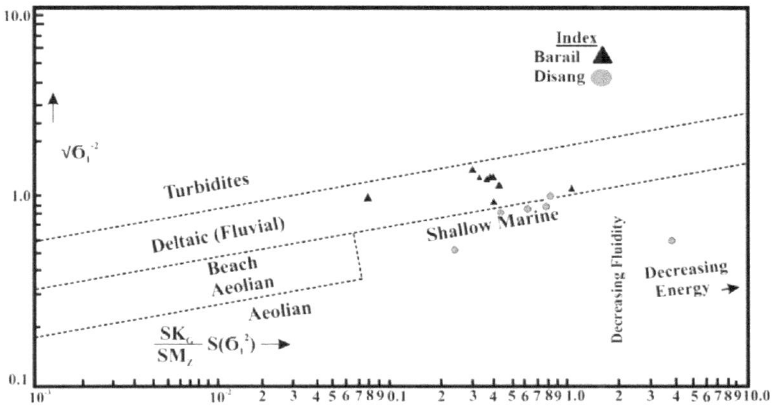


**Figure 2.9** Log–log plot [15].

**Figure 2.10** CHSNO Analyser at CoEES (Oil India Limited), Guwahati.

## 2.5 Evaluation of the Hydrocarbon Source Potential

### 2.5.1 Elemental Analysis

Elemental analysis was carried out on 20 shale samples from the Barail (9 nos) and Disang (11 nos) Groups to determine the weight% of the carbon (C), hydrogen (N), nitrogen (N) and sulphur (S) present in the samples.

Elemental analysis results (Table 2.3) indicate that the carbon content of samples SN-09, SN-10, SN-25, SN-29, SN-33 and SN-34 is more than or equal to 1%. Further, rock-eval pyrolysis was carried out for these samples. The total sulphur versus TOC plot (Fig. 2.11) shows that the Barail shale units are deposited in a freshwater lacustrine environment, whereas the Disang shale units are deposited in a marine environment [16].

**Table 2.3** Elemental analysis results

| Litho unit | Sample | N% | C% | H% | S% |
|---|---|---|---|---|---|
| | 03 | 0.02 | 0.1 | 0.62 | 0.03 |
| | 06 | 0.9 | 0.07 | 0.82 | 0.01 |
| | 09 | 0 | 0.87 | 0.23 | 0.28 |
| | 10 | 0.01 | 6.87 | 1.56 | 2.66 |
| Barail gp. | 11 | 0 | 0.35 | 0.46 | 0.19 |
| | 20 | 0.08 | 0.27 | 0.71 | 0.01 |
| | 21 | 0 | 0.33 | 0.74 | 0.17 |
| | 22 | 0 | 0 | 1.02 | 0.05 |
| | 23 | 0 | 0.82 | 0.71 | 0.16 |
| | 24 | 0.03 | 0.39 | 1.23 | 0.02 |
| | 25 | 0 | 0.99 | 0.57 | 0.74 |
| | 26 | 0 | 0.72 | 1.21 | 0.91 |
| | 28 | 0 | 0.58 | 0.91 | 0.23 |
| | 29 | 0.04 | 2.89 | 1.06 | 2.8 |
| Disang gp. | 31 | 0 | 0.32 | 0.91 | 0.12 |
| | 33 | 0.04 | 1.08 | 0.81 | 0.12 |
| | 34 | 0.05 | 1.09 | 0.86 | 0.14 |
| | 35 | 0 | 0.02 | 1.23 | 0.03 |
| | 37 | 0 | 0.33 | 0.89 | 0.26 |
| | 38 | 0.05 | 0.11 | 0.96 | 0.02 |

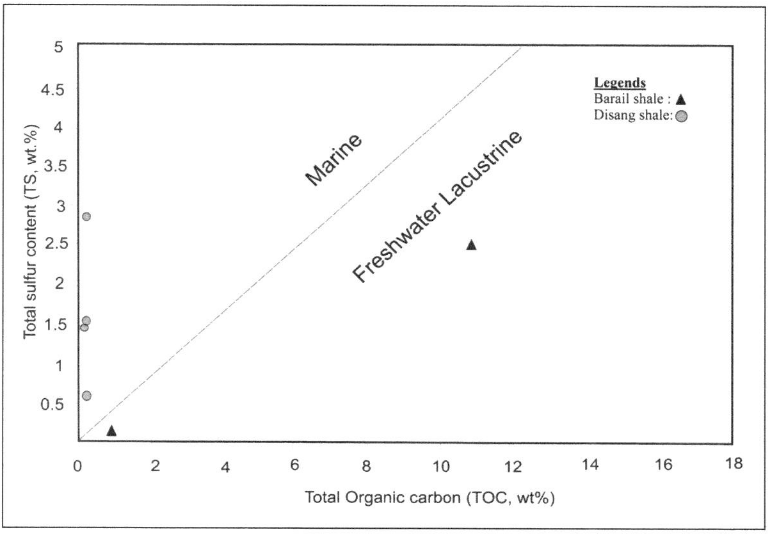


**Figure 2.11** Total sulphur vs. TOC [16].

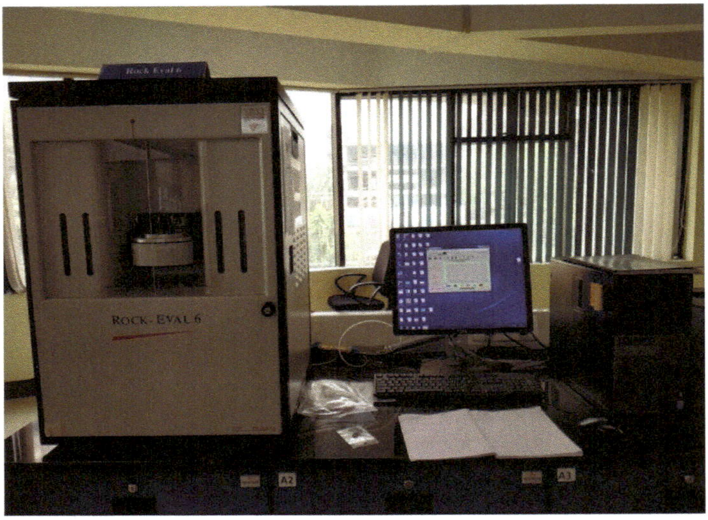


**Figure 2.12** Rock Eval 6 at CoEES (Oil India Limited), Guwahati.

### 2.5.2 Rock-Eval Analysis

For rock-eval analysis, approximately 60 mg of crushed shale rock samples were taken and placed into the Rock Eval 6 instrument. The abundance of free HCs was measured using a flame ionisation detector (FID) and recorded as the S1 peak in units of mg HC/g rock. The sample was then pyrolysed by increasing the temperature of the furnace from 300°C to 550°C or 600°C at the rate of 25°C/min. The amount of HCs (pyrolysate) generated during this period was measured using an FID and reported as the S2 peak. $T_{max}$ was measured in °C at which the maximum rate of pyrolysate generation occurs [11]. Rock-eval analysis was carried on 6 samples from the Barail (2 nos.) and Disang (4 nos.) Groups of rocks using the Rock Eval 6 analysis instrument at CoEES (OIL), Guwahati.

From the rock-eval data in Table 2.4, the TOC values of the Barail Group are 0.88% and 10.71%, which indicates fair and very good source rock potential, respectively. The TOC values of the Disang Group ranged between 0.05% and 1.53%, which is considered as having poor to good source rock potential. $T_{max}$ is the temperature at which the S2 (mg/g rock) peak reaches its maximum amount of HC

**Table 2.4** Rock-eval analysis results

| Litho unit | SN | S2 (mg HC/ g rock) | PI | $T_{max}$ °C | HI (mg HC/g C) | OI (mg $CO_2$/ g C) | TOC (wt%) |
|---|---|---|---|---|---|---|---|
| | 9 | 0.02 | 0.11 | 306 | 2 | 81 | 0.88 |
| Barail gp. | 10 | 0.01 | 0.18 | 313 | 7 | 7 | 10.71 |
| | 25 | 0.34 | 0.28 | 476 | 680 | 780 | 0.05 |
| Disang gp. | 29 | 0.07 | 0.17 | 496 | 140 | 240 | 0.05 |
| | 33 | 0.1 | 1.00 | 496 | 5 | 21 | 1.53 |
| | 34 | 0.2 | 0.34 | 417 | 110 | 225 | 1.44 |

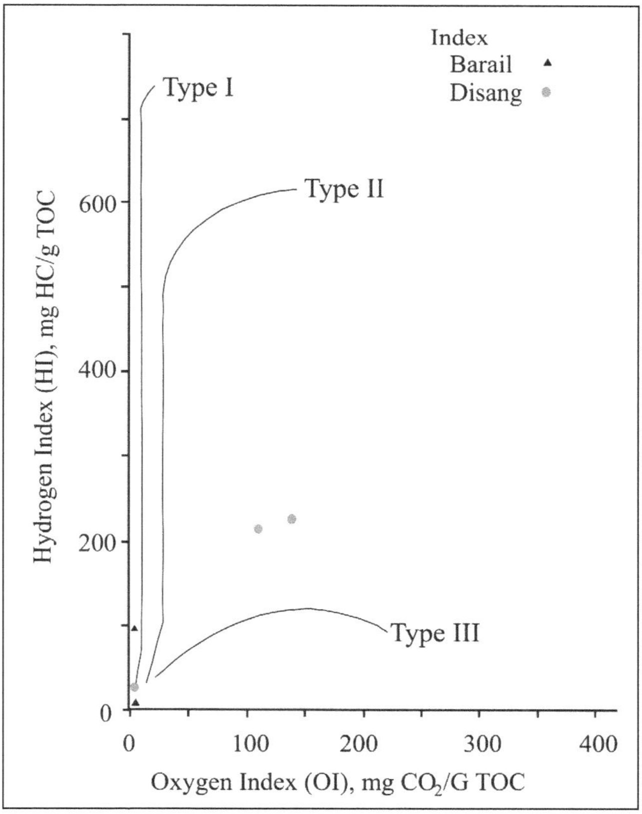


**Figure 2.13** HI vs. OI (Van Krevelen diagram) [18].

generation during rock-eval pyrolysis [17]. Its value increases with increasing maturation of organic matter. A low $T_{max}$ value indicates that the organic matter has not generated a substantial 306°C and 313°C, indicating that the organic matters are in the immature amount of HC [17]. The $T_{max}$ values of the Barail Group were stage. The $T_{max}$ values of the Disang Group ranged between 417°C and 496°C, indicating immature to over mature-source organic matter.

The plot in Fig. 2.13 (Van Krevelen diagram) between HI and OI indicates that the samples are type I and type II variety of organic matter, which are oil- and gas-prone [18]. The plot of production index (PI) versus $T_{max}$ plot (Fig. 2.14) of the shale samples considers that the Disang shale samples are an oil and dry gas zone [19].

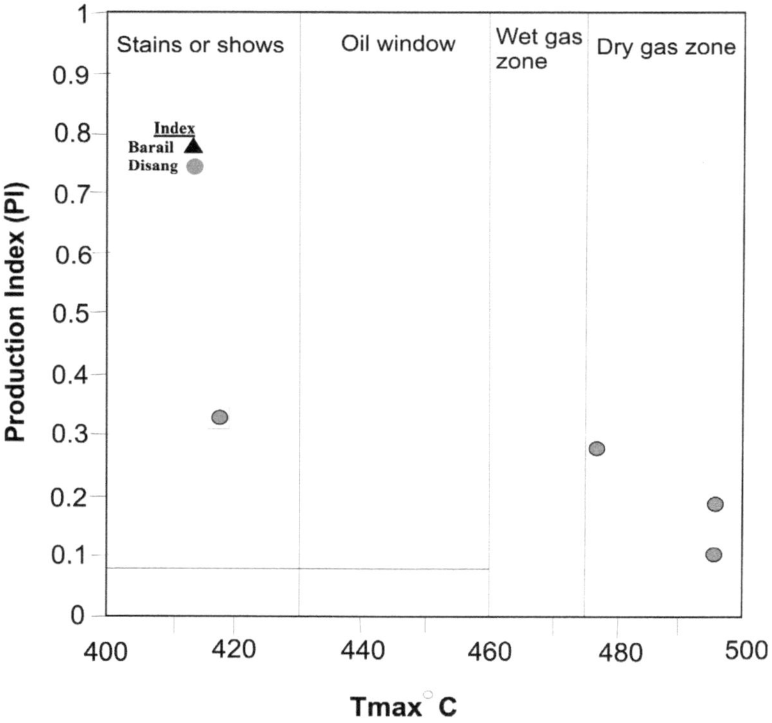


**Figure 2.14** PI vs. $T_{max}$ [19].

## 2.6 Conclusion

The traverse section consists mainly of alternate beds of sandstone where sandstones are dominant. By the granulometric analysis it is seen that the rocks are moderately and poorly sorted. The relationship between the textural parameters shows the Barail and Disang Groups of sediments deposited under river and beach environments, respectively, in high-energy conditions. The plot between total S and TOC also indicates a freshwater lacustrine and marine environment of the Barail and Disang Groups of rocks. The coal seams of the Barail Group indicate deltaic deposits. The S2 and HI values obtained are low, indicating poor HC potential. However, the plot of PI versus $T_{max}$ indicates dry gas potentiality of the Disang shale. The overall source rock analysis deciphers that the Barail and Disang shale units show type I and type II organic matter with some potential of HC gas generation.

## Acknowledgements

We would like to express our sincere gratitude to the Centre of Excellence for Energy Studies (Oil India Limited), Guwahati, for assisting us financially (Project Contract No. 6111264) and providing laboratory access.

# Chapter 3

# Characterisation of Barail Sandstones of Mizoram: A Sedimentological Approach

**Parakh Protim Phukan,[a] Pradip Borgohain,[a] Bubul Bharali[b] and Devojit Bezbaruah[c]**

[a] *Department of Petroleum Technology, Dibrugarh University, Assam, India*
[b] *Pacchunga University College, Mizoram, Assam, India*
[c] *Department of Applied Geology, Dibrugarh University, Assam, India*
pcube92@gmail.com

Mizoram is situated in the centre salient of the Chittagong–Mizoram–Tripura fold thrust belt of India. In the same belt prolific hydrocarbon (HC) prospects have been identified in Tripura but with very poor results in Mizoram. The present chapter is confined to the lithofacies analysis, granulometry, petrograpHy and porosity determination of the Barail sandstones exposed along the Champhai–Zokhawthar Road section (N23°28.435′, E93°20.007′ and N23°21.792′, E093°23.284′) of Mizoram. In the study area, the Barail sandstone represents alternate beds of sandstone with shale and siltstone. The sandstones are light grey to brown in colour. The lithofacies is characterised by a coarsening upward sequence, which suggests regressive effects of the sea during the deposition of the sediments. Granulometric analysis including various bivariate plots of statistical size parameters suggests

*Advances in Petroleum Technology*
Edited by Subrata Borgohain Gogoi

ISBN 978-981-4877-19-0 (Hardcover), 978-1-003-04993-7 (eBook)
www.jennystanford.com

shallow marine depositional setting for the Barail sandstones in the study area. A rock thin-section study indicates the sandstones are composed of mainly quartz (54.85%), followed by feldspar (7.22%) and rock fragments (1.67%). The constituent grains are mainly subhedral to anhedral in shape and cemented in siliceous, argillaceous and calcareous matter. They exhibit a moderate sorting texture. Certain diagenetic alterations like dissolution of framework grains by calcareous cement, surficial and peripheral replacement of quartz grains by argillaceous cement and fracturing of grains enhances the secondary porosity of the rock. However, quartz overgrowth and the formation of secondary minerals like authigenic mica, in certain cases, reduce the porosity. The average matrix concentration (14.47%) upholds the presence of wacke-type sandstones. Some of the rock fragments lead to the formation of phyllite clasts. Feldspars usually retain their shape, while some of them get dissolved and alter to clayey substance. The average porosity of 10.89%, obtained using a helium porosimeter, constitutes a moderate-quality reservoir rock for HC accumulation. From the overall analysis it can be concluded that the Barail sandstone in Mizoram deposited in a shallow marine environment, has undergone severe diagenetic alteration and bears a moderate reservoir quality.

## 3.1 Introduction

Mizoram covers an area of about 25,000 km$^2$. Almost 5000 m of a thick tertiary sediment sequence is exposed in Mizoram. Mizoram bears a mountainous terrain, and it is bounded by Tripura and the Chittagong Hill tracts (Bangladesh) in the west, in the north by Assam and Manipur, in the east by Chin Hills and in the south by Arakan Hill tracts (Myanmar). The Mizo fold belt is composed of a series of longitudinal folds arranged en-echelon. The anticlines are long, narrow and tight, but the intervening synclines are broad and gentle. The geological investigation in Mizoram is meagre owing to the inaccessibility of the terrain. Earlier geological investigations were carried out by different researchers [21–23]. Ganju (1975) along with several researchers provided valuable

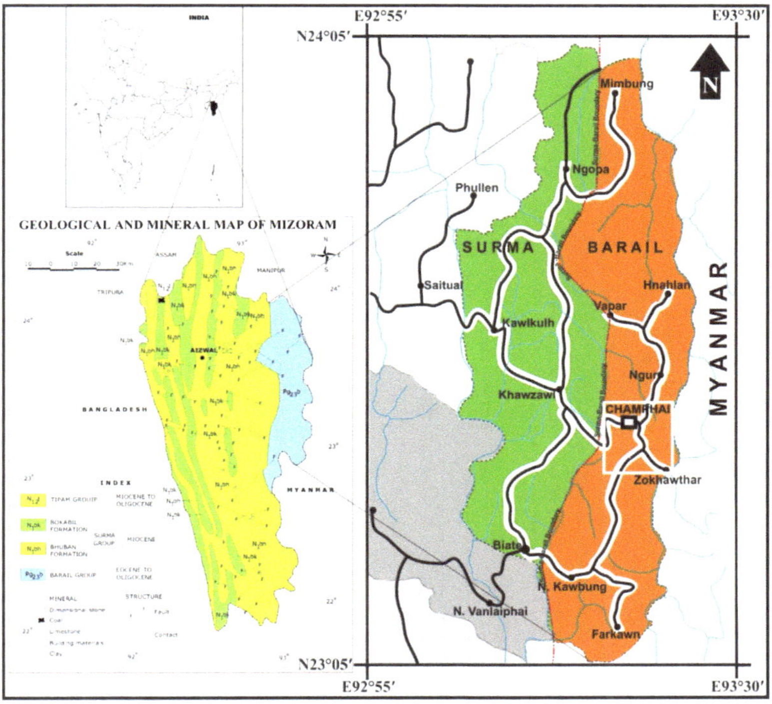


**Figure 3.1** Geological map of Mizoram, and the studied section is shown in a box.

information about the regional geological framework, including the tectonics, paleontology and magnetostratigrapHy of the region [24–29]. The geological map and stratigraphic succession of Mizoram is represented in Fig. 3.1 and Table 3.1 respectively.

Mizoram is known to house the Surma Group of rocks, although in the easternmost part of Mizoram, it is known to house the Barail Group of rocks in the Champhai district. Our study was carried out in the Champhai–Zokhawthar Road section of the Champhai district. Lithologically the Barails exposed in Nagaland and Upper Assam are quite different. And we know that the Barails of Upper Assam bears a good amount of hydrocarbon (HC). In this work we tried to decipher the depositional environment under which the Barails of Mizoram got deposited, calculate the average porosity of the Barail Group of rocks and comprehend the role of

**Table 3.1** Generalised geological succession of Mizoram [24, 31]

| Age | Group | Subgroup | Formation | Thickness (m) | Generalised lithology | Depositional environment |
|---|---|---|---|---|---|---|
| Recent Alluvium | | | | | Silt, clay and gravel | River deposits |
| Early Piliocene to late Miocene | Tipam | | | +900 | Friable sandstone with occasional clay bands | Stream deposits |
| Miocene to Upper Oligocene | SURMA | Bokabil | | +950 | Shales with siltstones and sandstones | Shallow marine |
| Miocene to Upper Oligocene | SURMA | Bhuban | Upper Bhuban | +1100 | Arenaceous with sandstones, shalesand siltstones | Shallow marine, near shore to lagoonal |
| Miocene to Upper Oligocene | SURMA | | Middle Bhuban | +3000 | Arenaceous with shalesand siltstones | Deltaic complex |
| Miocene to Upper Oligocene | SURMA | | Lower Bhuban | +900 | Arenaceous with sandstones and silty shales | Shallow marine |
| Oligocene | Barail | | | +3000 | Shales, siltstones and sandstones | Shallow marine |

various mineralogical constituents in controlling the petrophysical properties. The relation and intensity of the constituent minerals have a direct intervening effect on porosity. Petrological data will allow us to define the reservoir petrofacies and conclude on the reservoir quality of the Barail Group of Mizoram.

## 3.2 Field Characteristics

The Champhai–Zokhawthar Road section, N23°28.435, E93°20.007′ and N23°21.792′, E093°23.284′, bears an average elevation of 1300 m (MSL). This section consists of alternate beds of sandstone, shale and siltstone (Fig. 3.2a). Sandstones are fine- to medium-grained and attain a thickness of about 2 m at some places. Shales are grey in colour and alternate with siltstones. Shale clasts are also observed within the sandstone bed. Coarsening upward sequences are commonly observed throughout the section, where the fine part is dominated by shale at the base, which gradually changes to shale/siltstone and finally to sandstone (Fig. 3.3). Laminations within sandstone beds, showing cross-stratification, are observed

**Figure 3.2** Exposures showing alternate beds of sandstone, shale and siltstone.

**Figure 3.3** Presence of a coarsening upward sequence of sandstone, siltstone and shale

**Figure 3.4** Laminations showing cross-stratification.

**Figure 3.5** Presence of a ripple mark.

(Fig. 3.4). Ripples marks are also observed at places (Fig. 3.5). A representation of a geological map and a vertical lithocolumn of the studied section are shown in Figs. 3.6 and 3.7, respectively.

## 3.3 Methodology

Lithofacies analysis, depending on the field data and geological map and lithocolumn, was done (Figs. 3.6 and 3.7).

For grain size analysis 100 g of representative samples was crushed using an agate mortar and pestel. Then 100 g of each sample was sieved using standard sieve sets in a sieve shaker. The weight percentages thus obtained were used for calculating the data required for grain size analysis. Petrogarphic analysis of rock thin sections was done with a Leica DMLP polarising microscope fitted with a camera in the Department of Petroleum Technology, Dibrugarh University, Assam. The modal analysis was done following the Gazzi–Dickinson (1966) [30] method. A minimum of 500 counts were done for each of the rock thin slides. The outcrop samples collected from the field were utilised for preparation of core

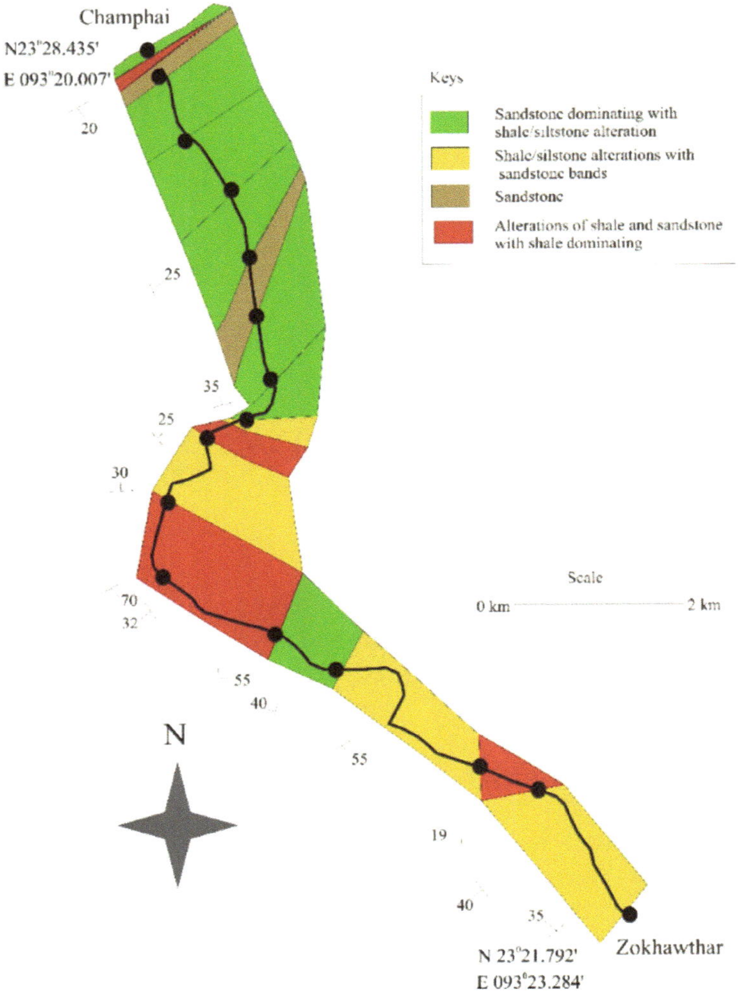


**Figure 3.6** Geological map of the Champhai–Zokhawthar Road section.

samples. A total of five representative field samples were selected for preparation of the core samples. Then, 1.5 in core samples were required for carrying out the experiments. The cores were obtained using the core-cutting machines. Porosity calculation was carried out using a helium porosimeter. The instrument used was TPI-2198.

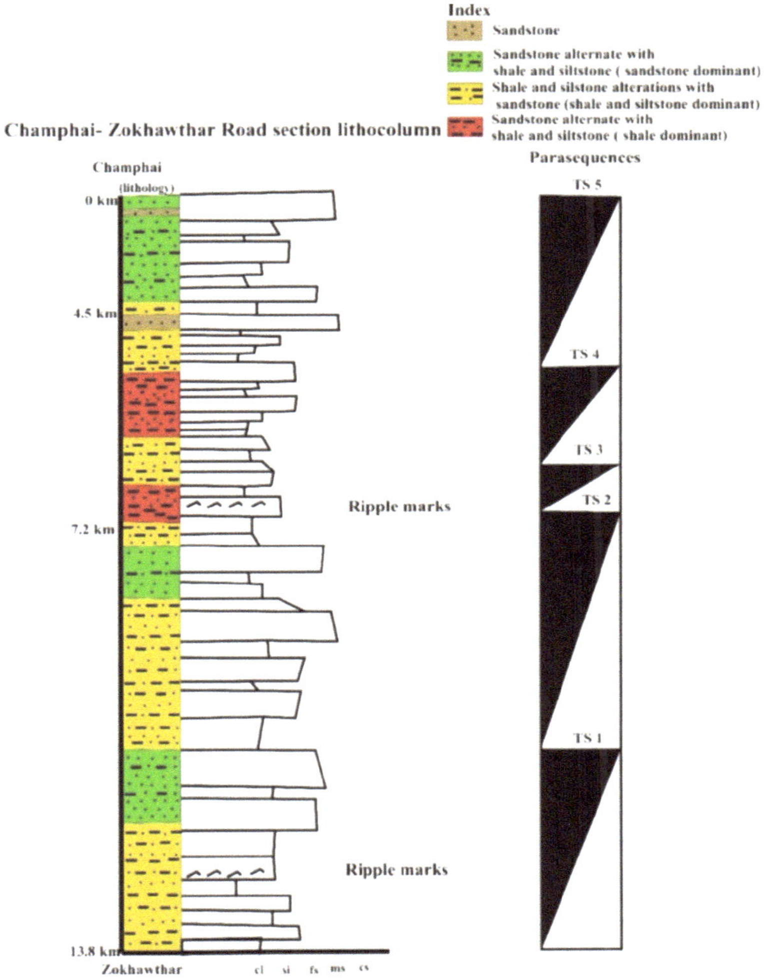


**Figure 3.7** Lithocolumn of the Champhai–Zokhawthar Road section.

## 3.3.1 Depositional Environment

### 3.3.1.1 Lithofacies analysis

The Champhai–Zokhawthar Road section consists of alternate beds of sandstone, shale and siltstone. The lithocolumn prepared for the section shows an overall coarsening upward sequence. For

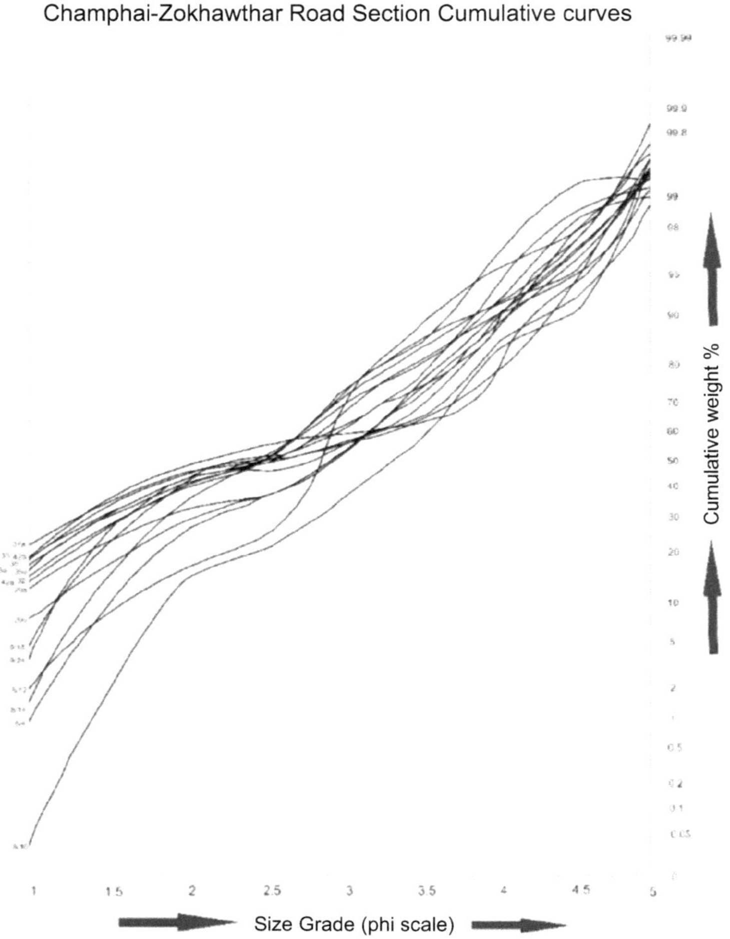


**Figure 3.8** Cumulative curves of the Champhai–Zokhawthar Road section.

convenience the lithocolumn (Fig. 3.7) was subdivided into five parasequences, namely TS 1, TS 2, TS 3, TS 4 and TS 5. Each of the five parasequences showed a coarsening upward sequence individually. In TS 1 it was seen that the base was dominated by shale and siltsone alteration, and gradually it was overlaid by sandstone dominated sequence. TS 2 bore a similar characteristic as TS 1. TS 3, TS 4 and TS 5 similarly showed the same lithological variations, which is indicative of the sequences to be coarsening upward sequences. The regressive nature of the sea was evident from such coarsening upward sequences. The coarsening upward sequences and regressive nature conformed with the deposition of sediments in marine conditions.

#### 3.3.1.2 Granulometric analysis

A total of 16 representative samples were taken into consideration for granulometric analysis. On the basis of granulometric data the cumulative wt.% versus grain size curves were prepared which showed an S-type slope, as represented in Fig. 3.8. On visual analysis of the curves it was evident that the tail ends of the curves showed a steep trend as compared to the central portion of the curves. So, better sorting was observed at the tail ends, whereas the central portion tended to poor sorting of the grains. The Barail Group of rocks shows a moderate nature of sorting. Univariate grain size parameters, graphic mean, graphic standard deviation, graphic skewness and graphic kurtosis, were calculated (Table 3.2). These univariate parameters were then utilised in various bivariate plots for deciphering the depositional environments of the Barail Group of rocks of Mizoram.

In a bivariate plot of skewness versus kurtosis [14] (Fig. 3.9), although showing scattered plots, most of the samples clustered in the beach field. A plot of skewness versus mean grain size [14], as shown in Fig. 3.10, also shows the clustering of the samples in the beach environment. A log–log plot of the mean $\phi$ deviation against the ratio of standard deviation of kurtosis to standard deviation of mean size times standard deviation (Sahu, 1964) showed the trend of the samples to be deposited in shallow marine conditions. To distinguish the sediments from eolian, beach, shallow-agitated marine and fluvio-deltaic conditions [15], (Fig. 3.11) proposed the linear discriminant functions Y1, Y2, Y3 and Y4 (Table 3.3). Y1 shows the sediments to deposited in beach conditions, Y2 in shallow-agitated marine conditions, Y3 in shallow marine conditions and Y4 in turbidity currents. On the basis of the results of the field evidence and granulometric analysis, it can be inferred that the sediments

**Table 3.2** Univariate grain size parameters, $M_Z$, $\sigma$, $S_{Ki}$ and $K_G$ of the Champhai–Zokhawthar Road section

| S. no. | $M_Z$ | $\sigma_1$ | $S_{Ki}$ | $K_G$ |
|---|---|---|---|---|
| CZ 8/12 | 2.68 | 1.20 | −0.032 | 3.255 |
| CZ 8/4 | 2.49 | 0.89 | 0.079 | 1.493 |
| CZ 8/10 | 3.11 | 0.94 | −0.942 | 1.418 |
| CZ 37 | 2.36 | 1.27 | −2.048 | 3.445 |
| CZ 42 | 2.50 | 1.29 | −1.065 | 3.688 |
| CZ 32 | 2.38 | 1.10 | −1.292 | 2.27 |
| CZ 29 | 2.68 | 1.19 | −2.904 | 2.744 |
| CZ 33 | 2.31 | 1.40 | 3.611 | 4.547 |
| CZ 43 | 2.44 | 1.41 | 1.857 | 4.420 |
| CZ 35 | 2.34 | 1.22 | −0.873 | 2.393 |
| CZ 9/15 | 2.49 | 1.14 | 1.439 | 2.783 |
| CZ 9/24 | 2.56 | 0.87 | −1.520 | 0.68 |
| CZ 8/14 | 2.79 | 1.006 | −0.622 | 2.082 |
| CZ 36 | 2.31 | 1.25 | −0.32 | 3.114 |
| CZ 42 | 2.43 | 1.23 | −0.24 | 2.909 |
| CZ 29b | 2.74 | 1.14 | −1.665 | 2.618 |

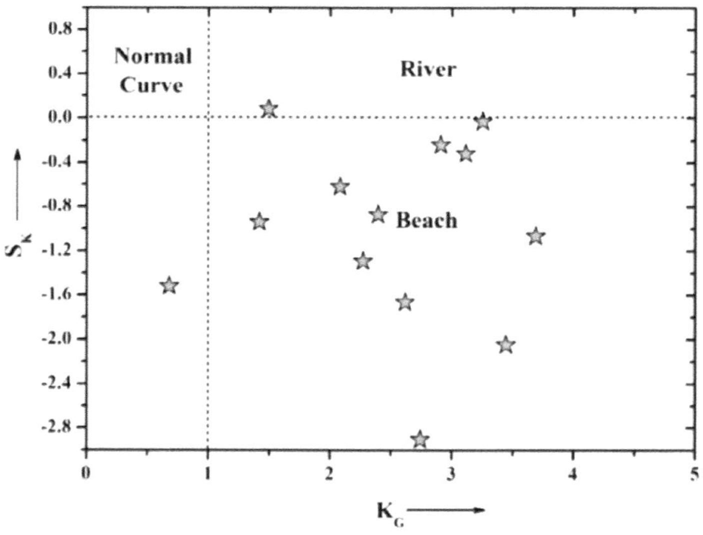


**Figure 3.9** Plot of $S_{Ki}$ vs. $K_G$ (Friedman, 1961).

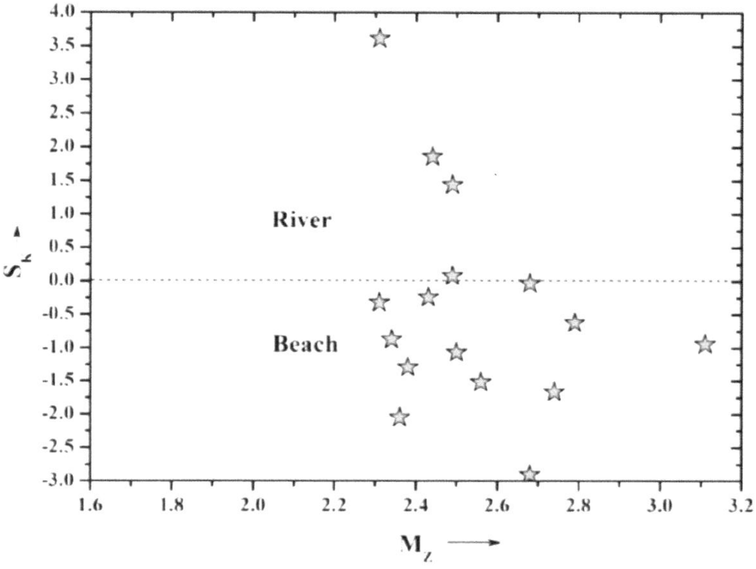


**Figure 3.10** Plot of $S_{Ki}$ vs. $M_Z$ grain size (Friedman, 1961).

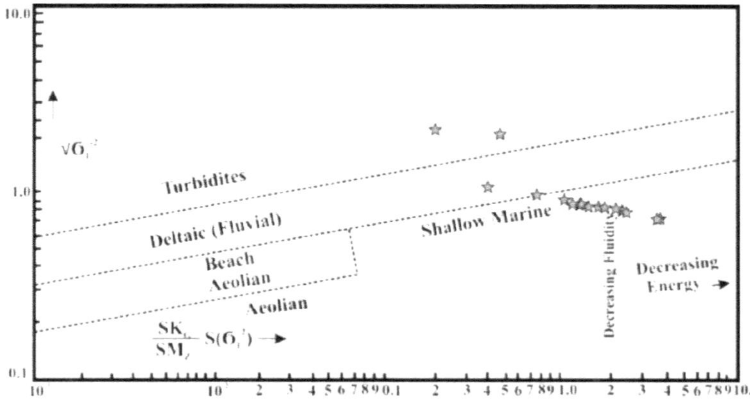


**Figure 3.11** Log–log plot of $M_Z$ $\phi$ deviation against $M_Z$ size times $\sigma$ (Sahu, 1964).

**Table 3.3** Linear discriminant functions Y1, Y2, Y3 and Y4 proposed by Sahu (1964)

| Sample | Y1 | Y2 | Y3 | Y4 |
|---|---|---|---|---|
| CZ8/12 | 5.9668 (B) | 196.2242 (SAM) | −11.5371 (F) | 18.3656 (FD) |
| CZ8/4 | −1.4698 (B) | 120.0825 (SAM) | −6.5435 (SM) | 8.4681 (TC) |
| CZ8/10 | 3.11 (B) | 115.9248 (SAM) | −2.1759 (SM) | 3.0510 (TC) |
| CZ37 | 12.5268 (B) | 169.5882 (SAM) | −3.2692 (SM) | 5.4985 (TC) |
| CZ42 | 10.9319 (B) | 197.4398 (SAM) | −8.4761 (F) | 13.4828 (FD) |
| CZ32 | 5.7358 (B) | 135.3734 (SAM) | −3.4898 (SM) | 4.5459 (TC) |
| CZ29 | 10.2513 (B) | 133.1945 (SAM) | 2.7008 (SM) | −3.6642 (TC) |
| CZ33 | 5.6794 (B) | 314.4729 (SAM) | −33.9617 (SM) | 49.2526 (FD) |
| CZ43 | 8.5567 (B) | 284.2444 (SAM) | −25.5942 (SM) | 36.8546 (FD) |
| CZ35 | 6.4219 (B) | 162.9036 (SAM) | −7.9845 (F) | 7.8767 (TC) |
| CZ9/15 | 1.6009 (B) | 201.9260 (SAM) | −17.5821 (F) | 25.69 (FD) |
| CZ9/24 | −1.0607 (B) | 74.8680 (SAM) | 1.5698 (SM) | −5.0918 (TC) |
| CZ8/14 | 1.5631 (B) | 137.4362 (SAM) | −4.9261 (SM) | 8.4371 (TC) |
| CZ36 | 7.8997 (B) | 190.6579 (SAM) | −11.3133 (F) | 15.3641 (FD) |
| CZ42 | 5.5303 (B) | 186.9323 (SAM) | −11.2459 (F) | 14.9242 (FD) |
| CZ29b | 6.6407 (B) | 146.5817 (SAM) | −2.3301 (SM) | 4.1003 (TC) |

B: beach; SAM: shallow-agitated marine; F: fluvial; SM: shallow marine; FD: Fluvio Deltaic; TC: turbidity current.

of the Barail Group of rocks were deposited in a shallow marine environment.

### 3.3.2 Porosity Determination

The five core samples prepared showed porosity in the range of 5.9% to 13.06%. Though the minimum percentage obtained was 5.9%, the remaining five samples showed porosity well above 10% (Table 3.4). On the basis of the results obtained, the average porosity

**Table 3.4** Results of porosity

| S. no. | Porosity (%) |
|---|---|
| CZ 42 | 11.6 |
| CZ 29 | 13.06 |
| CZ 37 | 11.5 |
| CZ 39 | 5.9 |
| CZ 33 | 12.4 |

calculated was 10.89%. Such percentage of porosity indicates the reservoir to be of moderate quality. Though the retention capacity of the reservoir is not that high, but it would act as a fair reservoir for HC accumulation.

## 3.3.3 Petrography

### 3.3.3.1 Observations

A thin slide study of 16 sandstone samples showed the presence of quartz, feldspar, rock fragments, micaceous minerals and others as framework minerals (Table 3.5). The matrix and cements constituted the background materials. A rock thin-section study showed that the framework grains are predominantly composed of quartz (average 54.85%), feldspar (average 7.22%) and rock fragments (average 1.67%). The other detrital minerals include muscovite, biotite, chlorite and heavy minerals. The cementing materials are mainly siliceous and argillaceous with a minor content of calcareous

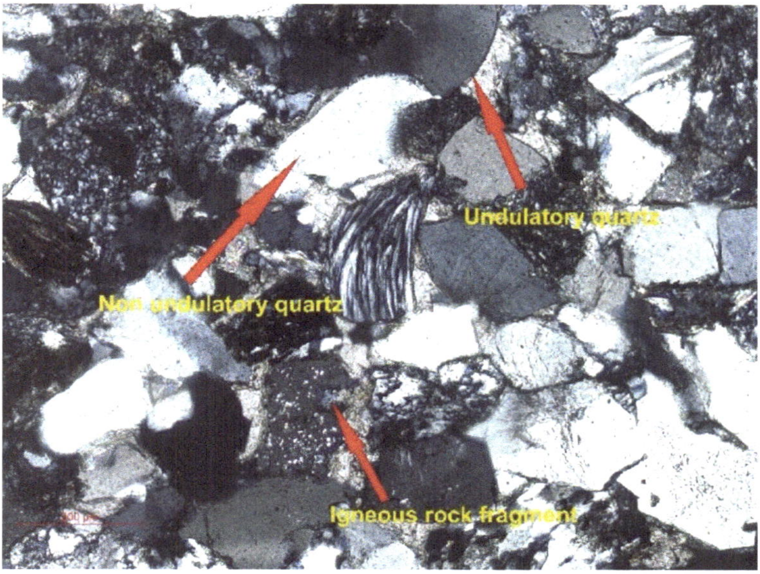

**Figure 3.12** Non-undulatory quartz, undulatory quartz and igneous rock fragment.

**Table 3.5** Modal count of petrographic study of thin section

| S. no. | $Q_{Mu}$ | $Q_{Mnu}$ | $Q_{P\text{-}2\text{-}3}$ | $Q_{P\geq 3}$ | $P_{Ca/Na}$ | $F_K$ | $RF_{Ig.}$ | $RF_{Sed.}$ | $RF_{Met.}$ | $C_{Sil}$ | $C_{Arg}$ | $C_{Fer}$ | $C_{Cal.}$ | Mi | Ma | Ch |
|---|---|---|---|---|---|---|---|---|---|---|---|---|---|---|---|---|
| CZ8 | 19.97 | 14.21 | 5.58 | 10.49 | 7.11 | 1.86 | 0 | 1.02 | 4.06 | 3.05 | 8.80 | 0 | 0 | 6.09 | 13.87 | 3.89 |
| CZ4 | 29.94 | 13.63 | 2.50 | 8.06 | 3.45 | 0 | 0 | 1.34 | 2.69 | 4.03 | 4.03 | 0 | 0 | 9.02 | 15.93 | 5.37 |
| CZ3 | 28.96 | 21.04 | 1.98 | 6.65 | 3.96 | 0 | 0 | 1.26 | 7.01 | 5.04 | 6.12 | 0 | 0 | 0.54 | 17.45 | 0 |
| CZ24 | 19.47 | 23.43 | 1.98 | 5.45 | 8.58 | 1.82 | 0 | 0 | 6.27 | 1.98 | 0 | 0 | 14.52 | 5.28 | 9.41 | 1.82 |
| CZ14 | 28.65 | 13.85 | 8.92 | 8.92 | 4.17 | 2.09 | 0 | 1.71 | 5.12 | 2.09 | 3.98 | 0 | 0 | 4.74 | 15.75 | 0 |
| CZ23 | 25.42 | 17.28 | 4.65 | 7.14 | 6.31 | 1.99 | 0 | 0 | 4.49 | 2.16 | 3.65 | 1.00 | 0 | 8.80 | 17.11 | 0 |
| CZ7 | 18.77 | 24.40 | 4.44 | 7.17 | 8.02 | 1.19 | 0 | 0 | 5.46 | 1.19 | 4.44 | 0.68 | 0 | 8.02 | 14.16 | 2.05 |
| CZ31 | 19.93 | 24.95 | 1.86 | 5.59 | 8.38 | 1.86 | 0 | 0 | 5.96 | 2.23 | 0 | 0 | 13.04 | 5.21 | 9.12 | 1.86 |
| CZ35 | 21.39 | 9.70 | 6.95 | 9.83 | 5.50 | 0 | 0.50 | 0 | 7.60 | 1.18 | 4.33 | 0 | 12.84 | 4.46 | 8.26 | 7.47 |
| CZ32 | 27.37 | 19.93 | 1.49 | 6.33 | 5.03 | 0.74 | 0 | 0 | 8.01 | 3.17 | 2.42 | 0 | 0 | 6.89 | 14.53 | 4.10 |
| CZ29 | 31.93 | 16.63 | 2.68 | 8.60 | 6.69 | 1.34 | 0 | 0 | 5.35 | 3.06 | 4.97 | 0 | 0 | 5.74 | 10.13 | 2.87 |
| CZ36 | 30.92 | 19.96 | 1.37 | 5.48 | 6.26 | 2.74 | 0 | 0 | 4.89 | 3.52 | 7.24 | 0 | 0 | 5.28 | 12.33 | 0 |
| CZ37 | 23.80 | 20.46 | 1.67 | 6.89 | 4.80 | 1.25 | 0 | 0 | 0 | 4.59 | 7.10 | 0 | 0 | 6.68 | 20.46 | 2.30 |
| CZ38 | 26.14 | 19.71 | 2.49 | 6.02 | 5.39 | 1.04 | 0 | 0 | 2.90 | 3.11 | 5.81 | 0 | 0 | 5.81 | 18.26 | 3.32 |
| CZ42 | 26.41 | 19.81 | 3.69 | 6.99 | 4.85 | 1.94 | 0 | 0 | 0 | 2.72 | 6.60 | 0.00 | 0 | 5.83 | 18.45 | 2.72 |
| CZ43 | 26.10 | 20.22 | 3.68 | 7.17 | 5.51 | 1.65 | 0 | 0 | 4.60 | 2.21 | 4.78 | 0 | 0 | 5.88 | 16.36 | 1.84 |

**Quartzs** are $Q_{Mu}$; $Q_{Mnu}$; $Q_{P\text{-}2\text{-}3}$ and $Q_{P\geq 3}$ **Feldspars** are $P_{Ca/Na}$ and $F_K$; Rock
**Rock fragments** are $R_{Fig.}$; $RF_{Sed.}$ and $RF_{Met.}$ **Cements** are $C_{Sil}$; $C_{Arg}$; $C_{Fer}$ and $C_{Cal.}$
**Mica**: Mi; **Matrix**: Ma and **Chert**: Ch

**Figure 3.13** Presence of polycrystalline quartz and calcareous cement.

and ferruginous. Amongst the different detrital varieties, quartz is the most abundant constituent and includes undulatory (average 24.98%) (Fig. 3.12), non-undulatory (average 18.70%) (Fig. 3.12) and polycrystalline varieties of two to three units (average 3.49%) and $\geq$3 units (average 7.29%) (Fig. 3.13). Feldspars (average 7.22%) are mostly subangular to subrounded, and both plagioclase and K-feldspar (Fig. 3.14) varieties were reported. Gneiss and schists constitute the metamorphic rock fragments (Fig. 3.15), whereas shale and chert (Fig. 3.16) constitute the sedimentary rock fragments. Amongst all the different varieties of rock fragments (average 1.67%), the metamorphic types dominate over others, while igneous rock fragments mark their minor presence (Fig. 3.3). Among micas, muscovite dominate over biotite and is observed as being elongated and subhedral in form. However, some chlorite flakes are also seen. Some of the mica flakes exhibit jagged corners with microfractures and kink bends (Fig. 3.17). The matrix (average 14.47%) is present as a complex mixture of finely constituted re-crystallised materials of less than 4 μm filling the intergranular pore

**Figure 3.14** Microcline and quartz overgrowth.

**Figure 3.15** Metamorphic rock fragment and siliceous cement.

Figure 3.16 Chert.

Figure 3.17 Concavo-convex and tangential contact and bending of mica.

spaces. All varieties of cements are recorded, although argillaceous cements are common and observed in most of the slides. The framework grains are mainly subhedral to anhedral in shape and exhibit normal packing with tangential (Fig. 3.17), concavo-convex (Fig. 3.17) and sutured contacts between the grains, depicting a different degree of compaction. Point and long contacts reflect early diagenesis, whereas concavo-convex and sutured contacts point to late diagenetic changes. Floating grains with corroded boundaries may be attributed to replacement of framework grains by cementing materials. Quartz overgrowths are also present (Fig. 3.14). Besides silica cements, iron oxide and calcite cement (Fig. 3.13) are also observed in the Barail sandstones. Silica cement is dominant over the others. Ferruginous cement is recorded as thin coatings around detrital grains and as scattered aggregates of pore-filling cement. Grains are poor to moderately well-sorted. Framework grains show dissolution effects where calcareous cement is present. Heavy mineral inclusions of zircon and rutile are observed in some of the quartz grains. Feldspars usually retain their shape, while some of the grains dissolve or alter at places to clayey substance. Some of the rock fragments leads to the formation of phyllite clasts. Quartz grains and other constituent framework minerals, due to the compaction effect, at many of the places get broken down. This leads to the formation of in situ matrix materials.

### 3.3.3.2 Porosity

Macroporosity is primarily intergranular. But enhancement in the porosity by secondary factors is also well-observed. Intense compaction dissolution of a newly formed pseudomatrix (Fig. 3.18) and quartzgrains leads to secondary porosity enhancement (Fig. 3.19). The presence of calcareous cements leads to a corrosive effect on the framework grains (Fig. 3.20). This leads to enhancement of porosity. Fracture porosity is another factor which contributes to porosity enhancement (Fig. 3.21). Bioturbation effects led to the formation of excavation pore spaces in some samples (Fig. 3.22). Pre-existing casts or moulds or their impressions led to the formation of mouldic porosity in a few samples (Fig. 3.23).

**Figure 3.18** Enhancement of porosity by dissolution of pseudomatrix.

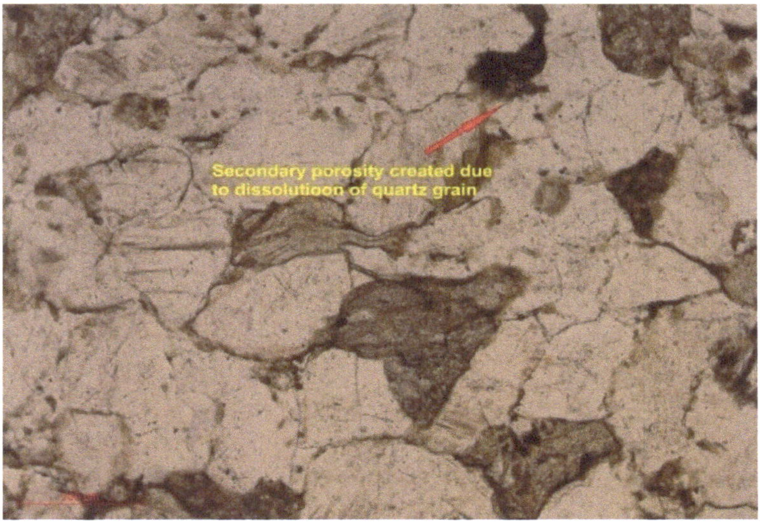


**Figure 3.19** Enhancement of porosity by secondary porosity created due to dissolution of quartz grains.

**Figure 3.20** Enhancement of porosity by calcareous cement dissolving quartz grains.

**Figure 3.21** Enhancement of porosity by fracturing of quartz grains.

**Figure 3.22** Enhancement of porosity by excavation of pore spaces.

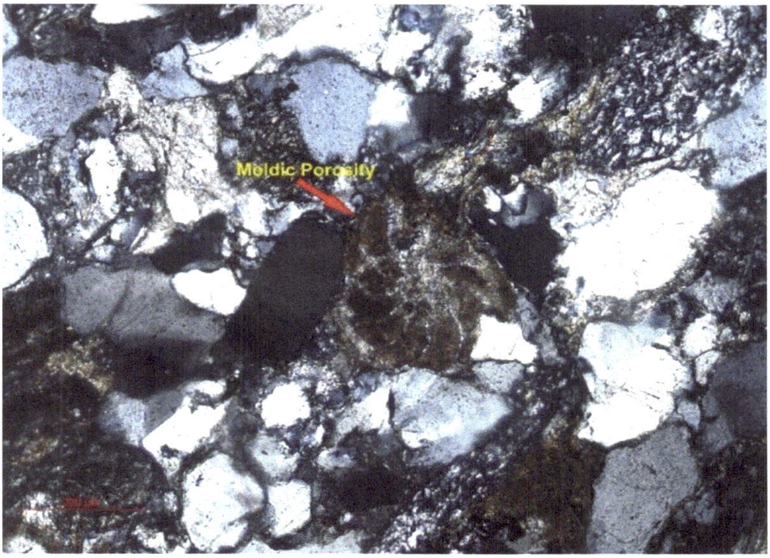


**Figure 3.23** Enhancement of porosity by mouldic porosity.

**Figure 3.24** Reduction of porosity by dissolution of argillaceous cement and intense packing of quartz grains.

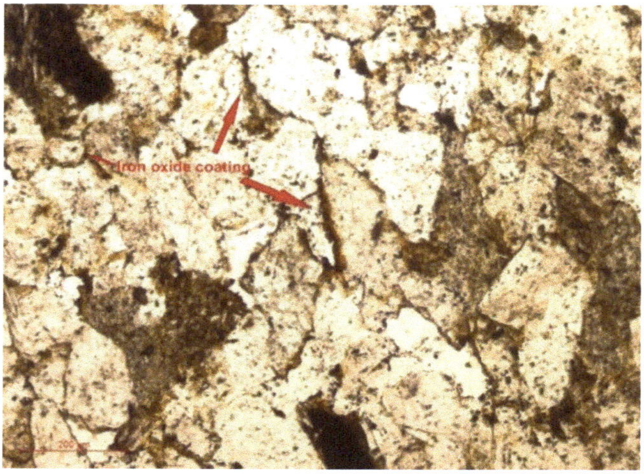


**Figure 3.25** Reduction of porosity by iron oxide coating.

Intense compaction reduces the pore spaces and also dissolves the framework grains, thereby creating more matrix content and reducing the porosity (Fig. 3.24). When compaction becomes intense

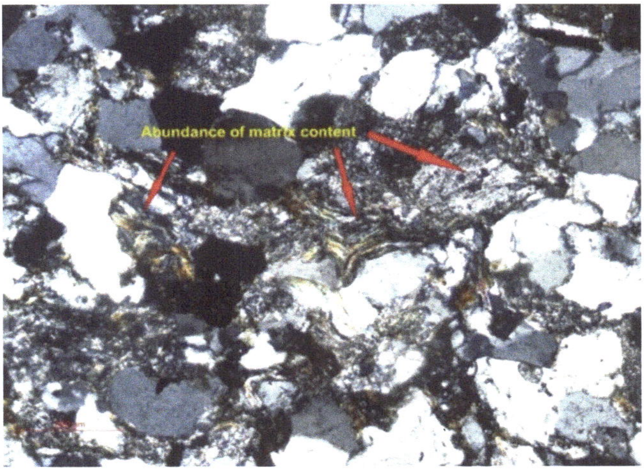


**Figure 3.26** Reduction of porosity by abundance of matrix content.

**Figure 3.27** Reduction of porosity by pore space filled by cementing material.

in mud intraclasts and low-grade metamorphic rocks, it leads to the formation of a pseudomatrix, which clogs porosity (Fig. 3.28). Coating of the grains by iron oxide is another factor which leads to the

**Figure 3.28** Reduction of porosity by abundance of pseudomatrix.

**Figure 3.29** Reduction of porosity by quartz overgrowth.

reduction of porosity (Fig. 3.25). Precipitation of secondary siliceous cements contributes to quartzovergrowth. This leads to the reduction in porosity (Fig. 3.29).

Although the primary macroporosity and secondary porosity contribute in enhancing the porosity, there are factors which also contribute to the reduction in porosity. So, the Barail Group of rocks of Mizoram bears a moderate reservoir quality. The rocks might act as good reservoir rocks.

## 3.4 Conclusion

The Barail sandstones of the Champhai–Zokhawthar Road section are mostly subarkosic to arkosic wacke, followed by sublitharenite to lithic greywacke. Granulometric analysis shows a moderate sorting nature of the grains. The Barail sediments are deposited in a shallow marine environment which could act as a favourable spot for HC accumulation. An average porosity of 10.89% was obtained, from which it can be inferred that the reservoir rocks bear a moderate reservoir quality. Diagenetic changes play an important role in modifying the reservoir characteristics. Primary macroporosity contributes to the maximum amount of porosity. Secondary porosity enhancement also marks the enhancement of porosity. The same intense compaction and formation of a pseudomatrix contribute to a reduction in porosity.

## Acknowledgements

On the very onset I would like to thank the DST-SERB for providing financial assistance in carrying out the field work (Project File no.: EMR/2016/003915) and costs involved in various laboratory analysis (Project File No. EMP/2016/003915). Also I would like to thank the faculty members of Mizoram University and Pachchunga University College, Mizoram, for their help and support. I also would like to thank the faculty members of the Department of Petroleum Technology and Department of Applied Geology, Dibrugarh University, in carrying out the work.

# Chapter 4

# Chemical Enhanced Oil Recovery with Special Reference to the Upper Assam Basin

**Kalpajit Hazarika**[a] **and Subrata Borgohain Gogoi**[b]

[a] *Department of Petroleum Engineering, Presidency University, Bengaluru, India*
[b] *Department of Petroleum Technology, Dibrugarh University, Assam, India*
kalpajithazarika@presidencyuniversity.in

This chapter presents results of the laboratory investigation of alkali-surfactant-polymer (ASP) flooding technology for improving recovery of crude oil from Upper Assam oilfields. The oil displacement efficiency of the two-component ASP formula consisting of a polymer-surfactant tensiometer was modified by added alkali and was tested by conducting core flood studies. During the course of the experiments, 11 pore volume (PV) of brine and different chemical slugs was injected into the core samples. This technology has potential to become an alternative to the other chemical recovery methods, especially in cases where application of these methods is restricted by geological, economical or ecological factors. The identified combination of surfactant, alkalis and polymer was sodium dodecyl sulphate (SDS)+$Na_2CO_3$+polyacrylamide (PAM). The efficiency of these slugs will be determined with respect to the mobility ratio, (MR) stability and recovery efficiencies. From the

*Advances in Petroleum Technology*
Edited by Subrata Borgohain Gogoi

ISBN 978-981-4877-19-0 (Hardcover), 978-1-003-04993-7 (eBook)
www.jennystanford.com

overall efficiency ($E_o$), ASP flooding was found to be most efficient over the others. The maximum recovery of 70.6% was achieved with SDS+$Na_2CO_3$+PAM as an ASP slug. In A, S and AS flooding the overall efficiency, $E_o$, is the microscopic displacement efficiency ($E_d$), while in P, SP and ASP flooding, $E_o$ is the product of $E_d$ and macroscopic sweep efficiency, $E_s$. In this research the novelty lies in the fact that $E_d$ plays a major role in contributing to the $E_o$ of ASP–enhanced oil recovery (ASP-EOR) technology. The reduction in MR, and hence the increase in areal sweep efficiency ($E_a$) in the case of P, SP and ASP flooding may be the factor for more $E_o$ over the others (A, S and AS). The optimised composition of the ASP slug for the Naharkatiya (NH) crude oilfield is 0.4 vol% SDS + 0.85 wt% $Na_2CO_3$ + 1500 ppm PAM. The slug optimised for ASP-EOR was 0.4 vol% SDS + 0.85 wt% $Na_2CO_3$ + 1500 ppm PAM, which was found to be and recovered 70.6% of crude oil.

## 4.1 Introduction

The demand for the chemical enhanced oil recovery (CEOR) process is increasing day by day due to the depletion of most reservoirs of the Upper Assam Basin. The residual oil saturation ($S_{or}$) left in the reservoir after secondary brine flooding can be recovered by proper selection of alkali-surfactant-polymer–enhanced oil recovery (ASP-EOR) slug on the basis of the reservoir petrophysical and fluid–fluid interactions. The feasibility of the ASP-EOR process depends on the selection of chemicals and their proportion. This chapter aims to study the comparison between different ASP-EOR slugs in terms of $E_o$ after secondary brine flooding. The $\eta$ of these slugs will be determined with respect to the mobility ratio (MR), stability and recovery of $S_{or}$ which remains in the reservoir after secondary oil recovery. Earlier researchers identified black liquor (BL) as an efficient surfactant for the Upper Assam Basin as it was cheap and easily available in the locality, although much reduction in interfacial tension (IFT) was not achieved [32–35]. Some polymers were found efficient for the Upper Assam Basin in improving the MR between the displacing fluid, slug and displaced fluid, and oil [36]. Gogoi (2011) studied alkali-surfactant (AS) flooding with sodium lignosulphonate (SLS) as a surfactant and NaOH as an alkali in oilfields of Upper Assam, and it was observed that there was EOR but there was the phenomenon of adsorption in the porous media [36, 37]. Das and Gogoi (2015) also observed that the surfactant in the form of micellar is more efficient in reducing the IFT between the oleic and the aqueous phase in the reservoir to $10^{-3}$ mN/m, thus leading to EOR [38, 39]. Phukan et al. (2018) and Gogoi and Kakoty (2017) worked on several chemical and gas EOR processes, including $CO_2$ miscible and immiscible EOR processes [40, 41]. Initial studies on sodium dodecyl sulphate (SDS) were done by Hazarika et al. (2018 and 2014) where SDS was taken alone as an EOR and along with alkali or/and polymer; further, their research proved that an anionic surfactant, mainly SDS, is the most appropriate surfactant for CEOR for the Upper Assam Basin [42, 43]. The effect of alkali concentration on alteration of wettability, reduction of IFT and reduction of adsorption of the surfactant on porous media, rheology of the CEOR slug and the propagation of surfactants as well as the CEOR slug in cores are the most contributory factors in recovery.

## 4.2 Experimental

### 4.2.1 Material

The core plugs samples were collected from the Upper Assam Basin. The chemicals used for the different experiments are tabulated in Table 4.1.

**Table 4.1** List of materials

| S. no. | Material | Source |
|---|---|---|
| 1 | Sodium dodecyl sulphate | Sigma-Aldrich Co., Mumbai |
| 2 | Sodium carbonate ($Na_2CO_3$) | Loba Chemie, Mumbai |
| 3 | Liquid paraffin (light) | Fisher Scientific, New Delhi |
| 4 | Polyacrylamide (PAM) | Himedia, Mumbai |
| 5. | Core Sample | Collected from Upper Assam Basin |

**Table 4.2** Slug for core flooding in the NHP core plug

| S. no. | Core | Alkali (0.85 wt%) | Surfactant (0.4 vol%) | Polymer (1500 ppm) |
|---|---|---|---|---|
| 1 | NHP | $Na_2CO_3$ | | |
| | | – | – | |
| 2 | | – | SDS | – |
| 3 | | $Na_2CO_3$ | SDS | – |
| 4 | | – | – | PAM |
| 6 | | – | SDS | PAM |
| 7 | | $Na_2CO_3$ | SDS | PAM |

## 4.2.2 Methodology

### 4.2.2.1 Core flooding

In the core-flooding experiment, the core samples were first saturated with 3000 ppm brine, and then the brine was displaced by oil through a drainage process until irreducible water saturation was achieved in the core. Seven core-flooding experiments were done in the core-flooding instrument. Total 11 PV comprising 4 PV of brine, 3 PV of polymer and 2 PV of SP and ASP slug was flooded. The fluids constituting the slugs are in Table 4.2. The flow diagram of the core-flooding apparatus is shown in Fig. 4.1.

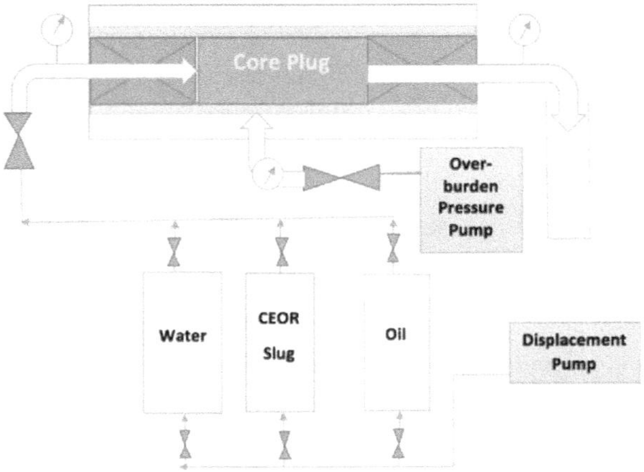


**Figure 4.1** Core-flooding instrument.

#### 4.2.2.2 Selection of a CEOR slug

The ASP slug selected for core-flooding experiments consisted of 0.4% by volume of SDS, 0.85% by weight of $Na_2CO_3$ and 1500 ppm polyacrylamide (PAM). The SDS concentration was selected from IFT experiments where 0.4% by volume of SDS observed was the lowest IFT between the surfactant slug and the oil. Again 0.85% by weight of $Na_2CO_3$ as alkali was selected on the basis of IFT and alteration of wettability. Initially the core samples were intermediate-wet but after addition of an alkali they shifted to strongly water-wet, which is a prerequisite for CEOR processes. Then 1500 ppm of PAM was selected on the basis of its ability to lower the MR between the displacing and the displaced fluids [44].

#### 4.2.2.3 Recovery $\eta$

After secondary brine flooding or the imbibition process, the core plugs were subjected to ASP flooding. Two sequences of flooding were adapted in this chapter:

(1) The first sequence of flooding was as follows and is exhibited in Table 4.2:

   (a) A as $Na_2CO_3$
   (b) S as SDS
   (c) AS is [$Na_2CO_3$+SDS]
   (d) P as PAM

(2) The second sequence of flooding was as follows and is exhibited in Table 4.2:

   (a) SP
   (b) ASP

#### 4.2.2.4 Areal displacement efficiency ($E_d$)

$E_d$ was calculated from the results of core-flooding experiments to determine the efficiency of A, S and AS flooding by using the following Eq. 4.1:

$$E_d = \frac{S_o - S_{or}}{S_o}$$

$$E_d = 1 - \frac{S_{or}}{S_o} \tag{4.1}$$

#### 4.2.2.5 Areal sweep efficiency ($E_a$)

The MR of the plugs was calculated from the results of the core-flooding experiments. $E_a$ can be calculated by taking the MR and inverse of the MR ($MR^{-1}$) from Dyes et al. (1954) [45]. The determination of $E_a$ was done to find out the efficiency of P, SP and ASP flooding using the following Eq. 4.2:

$$MR = \frac{\lambda_w}{\lambda_o} = \frac{\frac{k_{rw}}{\mu_w}}{\frac{k_{ro}}{\mu_o}} = \frac{k_{rw}\mu_o}{k_{ro}\mu_w} \tag{4.2}$$

where

| | |
|---|---|
| $k_{rw}$ | relative permeability of water or brine |
| $k_{ro}$ | relative permeability of oil |
| $\mu_o$ | apparent viscosity of oil |
| $\mu_w$ | apparent viscosity of water or brine |
| $\lambda_w$ | mobility of the displacing fluid or water or brine |
| $\lambda_o$ | mobility of the displaced fluid or oil |

#### 4.2.2.6 Overall efficiency ($E_o$)

$E_o$ was found out for individual slugs as a product of $E_d$ and $E_v$, where $E_v$ is the product of $E_a$ and invasion or vertical sweep efficiency ($E_i$) by Eq. 4.3 and Eq. 4.4.

$$E_o = E_d \times E_v, \tag{4.3}$$

where

$$E_v = E_a \times E_i. \tag{4.4}$$

#### 4.2.2.7 Thermogravimetric analysis

The thermal stability of the formulated slug was determined by thermogravimetric analysis (TGA) under reservoir conditions. The principle behind this analysis was the changes in physical and chemical properties of materials as a function of increasing temperature (with a constant heating rate). There are two temperatures in the reaction, starting of decomposition temperature ($T_i$) and final temperature ($T_f$), representing the lowest temperature at which the onset of a mass change is seen and the lowest temperature at which the process has been completed, respectively [46].

## 4.3 Results and Discussion

### 4.3.1 Core Flooding

$\eta$ was found to be highest in the case of ASP, followed by SP, P, S and A in descending order. $E_o$ in terms of $E_d$ and $E_v$ was found in the case of ASP, SP and P, while in the case of A, S and AS $E_o$ was only $E_d$. This proves that a polymer as a slug alone will contribute not only to $E_v$ but also to $E_d$. This is the novelty of this research. Several researchers in the past have claimed that a polymer does not contribute to $E_d$ [47, 48]. It was also observed that the MR is inversely proportional to $E_o$, which is in line with several researchers [49–51]. For the NHP-1 core plug the $E_o$ was found to be 68%. The core flooding was mainly conducted to compare the different CEOR processes. The results of core-flooding experiments show that the recovery is highest in the case of ASP flooding, followed by SP, P, S and A, as in Tables 4.3–4.14. During secondary brine flooding or the imbibition process the recovery was approximately 24% of the

**Table 4.3** Brine flooding

| Core | $S_{oi}$ (%) | Secondary recovery (%) | $S_{or}$ (%) | $E_d$ (%) |
|---|---|---|---|---|
| NHP-1 | 82.36 | 45.3 | 37.06 | 55.00 |
| NHP-2 | 89.16 | 49.54 | 39.62 | 55.56 |
| NHP-3 | 90.8 | 47.1 | 43.7 | 51.87 |

**Table 4.4** Alkali flooding

| Core | $S_o$ (%) | Slug | Recovery (%) | $S_{or}$ (%) | $E_d$ (%) |
|---|---|---|---|---|---|
| NHP-1 | 37.06 | $Na_2CO_3$ | 5.4 | 32.61 | 12.00 |
| NHP-2 | 39.62 | $Na_2CO_3$ | 4.4 | 35.70 | 9.90 |
| NHP-3 | 43.70 | $Na_2CO_3$ | 3.8 | 40.25 | 7.90 |

**Table 4.5** Surfactant flooding (SDS)

| Core | $S_o$ (%) | Recovery (%) | $S_{or}$ (%) | $E_d$ (%) |
|---|---|---|---|---|
| NHP-1 | 32.61 | 8.0 | 26.02 | 20.20 |
| NHP-2 | 35.70 | 7.1 | 29.37 | 17.73 |
| NHP-3 | 40.25 | 6.6 | 34.26 | 14.89 |

**Table 4.6** AS flooding by SDS+$Na_2CO_3$ slug

| Core | $S_o$ (%) | Recovery (%) | $S_{or}$ (%) | $E_d$ (%) |
|---|---|---|---|---|
| NHP-1 | 26.02 | 1.1 | 25.11 | 3.48 |
| NHP-2 | 29.37 | 2.2 | 27.40 | 6.67 |
| NHP-3 | 34.26 | 4.3 | 30.35 | 11.35 |

**Table 4.7** MR(bo) of brine

| Core | $\mu_o$ (cp) | $K_{rw}$ (mD) | $K_{ro}$ (mD) | MR (bo) | $E_a$ (%) |
|---|---|---|---|---|---|
| NHC-1 | 4.8 | 0.30 | 0.21 | 6.80 | 45 |
| NHC-2 | 5.28 | 0.35 | 0.27 | 6.79 | 45 |
| NHC-3 | 4.42 | 0.38 | 0.30 | 15.55 | 46 |

**Table 4.8** MR(po)$^{-1}$ of PAM

| Core | $\mu_p$ cp | $\mu_o$ cp | $K_{ro}$ mD | $K_{rp}$ mD | MR (po) | MR (po)$^{-1}$ | $E_a$ (%) |
|---|---|---|---|---|---|---|---|
| NHC-1 | 9.11 | 4.80 | 0.21 | 0.20 | 0.50 | 2.00 | 78 |
| NHC-2 | 9.11 | 5.28 | 0.27 | 0.25 | 0.54 | 1.85 | 76 |
| NHC-3 | 9.11 | 4.42 | 0.30 | 0.26 | 0.42 | 2.38 | 79 |

**Table 4.9** MR(spo)$^{-1}$ of PAM

| Core | $\mu_{spo}$ cp | $\mu_{spo}$ cp | $K_{ro}$ mD | $K_{rsp}$ mD | MR (spo) | MR (spo)$^{-1}$ | $E_v$ (%) |
|---|---|---|---|---|---|---|---|
| NHC-1 | 4.80 | 9.15 | 0.21 | 0.21 | 0.52 | 1.91 | 78 |
| NHC-2 | 5.28 | 9.15 | 0.27 | 0.25 | 0.53 | 11.87 | 78 |
| NHC-3 | 4.42 | 9.15 | 0.30 | 0.26 | 0.42 | 2.39 | 80 |

original oil in place (OOIP). Therefore, polymer flooding increased the recovery up to 61% of OOIP and after ASP flooding only 30% of OOIP was left in the core sample. The IFT values between aqueous

**Table 4.10** Inverse MR $(aspo)^{-1}$ of ASP

| Core | $\mu_{aspo}$ cp | $\mu_{aspo}$ cp | $K_{ro}$ mD | $K_{rasp}$ mD | MR (aspo) | MR $(aspo)^{-1}$ | $E_v$ (%) |
|---|---|---|---|---|---|---|---|
| NHC-1 | 4.80 | 9.15 | 0.21 | 0.18 | 0.45 | 2.22 | 78 |
| NHC-2 | 5.28 | 9.15 | 0.27 | 0.22 | 0.47 | 2.13 | 78 |
| NHC-3 | 4.42 | 9.15 | 0.30 | 0.23 | 0.37 | 2.70 | 82 |

**Table 4.11** Efficiencies of Brine slug

| Core | $E_d$ (%) | $E_a$ (%) | $E_o$ (%) |
|---|---|---|---|
| NHC-1 | 55.0 | 45.0 | 24.7 |
| NHC-2 | 55.6 | 45.0 | 25.0 |
| NHC-3 | 51.9 | 46.0 | 23.9 |

**Table 4.12** Efficiencies of PAM slug

| Core | $E_d$ (%) | $E_a$ (%) | $E_o$ (%) |
|---|---|---|---|
| NHC-1 | 78.2 | 78.0 | 61.0 |
| NHC-2 | 75.1 | 76 | 57.1 |
| NHC-3 | 71.2 | 79.0 | 56.2 |

**Table 4.13** Efficiencies of SDS+PAM slug

| Core | $E_d$ (%) | $E_a$ (%) | $E_o$ (%) |
|---|---|---|---|
| NHC-1 | 79.6 | 78.0 | 62.1 |
| NHC-2 | 79.1 | 78.0 | 61.8 |
| NHC-3 | 76.4 | 80.0 | 61.1 |

**Table 4.14** Efficiencies of ASP slug

| Core | $E_d$ (%) | $E_{va}$ (%) | $E_o$ (%) |
|---|---|---|---|
| NHC-1 | 88.2 | 78.0 | 68.8 |
| NHC-2 | 85.8 | 78.0 | 66.9 |
| NHC-3 | 86.1 | 82.0 | 70.6 |

and oleic phases for NHC-1, NHC-2 and NHC-3 were found to be lowest in the range of $10^{-3}$ mN/m for the SDS + $Na_2CO_3$ slug. The justification of taking PAM may be because PAM is hydrolysed in a reservoir environment to generate $-COO^-$ ions before getting converted to hydrolysed polyacrylamide (HPAM) with the liberation of $NH_3$, leading to the formation of an alkaline environment. If $-COO^-$ ions increase it will increase the repulsive force between the ions, leading to an increase in viscosity. During hydrolysis all the amide ($-NH_2$) ions are not replaced by $-ONa$ ions. The stability of HPAM depends on the number of $-NH_2$ ions replaced by $-ONa$. The $-NH_2$ of PAM will get replaced by $-ONa$. Therefore, $-ONa$ will increase in the molecule of HPAM with the decrease of $-NH_2$, leading to instability. If the $Na^+$ is being replaced by the divalent ions of the formation water (FW) there will be precipitation in the reservoir.

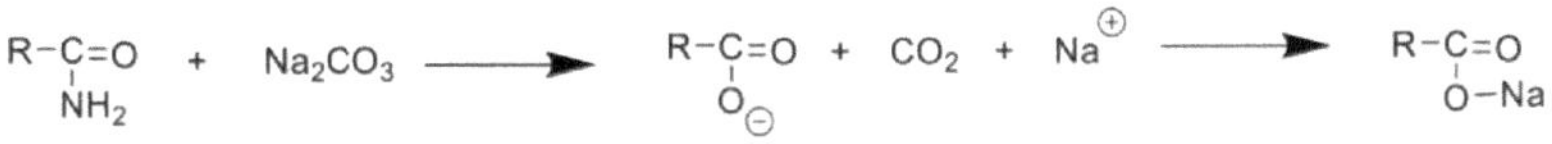


## 4.3.2 Mobility Ratio

The MR between the displacing fluid and the displaced fluid was determined in order to study the effectiveness of P, SP and ASP

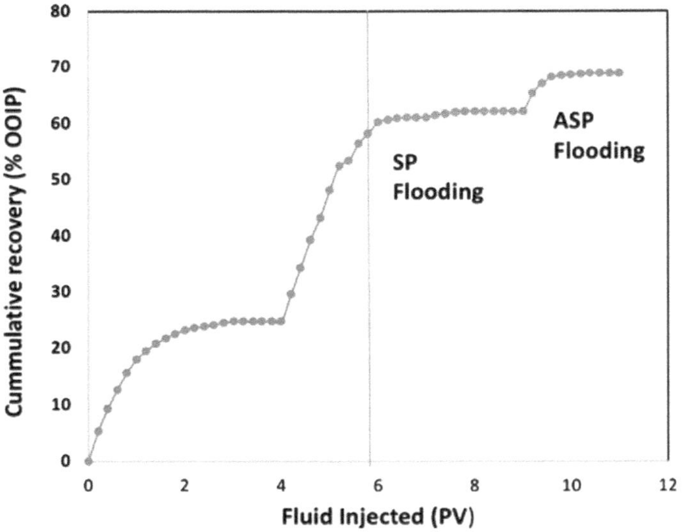


**Figure 4.2** Cumulative oil recovery during different EOR slug floodings.

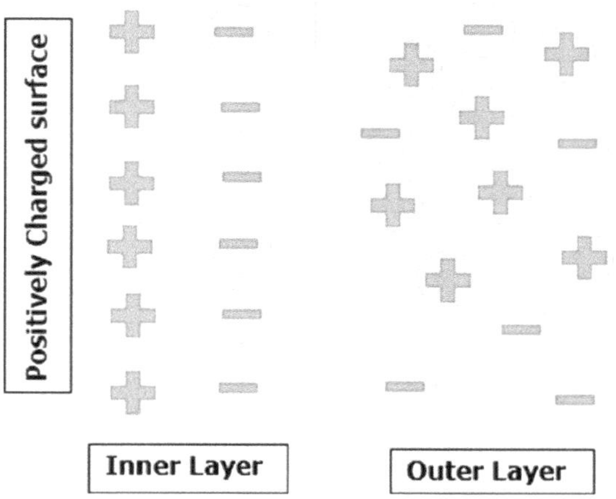


**Figure 4.3** Electrical double layer at a charged surface.

flooding. The MRs of oil water ($MR_{ob}$), oil polymer ($MR_{op}$), oil SP ($MR_{osp}$) and oil ASP ($MR_{oasp}$) are tabulated in Tables 4.7–4.10. $E_v$ of brine against the MR was taken from Dyes et al. (1954) [45]. $MR^{-1}$ were calculated for $MR_{op}$, $MR_{osp}$ and $MR_{oasp}$. Sweep efficiency of P, SP and ASP flooding was taken from from Dyes et al. (1954) [45] against $MR^{-1}$. In all the cases the MR of OW was more than 1; therefore, the fingering effect will occur and brine will bypass the oil [52]. The MR of OP was found to be less than 1, which signifies the mobility of the polymer solution inside the porous media will be less than the mobility of oil; therefore, there is less chance of the fingering effect to occur, which will enhance the recovery. $MR_{oasp}$ was comparatively less than $MR_{op}$ and $MR_{osp}$; therefore, $E_v$ of ASP flooding was more than that of P and SP flooding, as shown in Tables 4.11–4.14. Figure 4.4 shows the comparison of recovery efficiency for different CEOR slugs. During the secondary brine flooding or the imbibition process, 4 PV of 3000 ppm brine was flooded and the maximum recovery for the NHP-1 sample was found to be 24% of OOIP. After that 3 PV of polymer followed by 2 PV of SP slug followed by 2 PV of ASP slug was flooded and $E_o$s were found to be 61% of OOIP for polymer flooding, 62% of OOIP for SP flooding and 68% of OOIP for ASP flooding.

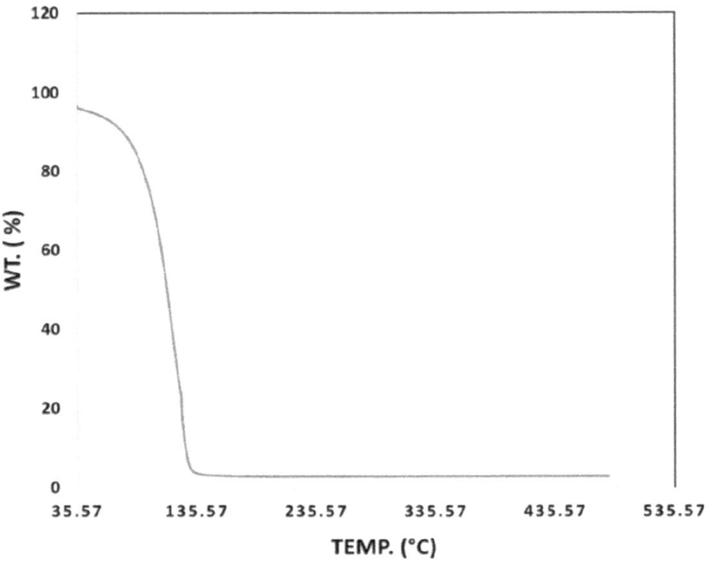


**Figure 4.4** Thermal stability of ASP ($Na_2CO_3$+SDS+PAM) slug.

### 4.3.3 Effect of Polymer on the CEOR Slug

In polymer flooding, the viscoelasticity properties, more specifically MR of PAM, are believed to contribute more oil recovery in general, and some researchers claim that it might reduce $S_{or}$ due to only piston-like effect mechanisms [53–56]. HPAM shear-thickening behaviour may, in some cases, contribute to improving front stability and oil recovery [57]. On the other hand, the significant pressure gradient associated with shear-thickening phenomena can limit polymer injection and cause wellbore damage or fracturing.

### 4.3.4 Effect of Alkali on the CEOR Slug

The effect of alkali does not only have an effect with regard to the charge of the head group of the surfactant; it also alters the aqueous phase solubility. As the concentration of salt increases, the solubility of the surfactant in the aqueous phase decreases. As the solubility decreases, the surfactant starts accumulating at the interface instead of in the bulk. The salt concentration where the solubility of the surfactant is equal in both the oil and the water phase is where the surfactant has its highest affinity to the interface. Increasing salinity will increase the surfactant's affinity to the interface, resulting in a higher concentration of monomers at the interface and thus a lower IFT. When a charged particle is present in a solution containing excess ions, those ions will orient themselves around the charged particle to electrostatically neutralise the particle. Closest to the particle, a layer consisting of only ions with the opposite charge of the particle will accumulate. Further out from the charged particle, both ions of the same charge and the opposite charge will accumulate in a layer larger than the inner one. These two layers are what is called the electrical double layer (EDL) [58, 59]. An illustration of the EDL can be seen in Fig. 4.3 [60]. When two surfactant monomers with the same charge of the head group are present at an interface, their EDLs will interact and repel each other, as the two monomers have the same charge. However, when a salt is introduced to the system, positive and negative ions will interact with the double layer, decreasing the size of the EDL and hence the repulsion between the two monomers [60]. This is the reason the critical micelle concentration (CMC), IFT as well as $E_{\mathrm{d}}$ decrease in a surfactant-containing system when an alkali is introduced to the system. The main function of a polymer is to give the viscosity of the ASP slug. The viscoelasticity of the ASP slug depends on the concentration of the alkali. The increase in $Na_2CO_3$ concentration will increase the degree of hydrolysis of PAM due to the formation of $-COO^-$ ions. The increase of $-COO^-$ ions increases the repulsive force between the negative ions, and hence stretching of the polymer chain takes place. The more that the molecules of the polymer are stretched, the higher the viscosity of its solution is. But this mechanism continues till a certain concentration of alkali. A further increase in the concentration of the alkali will form EDL, which reduces the viscosity of the ASP slug due to compression of the polymer chain. In addition to this, the high concentration of the alkali will degrade the polymer and lead to polymer weight loss due to which viscosity will be reduced [61–63].

### 4.3.5 Stability of ASP Slugs

The results of TGA, as observed from Fig. 4.4, indicates that there was degradation of only 10% below 100°C; from 100°C to 150°C the degradation suddenly rose to 97%, and after 150°C the degradation was constant. Since the reservoir temperature under study was below 100°C, it can be assumed that the slugs prepared were thermally stable.

## 4.4 Conclusion

This chapter gave an overview of different CEOR processes, more specifically it helped to compare different CEOR processes in terms of recovery $\eta$. The $E_o$ of ASP-EOR was found to be more efficient over others (A, S, AS). This is due to the combined effect of alkali, surfactant and polymer. A surfactant and alkali reduce the IFT, an alkali enhances the viscoelastic property of a polymer and alters the wettability from intermediate-wet to strong water-wet, and the polymer reduces the MR. In the NH oilfield the maximum recovery of 70.6% was achieved in NHP-3 with SDS+$Na_2CO_3$+PAM as the ASP slug. In A, S and AS flooding the $E_o$ was the $E_d$, while in P, SP and ASP flooding the $E_o$ was the product of $E_d$ and $E_s$. In the case of polymer flooding finding out $E_d$ was the novelty of this research. The reduction in the MR and hence the increase in $E_a$ in the case of P, SP and ASP flooding may be a factor for more $E_o$ over the others. The optimised composition of the ASP slug for the NH oilfield is 0.4 vol% SDS + 0.85 wt% $Na_2CO_3$ + 1500 ppm PAM. All the slugs used in the flooding operation were found to be thermally stable because only 10% of the slug was degraded below 100°C, which was the reservoir temperature under study. But 97% of the slug was degraded at about 150°C. Therefore, these slugs can be efficiently used for reservoirs which have temperatures below 100°C.

## Acknowledgements

We would like to acknowledge the following:

(a) DST Project no SB/S3/CE/057/2013 entitled 'Identification of Competent Alkali-Surfactant-Polymer Formulations for Enhanced Oil Recovery of Assam Crude Oil'
(b) 21st Century Knowledge initiative programme under University Grants Commission Indo-US Project F no. 194-1/2009 (IC) dated February 2015, entitled 'Foam-Assisted $CO_2$ Flooding for the Depleted Reservoirs of Upper Assam Basin and in Candidate Reservoirs in Louisiana'
(c) Dibrugarh University for laboratory facilities

## Chapter 5

# Formulation of a Stable Microemulsion Slug for Enhanced Oil Recovery in the Upper Assam Basin

**Shilpi Sarmah,[a] Subrata Borgohain Gogoi[a] and Fan Xianfeng[b]**
[a] *Department of Petroleum Technology, Dibrugarh University, Assam, India*
[b] *School of Engineering, University of Edinburg, UK*
ta_shilpisarmah@dibru.ac.in

A microemulsion is a multiphase system of a transparent solution comprising water, oil, surfactant and alcohol/co-surfactant. The microemulsion-flooding application is an integral part in oilfields of the Upper Assam Basin because the basin faces various issues such as decrement in the oil production rate, depleted oil reservoirs and limitation of primary and secondary recovery methods. It is an important yet challenging aspect in the enhanced oil recovery (EOR) process. It is important because of its characteristic property of having ultralow interfacial tensions (IFTs) as well as enhancing viscosity, thereby inhibiting the fingering effect. However, the challenge lies in formulating the Winsor type III microemulsion, where a surfactant-rich middle phase coexists with excess water and oil. The study focuses on formulating a stable microemulsion slug for EOR, where different anionic surfactants such as sodium dodecyl

*Advances in Petroleum Technology*
Edited by Subrata Borgohain Gogoi

ISBN 978-981-4877-19-0 (Hardcover), 978-1-003-04993-7 (eBook)
www.jennystanford.com

sulphate (SDS); sodium dodecyl benzene sulphonate (SDBS); non-ionic surfactants namely Tergitol 15-s-7, Tergitol 15-s-9, Tergitol 15-s-12, Triton X-100 (TX-100) and Triton X-405 (TX-405); and co-surfactant propan-2-ol were studied by determining their surface tension (ST) and IFT. The measurements were carried out by the Du Nouy Ring method to determine the critical micelle concentration (CMC) of surfactants. The phase behaviour study was done by using the Winsor classification method to obtain the optimum salinity and temperature at which water, oil and surfactant coexist. Thereafter, characterisation of the microemulsion was completed by examining its density, viscosity, particle size analysis, microscopic study and wettability. The efficiency of surfactants was examined through a core-flooding experiment. Phase behaviour studies suggest that at an optimum salinity of 6000 ppm, an anionic surfactant (SDBS) of 1.4 mM, a co-surfactant (propan-2-ol) of 0.004% mole fraction and Upper Assam crude oil of density 0.866 g/cc, microemulsion stage III was observed, and this mixture when flooded in the core flooding resulted in a crude oil recovery of 59.92% of residual oil saturation, whereas at an optimum salinity of 4250 ppm a non-ionic surfactant of a mixture of Tergitol 15-S-7 and TX-405 with the same Upper Assam crude oil, microemulsion stage III was observed. The slugs were further flooded in the core flooding to obtain a recovery of 54.47% residual oil saturation. This suggests the efficiency of the above two formulated slugs in EOR application.

## 5.1 Introduction

Currently, the demand for oil and gas is increasing worldwide. However, new exploration of oil and gas fields is reducing constantly. Oil recovery techniques play a vital role in solving this issue. The oil recovery process can be classified into three important stages, which are primary recovery, secondary recovery and tertiary recovery [64]. Enhanced oil recovery (EOR), also known as tertiary oil recovery, is a method for recovering the amount of residual oil left over in the reservoirs after primary and secondary oil recovery. EOR is a technique which improves oil production through injection of chemicals and gas to ease the mobility of hydrocarbons (HCs). There are three main types of EOR techniques which are used at present, which are thermal, miscible and chemical [65]. In recent years, microemulsion flooding is favoured among the EOR techniques. There are various companies which have implemented the application of microemulsion flooding, which is a better alternative method compared to other EOR techniques.

The exploration of microemulsions commenced in 1943. In 1959, Schulman and his colleagues constructed the term 'microemulsion'. Moreover, microemulsions contribute a significant part in our daily-use products like cosmetic products and pharmaceutical products [66]. Furthermore, the debate about the development of microemulsions like the oil-dispersed-in-water (o/w) or water-dispersed-in-oil (w/o) systems has been discussed in the past years. The common elements that the dispersions have are oil phase, aqueous phase, surfactant and co-surfactant. The critical micelle concentration (CMC) acts as an important parameter for a stable microemulsion. The criteria which influence the CMC are surfactant structure, electrolyte presence, presence of organic additives, presence of a second liquid phase and temperature [65]. Winsor nominated a strategy for a stable microemulsion formulation with oil, water and surfactants. There are four types of Winsor classification: Winsor type I (an oil phase and a water phase with dissolved surfactants), Winsor type II (an oil phase with dissolved surfactants and a water phase), Winsor type III (a middle phase with oil, surfactants and water; an excess water phase; and an excess oil phase) and Winsor type IV (mixture of oil, water and surfactants) [64]. The middle phases of Winsor type III systems and Winsor type IV systems were considered as microemulsions due to their thermodynamic stability. Practically, it is not simple to identify the microemulsion on the basis of the Winsor classification strategy because the system is mixed. Centrifugation is the solution for this issue because it can aid in accelerating the equilibration [67].

Microemulsions pass through the porous media more freely due to their ultralow interfacial tension (IFT) property, which upgrades the oil withdrawal performance. The stability of surfactants, the IFT between oil and water and minimum surfactant loss by adsorption on the reservoir rocks are vital for the effectiveness of a microemulsion. Surfactant retention plays a key role in the effectiveness of the oil recovery process by microemulsion flooding [64]. Hence, a combination of anionic surfactant sodium dodecyl benzene sulphonate (SDBS) and co-surfactant (propan-2-ol), a combination of non-ionic surfactants Tergitol 15-S-7 and TX-405 and a combination of anionic surfactant SDBS, co-surfactant (propan-2-ol) and non-ionic surfactants Tergitol 15-S-7 and TX-405 were tested to choose an optimum slug for EOR in the Upper Assam Basin.

## 5.2 Materials and Methods

### 5.2.1 Materials

The chemicals and instruments used are listed in Table 5.1.

**Table 5.1** Materials

| S. no. | Material | Make |
|---|---|---|
| 1 | SDBS | Loba Chemie, Mumbai |
| 2 | Propan-2-ol | Thermo Fisher Scientific, India |
| 3 | Liquid paraffin light | |
| 4 | Tergitol 15-S-7 | Sigma-Aldrich Chemie, USA |
| 5 | TX-405 | |
| 6 | Sodium chloride (NaCl) | RFCL Limited, New Delhi |
| 7 | Distilled water (DW) | Distilled in laboratory |

### 5.2.2 Methods

#### 5.2.2.1 Experiment on IFT measurements

The CMC of SDBS was determined by varying its concentration in distilled water. The co-surfactant propan-2-ol was added to the above to observe a further decrease in the IFT. Another set of non-ionic slugs was also taken and compared [67]. The Kruss Easy Dyne tensiometer was used to conduct IFT experiments. The IFT was measured through a tensiometer with a probe suspended from a force sensor. The probe was a platinum ring.

#### 5.2.2.2 Experiment on phase behaviour

Phase behaviour analysis was done for three sets of slugs, viz. (SDBS+propan-2-ol), (Tergitol 15-S-7+TX-405) and (SDBS+propan-2-ol+Tergitol 15-S-7+TX-405) by varying salinity from 1000 to 5000 ppm. The samples were then mixed with crude oil in a 1:1 ratio and rotated in a rotaspin at 10 rpm for 1 hour [67]. All the prepared mixtures were poured into closed-end pipettes for phase behaviour study. The Winsor classification method has been used for the favourable formulation of microemulsions. Moreover, all the closed-end pipettes were placed in an oven at 67°C until a Winsor type III microemulsion was formed.

#### 5.2.2.3 Characterisation of microemulsions

The three sets of slugs designed were characterised on the basis of density, kinematic viscosity in Cannon Fenske routine viscometer, particle size distribution in Zetasizer 7.11 (Malvern instruments Ltd.), wettability by the sessile drop method and capillary force executed by the slugs on reservoir rock.

#### 5.2.2.4 Core flooding

The slugs were further studied in terms of recovery efficiencies in core-flooding apparatus. The combinations and concentrations taken for the three slug sets are shown in Table 5.2.

**Table 5.2** Sample preparation

| S. no. | SDBS (mM) | Propan-2-ol (mF) | Tergitol-15-S-7 (%) | TX-405 (%) | Salinity (ppm) |
|---|---|---|---|---|---|
| 1 | 1.4 | 0.05 | | – | 2000 |
| 2 | – | – | 0.004 | 0.0024 | 4250 |
| 3 | 1.4 | 0.05 | 0.004 | 0.0024 | 6000 |

## 5.3 Results and Discussion

### 5.3.1 IFT Measurements

From Fig. 5.1 it can be observed that at 1.4 mM, the CMC is attained for SDBS. Addition of a co-surfactant, propan-2-ol, to SDBS further

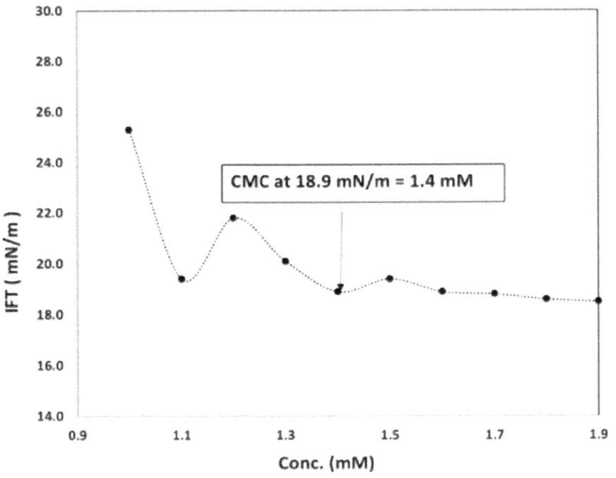


**Figure 5.1** IFT for varying concentrations of SDBS in DW.

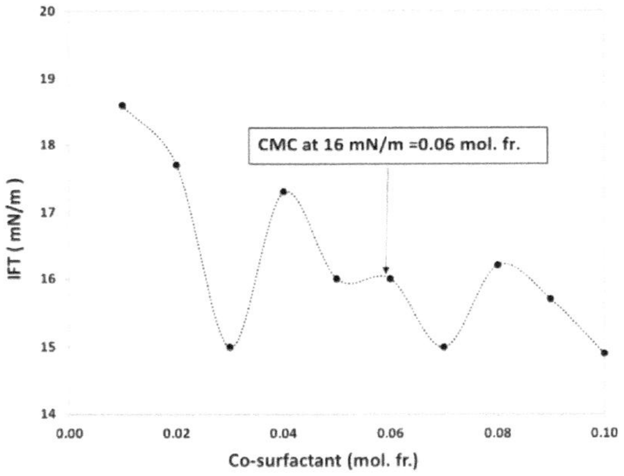


**Figure 5.2** IFT for varying mFs of propan-2-ol in the CMC of SDBS.

reduces the IFT between the aqueous and the oleic phase, as seen in Fig. 5.2. Addition of a 0.06 mole fraction of propan-2-ol to SDBS shows much reduction of IFT. However, further increasing the concentration did not much decrease the IFT. So, the mole fraction of 0.06 propan-2-ol is considered as the optimum condition for achieving ultralow IFT. So, a combination of an anionic surfactant SDBS with a co-surfactant propan-2-ol is the first set of chemical

**Table 5.3** ST and IFT for optimum conc. of SDBS+Propan-2-ol+Tergitol 15-S-7+TX-405 in DW

| S. no. | SDBS(mM) + Propan-2-ol (mF) + Tergitol-15-S-7 (%) + TX-405 (%) | ST (mN/m) | IFT (mN/m) |
|---|---|---|---|
| 1 | 1.4 mM + 0.05 mF + 0.004% + 0.0024% | 28.9 | 5.9 |

slugs designed. The second set includes a mixture of two non-ionic surfactants Tergitol 15-S-7 and TX-405 taken from the study by Sarmah (2019) [67]. The third set of chemical slugs corresponds to the mixture of the above two. Table 5.3 presents results of the surface tension (ST) and IFT values for the combination of SDBS, propan-2-ol, Tergitol 15-S-7 and TX-405. It can be concluded that ST and IFT values for the third set of chemical slugs is the lowest having 28.9 mN/m for ST and 5.9 mN/m for IFT at 1.4 mM of SDBS, 0.06 mole fraction of propan-2-ol, 0.004% of Tergitol 15-S-7 and 0.0024% of TX-405. Overall, the combination of SDBS, propan-2-ol, Tergitol 15-S-7 and TX-405 has the lowest ST and IFT results at optimum concentrations of surfactants.

### 5.3.2 Phase Behaviour

Phase behaviour helps in the evaluation of favourable microemulsion formulations. In this experiment, the challenge was formulating Winsor type III microemulsions. So here the optimum salinity concentration at which oil, water and surfactant co-existed was determined. Figures 5.3 and 5.4 show a Winsor type III formulation at a salinity of 6000 ppm for the third set of chemical slugs.

### 5.3.3 Characterisation of Microemulsions

Crude oil characterisation was done on the basis of density, viscosity and particle size analysis, as seen in Table 5.4. The viscosity of crude oil for the three sets of slugs ranged from 0.400 to 0.568, the combination of anionic and non-ionic slugs being the maximum. The viscosity value changed for each surfactant mixture due to the different concentrations and masses of surfactants. However,

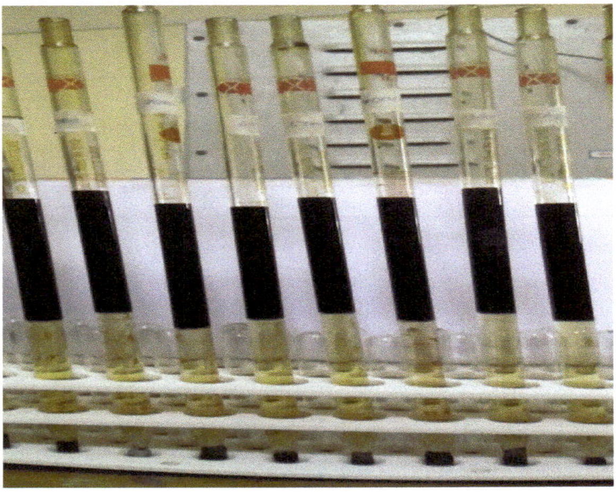

**Figure 5.3** Samples placed in pipettes on day 1.

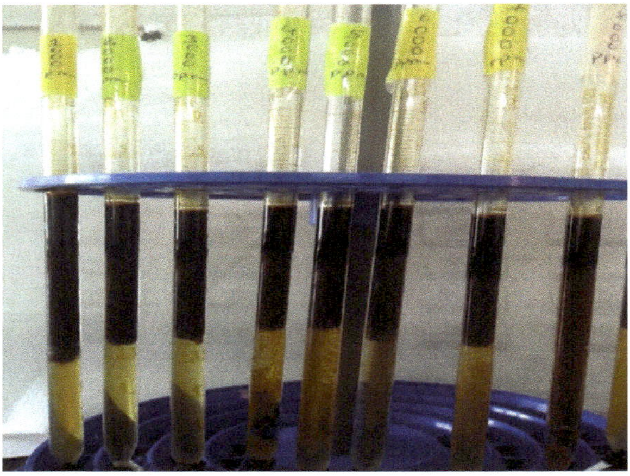

**Figure 5.4** Formulated microemulsion on day 16 (SDBS+propan-2-ol+Tergitol 15-s-7+TX 405+NaCl+DW+crude oil).

particle size for the same was found to be lower than the slug comprising only an anionic surfactant. The average particle size varies according to the various concentrations of surfactants and hence plays an integral role in the oil recovery process.

**Table 5.4** Density, viscosity and particle size of various surfactants and co-surfactant in optimum salinity brine

| S. no. | Sample | Density (g/ml) | Viscosity (cSt) | Particle size (d.nm) |
|---|---|---|---|---|
| 1 | A=[SDBS (1.4 mM)+ Propan-2-ol (0.05 mF)+DW] | 1.0084 | 0.520 | 388.9 |
| 2 | B=[Tergitol 15-S-7 (0.004%)+TX-405 (0.0024%)+DW] | 1.0366 | 0.400 | 8.361 |
| 3 | (A+B) | 0.9995 | 0.568 | 185.6 |

### 5.3.4 Core Flooding

The inlet pressure was maintained at 550 psi, and the overburden pressure was maintained at 600 psi. The overburden pressure plays a vital role in keeping the core seal tight. This was done because a higher overburden pressure ensures the injected fluid will only flow into the core. The higher the inlet pressure, the higher the overburden pressure. Different combinations of surfactants were used for the core-flooding experiment. On the basis of the results obtained from Table 5.5, it can be said that the third set of formulated chemical slugs had the highest oil recovery volume compared to the first and second sets.

**Table 5.5** Oil recovered for various surfactants and co-surfactants in optimum salinity

| S. no. | Sample | Oil recovery (%) |
|---|---|---|
| 1 | C=[SDBS (1.4 mM)+Propan-2-ol (0.05 mF) +NaCI (2000 ppm)] | 31.12 |
| 2 | D=[Tergitol 15-S-7(0.004%)+TX-405 (0.0024%)+NaCI (2000 ppm)] | 54.47 |
| 3 | (C+D) | 59.92 |

## 5.4 Conclusion

A comprehensive study of microemulsions like the ST and IFT of surfactants, phase behaviour study, characterisation of microemulsions and oil recovery rate is vital for the precise prediction of microemulsion flooding in the Upper Assam Basin, which consists of heterogeneous multilayered reservoirs. It can be concluded that a combination of anionic surfactant (SDBS), co-surfactant (propan-2-ol) and non-ionic surfactants Tergitol 15-S-7 and TX-405 at 6000 ppm is the optimum slug for microemulsion flooding due to its ultralow IFT, microemulsion Winsor type III formulation, smaller particle size and high oil recovery rate.

## Acknowledgements

We would like to acknowledge the following:

(a) Indo-US Project F. No. 194-1/2009(IC) dated February 2015, entitled 'Foam-Assisted $CO^2$ Flooding for the Depleted Reservoirs of Upper Assam Basin and in Candidate Reservoirs in Louisiana', of the 21st Century Knowledge Initiative programme under University Grants Commission
(b) Dibrugarh University for laboratory facilities

# Chapter 6

# Systematic Study of Low-Cost, High-Performance Chemical Flooding as EOR for Depleted Oilfields of the Upper Assam Basin

**Miranda Kakoty and Subrata Borgohain Gogoi**

*Department of Petroleum Technology, Dibrugarh University, Assam, India*

rs_miranda@dibru.ac.in

In this chapter, we present a systematic laboratory study with the aim to develop an alkali-surfactant (AS) enhanced oil recovery (EOR) slug for a depleted oilfield of Barail formation of Upper Assam. A series of experiments such as interfacial tension (IFT), aqueous stability, foam stability and thermal stability have been carried out to optimise the AS formulation. Depending on the anionic nature of most reservoirs of Upper Assam, a natural surfactant, black liquor (BL), and synthetic surfactants sodium dodecyl sulphate (SDS), sodium dodecyl benzene sulphonate (SDBS) and alpha olefin sulphonate (AOS) were screened. A surfactant at its critical micelle concentration (CMC) was mixed with alkalis: $Na_2CO_3$ and $NaHCO_3$ and corresponding CMC values at the lowest IFT were recorded. IFT experiments conducted separately with formation water (FW; containing $K^+$, $Ca^{2+}$, $Mg^{2+}$) and with laboratory-prepared brine

*Advances in Petroleum Technology*

Edited by Subrata Borgohain Gogoi



ISBN 978-981-4877-19-0 (Hardcover), 978-1-003-04993-7 (eBook)

www.jennystanford.com

(containing $Na^+$ only) established that IFT values are lower with FW than with the latter. This study conducted in the temperature range from 30°C to 90°C showed that temperature does not appear to have a significant effect on the IFT. On the basis of IFT-reducing capabilities, a slug comprising BL, AOS, SDBS and $Na_2CO_3$ was selected for the EOR application in the depleted oilfield under study. Thermal stability tests revealed that both the surfactant system and the alkali did not undergo any thermal degradation upto 90 days. The best chemical formulation was validated in core-flooding experiments. The AS formulation thus developed in the laboratory could lower the oil water IFT to the range $10^{-2}$–$10^{-3}$ mN/m, and the components were very much compatible with each other at reservoir temperature. The thermal stability study also produced expected results. The results of the core-flooding experiments were encouraging as 50%–76% additional recovery has been achieved.

## 6.1 Introduction

Chemical enhanced oil recovery (CEOR) is one of the very promising EOR techniques which is based on injection of different exotic fluids such as polymer, surfactant, micellar, emulsion and alkali into a reservoir. Around 50%–75% of the original oil is still in place after economic recovery through conventional techniques, that is, primary and secondary recovery [41, 68–71]. These reservoirs are viable candidates for CEOR which interplays with the properties of oil/aqueous/formation interfaces, viz. capillary forces, contact angle, wettability, viscous forces and interfacial tension (IFT) to improve displacement as well as sweep efficiency ($E_S$). However, designing these aqueous chemical mixtures is crucial and must be tailored to the reservoir rock and fluid (i.e., crude oil and formation brine) properties of the specific application. A strategic design methodology can help provide a low-cost, well-performing chemical formulation. Therefore, this chapter emphasises on the formulation of an alkali-surfactant (AS) slug through a series of extensive laboratory experiments which complement the necessary characteristics of the depleted field under study. In an approach to alleviate the high cost involved in CEOR, in this study the efficiency of black liquor (BL) as the primary surfactant assisted by other synthetic surfactants is explored. The design strategy first uses IFT experiments to quickly select and optimise concentrations of injected chemicals. Assessment of formulation optimisation strategies are carried out through determining the CMC value of all surfactants, varying different surfactant combination, selecting a suitable alkali, and utilising formation brine in the injection mixture. Formulations performing well in this experiment were subjected to aqueous stability, foam stability and thermal experiment respectively. Finally, the selected formulation is validated in core-flooding experiment which adheres to necessary design criteria such as pressure and salinity gradients, surfactant adsorption and capillary effects.

Two essential criteria that must be met for successful recovery of residual oil by surfactant flooding are (1) very low IFT between the chemical bank and residual oil and between the chemical bank and the drive fluid and (2) small surfactant retention losses to reservoir rock [40]. The decrease in the IFT between crude oil and formation water (FW) lowers the capillary forces and facilitates oil mobilisation and EOR [72, 73]. Therefore, the IFT of the reservoir fluid system is an important thermophysical property in investigating the capillary forces acting on trapped oil within the reservoir rocks. Crude oil contains surface-active molecules such as low-molecular-weight fatty acids, naphthenic acids and asphaltene. Depending on pH and salinity, these natural surfactants will interact with the added synthetic surfactants in order to enhance the oil recovery [74, 75]. An alkali reacts with naturally occurring organic acid (naphthenic acids) in the crude oil and forms foam, which interacts synergistically with the added surfactant to produce an ultralow IFT. The ultralow IFT is obtained through the distribution of the surfactant between oil and water phases and arrangement of the surfactant at the oil/water interface. In addition, the alkali injected with the surfactant can reduce surfactant adsorption and aid in lowering the IFT [72, 76–78].

## 6.2 Experimental

### 6.2.1 Apparatus

The surface tension (ST) of the crude oil sample and water was measured by the Easy Dyne tensiometer K20 (Du Nouy ring method); the IFTs were measured by the M6500 Spinning Drop tensiometer. A thermal stability test was carried out using a PerkinElmer STA 8000 model thermogravimetric analysis (TGA) instrument. Porosity of the core sample was measured with a coretest Systems, Inc. TPI-219 model Teaching Helium Porosimeter, and air permeability was measured with an OFITE Model 360 Air Permeameter. The core-flooding experiment was carried out using a Gon Engineering core-flooding instrument. FW analyses were carried out in a SYSTRONICS water analyser and a SYSTRONICS Flame Photometer 128.

### 6.2.2 Materials

#### 6.2.2.1 Surfactants

Sodium dodecyl sulphate (SDS) was procured from Merck Specialities Pvt. Ltd., Mumbai, with purity more than 90%, whereas sodium dodecyl benzene sulphonate (SDBS) was purchased from Sisco Research Laboratories Pvt. Ltd., India. Alpha olefin sulphonate (AOS), a 32% transparent yellowish liquid, was procured from Aman Enterprises, India. BL, obtained as an effluent of Nagaon paper mill, Assam, India, is a natural anionic hydrophilic surfactant. Its component sodium lignosulphonate ($C_{20}H_{24}Na_2O_{10}S_2$) possesses very good surface-active properties to lower the IFT between oleic and aqueous phases. Detailed specifications and properties of BL are discussed in [33].

#### 6.2.2.2 Alkalis

Sodium carbonate ($Na_2CO_3$) and sodium bicarbonate ($NaHCO_3$). Anhydrous sodium carbonate ($Na_2CO_3$) was procured from Loba Chemie Pvt. Ltd., Mumbai, India, with purity 99.5% (molecular weight [MW] 105.99 g/mol) and sodium bicarbonate ($NaHCO_3$)

from Avantor Performance Materials India Ltd. with purity 99% (MW 84.01 g/mol).

#### 6.2.2.3 Crude oil, formation water and synthetic soft brine

A crude oil (CO) sample was obtained from the Naharkatiya (NH) field of Barail formation of Upper Assam, and the corresponding FW was collected from Oil India Limited, Duliajan, Assam. Synthetic soft brine which mimics saline formation brine was prepared by sodium chloride (NaCl) salt procured from Merck Specialities Pvt. Ltd., Mumbai, with purity 99% (MW 58.44 g/mol).

A core sample from the NH oilfield of Upper Assam was used.

### 6.2.3 Procedure

#### 6.2.3.1 IFT determination

Aqueous solutions of SDS, SDBS and AOS at 0.1%–1.2% (w/w) in 2000 ppm brine and BL at 1%–10% (w/w) in 2000 ppm brine were prepared; their IFTs with the oil sample were measured at 70°C and corresponding CMC values were recorded. Then, BL at its CMC value was added to SDS, SDBS and AOS separately and IFTs at respective CMC values of the surfactant mixtures were recorded at 70°C. It was observed that among different surfactant mixtures, SDBS+BL showed the highest IFT reduction to 0.132 mN/m, followed by AOS and SDS. Then another surfactant mixture comprising BL, SDBS and AOS was tested for its IFT-reducing capability, and it was found that this mixture showed better IFT reduction compared to previous findings. Even though this surfactant mixture lowers the IFTs to a low value of 0.102 mN/m, the required ultralow IFTs for EOR was not obtained. Therefore, further experiments were conducted to examine the behaviour of the surfactant system as an AS slug for the oilfield under study by adding an alkali at its optimum value. To select the best alkali for the AS slug, two species of alkalis, $Na_2CO_3$ and $NaHCO_3$, were taken under consideration. Both these alkalis at different concentrations (1 wt% and 2 wt%) were added to the surfactant system, and measurement of their respective IFTs was done.

All IFTs so far in this study were measured in brine solutions prepared in distilled water for meaningful comparison of surfactant slugs meant for AS EOR. In the absence of any suitable surfactant which can impart an ultralow IFT to the reservoir crude–oil system, an attempt was made to at least measure the base IFT of the oil sample with its corresponding FW and laboratory-prepared brine. This compares the IFT behaviour between the brine and actual FW. The analysis of the FW was also carried out to validate the effect of the various ions present in the FW for the reduction of the IFT. This was accompanied by measuring the IFT at the various temperatures upto 90°C to examine the effect of temperature on the IFT.

The IFT at the crude oil/water interface was measured with a Spinning Drop tensiometer at 30°C, 60°C and 90°C by using Eq. 6.1.

$$\gamma = 1.44 * 10^{-7} * d^3 * \theta^2 * \Delta\rho, \tag{6.1}$$

where $\gamma$ is the IFT (mN/m), $d$ is the diameter of the drop (LCD read directly from instrument) (mm), $\theta$ is the rotation (read directly from instrument) (rpm) and $\Delta\rho$ is the density difference between crude oil and tested solution (g/cc).

#### 6.2.3.2 Aqueous stability

Aqueous stability assessment of the selected AS formulation from the IFT experiment was carried out at salinities ranges from 2000 to 10,000 ppm. This test determines the clarity and homogeneity of all the dispensed aqueous solutions and also provides an upper value of salinity below which all chemical constituents can coexist in the aqueous phase. The AS formulation was dispensed in an array of glass vials having salinity from 2000 to 10,000 ppm, agitated and then allowed to settle at 70°C. Aqueous stability was assessed by visually inspecting any cloudiness and/or phase separation.

#### 6.2.3.3 Foam stability

The foam stability test was carried out to determine the ability of surfactants to stabilise foam. An equal volume of the surfactant solution was dispensed in separate sealed test tubes and mixed with the oil sample in a 10:1 ratio. The test tubes were then shaken with equal intensity, resulting in the formation of foam at the top of the liquid column. The height of the foam was measured with time. Surfactants with the best foaming ability in the presence of oil will exhibit the lowest rate of foam destruction, that is, the height of the foam column will decrease slowest with time. The foam stability test is only a preliminary indicator of a surfactant's ability to provide stability to foam. The actual ability of a surfactant to provide foam stability in porous media may differ from results of the foam stability test due to adsorption, the presence of capillary forces and foam interactions with microemulsions.

#### 6.2.3.4 Thermal stability

The thermal stability of the samples prepared for the aqueous stability test was carried out by TGA. The Simultaneous Thermal Analyzer (STA) 8000 determines the mass loss of a surfactant with temperature. This test was conducted for a temperature range of 25°C to 500°C with a heating rate of 20°C/min in a $N_2$ environment.

#### 6.2.3.5 Core flooding

A sandstone core was initially drilled and cut to about 3.5 inch (8.8 cm) length and 1.5 inch (3.8 cm) diameter and then dried. Before commencing fluid injection, porosity and air permeability of the core were determined. Then the core was wrapped with Teflon or polytetrafluoroethylene (PTFE) tape and weighed, and inserted into the core holder. The displacing fluids were inserted into the appropriate chamber of the core-flooding instrument. Once all the flow tubing connections were made and the system was checked for leaks, fluid injection was initiated.

Core data and residual oil saturation for chemical flood design and oil recovery were calculated by the Johnson, Bossler and Neumann (JBN) method from a sequence of prechemical floods (brine, oil and water floods). Details of the JBN method are discussed in [79–81]. Brine flooding provides absolute permeability ($k_{abs}$) determined using Darcy's law. After brine saturation, the core was saturated with light paraffin oil. Oil saturation continued till a 100% oil cut in the effluent and stable pressure drop across the core were achieved. From this flooding, irreducible water saturation or connate water saturation ($S_{wi}$) and initial oil saturation ($S_{oi}$), effective permeability ($k_{eff}$) at $S_{wi}$ and relative permeability with respect to oil ($k_{ro}$) at $S_{wi}$ were determined. The oil flood was followed by a secondary waterflood at a frontal velocity (1 to 2 ft/day) like field-scale water flooding until no further oil was recovered. Waterflood data provided residual oil saturation ($S_{or}$) and relative permeability with respect to water ($k_{rw}$) at $S_{or}$. The eaterflood was followed by the formulated AS slug and recovery efficiency was determined.

## 6.3 Results and Discussion

### 6.3.1 Chemical Formula Design

The fundamental requirement of CEOR processes is to produce a low IFT. The ST and IFT of water and oil samples were measured and are presented in Table 6.1.

The principal limitation of most chemical flooding is adsorption of surfactants on the reservoir rock surface. As the clay of porous media of the Upper Assam Basin is negatively charged [39], an anionic surfactant is preferred. The porous media repels the anionic surfactant and thus shows low adsorption effects. This led to the selection of anionic surfactants for the porous media of Upper Assam. BL is a highly toxic compound and can cause severe health hazards to humans and poses a threat to aquatic life if disposed of without treatment. Thus, prior to discharge, it must undergo proper treatment, which, in turn, has high financial implications. Therefore, proper strategic use of BL for surfactant EOR may not only provide economic recovery of residual oil but also save the environment from its hazards. Figure 6.1 shows the IFT behaviour of crude oil

**Table 6.1** ST and IFT of water and oil samples at 30°C

| S. no. | Sample | ST or IFT mN/m |
|---|---|---|
| 1 | Brine | 68 |
| 2 | Oil | 24.3 |
| 3 | Oil/brine | 13.8 |

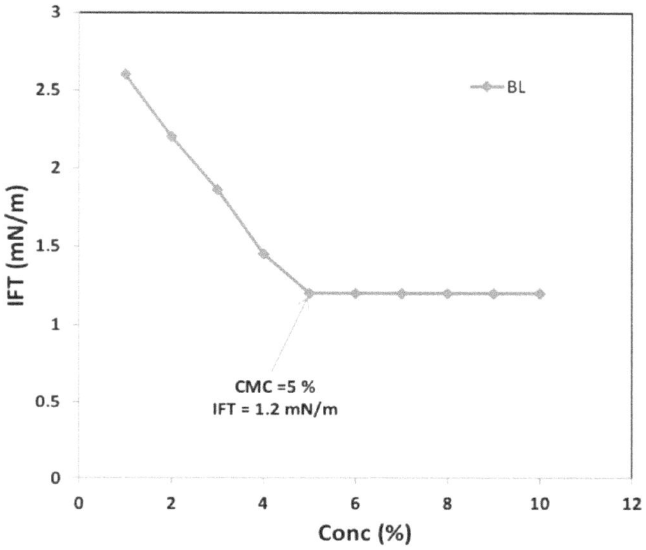


**Figure 6.1** CMC of BL at 70°C for the oil sample.

with BL, and Fig. 6.2 shows the IFT behaviour of the oil sample with different surfactant mixtures, each containing BL. The reservoir temperature of the Barail sand block is reported as 70°C–90°C. The CMC is the concentration at which the surfactant solution begins to form micelles in a large amount and all additional surfactants added to the system go to micelles.

It is seen from Fig. 6.1 that BL could lower the IFT of oil/brine to 1.2 mN/m from a value of 13.8 mN/m (Table 6.1). Figure 6.2 reveals that if other surfactants, viz. SDS, SDBS and AOS, were added to BL, there is a further reduction of the IFT to 0.239 mN/m, 0.132 mN/m and 0.153 mN/m, respectively. When a surfactant solution consisting of BL, SDBS and AOS at their corresponding CMC values was tested for its IFT-reducing capability, an encouraging reduction of the IFT to a value of 0.102 mN/m was obtained. Although this surfactant system was found to be more effective in lowering the IFT compared to BL SDS, SDBS and AOS alone, the required ultralow IFT for efficient EOR was not achieved [82–84] widely address the effectiveness of an alkali injected in conjunction

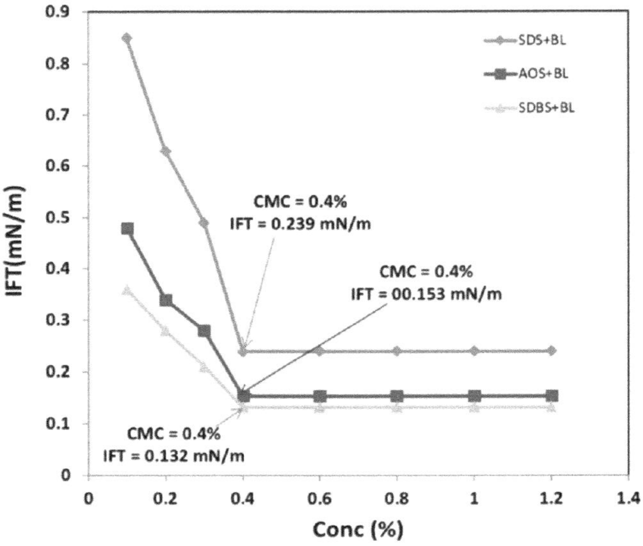


**Figure 6.2** CMC of different surfactant combinations at 70°C for the oil sample.

with a surfactant in recovering residual oil. Besides helping the surfactant to produce an ultralow IFT, the alkaline solution reacts with a divalent to form precipitates. The precipitates preferentially reduce high-permeability channels, and thus sweep efficiency is improved. This process is known as mobility-controlled caustic flood (MCCF) [85, 86]. Also, addition of alkalis increases pH, thus providing an optimal salinity for the surfactant slug, and lowers the surfactant adsorption or retention on reservoir rock [87], optimising the economic use of the surfactant solution.

Therefore, further experiments to examine the effects of an alkali in the interfacial behaviour of reservoir fluid at reservoir temperature were conducted. $Na_2CO_3$ and $NaHCO_3$ were added at different concentrations (1 wt% and 2 wt%) to the formulated surfactant system to study the change in the IFT. Table 6.2 shows the IFT for the surfactant system and AS slug recorded during this process.

It is evident from Table 6.2 that addition of both the alkalis to the surfactant mixture results in lowering of the IFT to lesser

**Table 6.2** Effect of alkalis on the IFT

| $Na_2CO_3$ conc. (wt%) | $NaHCO_3$ conc. (wt%) | IFT (mN/m) |
|---|---|---|
| 0 | 0 | 0.102 |
| 1 | 0 | 0.034 |
| 2 | 0 | 0.008 |
| 0 | 1 | 0.09 |
| 0 | 2 | 0.065 |

values compared to the surfactant mixture without an alkali. Adding $Na_2CO_3$ at 1 wt% concentration to the surfactant system brings down the IFT value from 0.102 mN/m to 0.034 mN/m. Increasing the $Na_2CO_3$ concentration to 2 wt% introduces further IFT reduction to a value of 0.008 mN/m. $NaHCO_3$ also reduces the IFT compared to the surfactant system alone, but the reduction is not much pronounced as found in the case of $Na_2CO_3$. Thus, it may be summarised that using $Na_2CO_3$ (2 wt%) with an aqueous surfactant system causes IFT reduction by a factor of $10^{-1}$–$10^{-2}$ compared to a surfactant mixture alone.

The oil/water IFT depends on multiple parameters such as reservoir temperature, pressure, crude composition, FW salinity and composition, to mention a few. All IFTs to this point were measured in brine solutions prepared in distilled water. The base IFT of the oil sample with its corresponding FW and brine having almost the same salinity as that of FW was calculated at 30°C, 60°C and 90°C. This compares the IFT behaviour with respect to the laboratory-prepared brine and actual FW and establishes the effect of temperature on the IFT. The outcome of this investigation is presented in Table 6.3.

**Table 6.3** Base IFTs at various temperatures in formation water and brine

| Temp °C | IFT in formation water (mN/m) | Formation water salinity as NaCl in (ppm) | IFT in 2000 ppm NaCl in DW (mN/m) |
|---|---|---|---|
| 30 | 4.34 | | 10.21 |
| 60 | 5.57 | 1850 | 13.12 |
| 90 | 5.92 | | 14.47 |

**Table 6.4** Formation water analysis data

| S. no. | Characteristics | Unit | Formation water |
|---|---|---|---|
| 1 | Appearance | – | Turbid |
| 2 | pH | – | 8.4 |
| 3 | salinity as NaCl | mg/l | 1850 |
| 4 | Total dissolved solids | mg/l | 3140 |
| 5 | Turbidity | NTU | 179 |
| 6 | Chloride | mg/l | 516 |
| 7 | $Na^+$ | mg/l | 2667 |
| 8 | $K^+$ | mg/l | 40 |
| 9 | $Ca^{2+}$ | mg/l | 16 |
| 10 | $Mg^{2+}$ | mg/l | 5 |

It is evident from Table 6.3 that crude IFTs in FW increase with temperature. However, within the range of 30°C–90°C, temperatures do not appear to have a significant effect on the IFT. The same is the case with the laboratory-prepared brine. Crude IFTs are higher in brine than in FW. This may be due to the fact that in the laboratory-prepared brine, many of the other ions present in the FW are absent. Table 6.4 shows the details found, while characterising the FW under consideration.

Another aspect of this analysis is that the concentration of divalent cations such as $Ca^{2+}$ and $Mg^{2+}$ will let us know the scaling that can cause severe formation plugging. In the formation water under study, the concentration of $Ca^{2+}$ and $Mg^{2+}$ is $<20$ mg/L (ppm); hence, scale formation should not be a matter of concern [88].

## 6.3.2 Aqueous Stability

Aqueous stability is important to ensure a stable surfactant slug for injection, one where the alkali and surfactant are compatible with the electrolyte concentration at the selected salinity. An experiment containing any phase separation or precipitation at the aqueous phase beyond the selected salinity fails this screening.

Observations of aqueous solutions were recorded first at room temperature, and next the ampules were equilibrated at 70°C in an oven. The aqueous solutions were monitored after 70°C also. It was observed that the aqueous solutions with the formulated AS slug yield a homogeneous and aqueous stable mixture with up to 10,000 ppm salinity. The aqueous stability did not change after 60 days, demonstrating the stability of the AS system under the condition of high temperature and salinity.

### 6.3.3 Foam Stability

This test was conducted both in the presence and in the absence of crude oil and at two different temperatures, that is, at room temperature and at reservoir temperature. Figure 6.3 shows the rate of foam destruction at 30°C and 70°C in the absence of crude oil. Figure 6.4 compares the rate of foam destruction at 70°C for different surfactants in the absence of oil.

It is clear from Fig. 6.4 that at 30°C, foam stability in the absence of oil is the highest for SDS, followed by AOS and SDBS. However, in

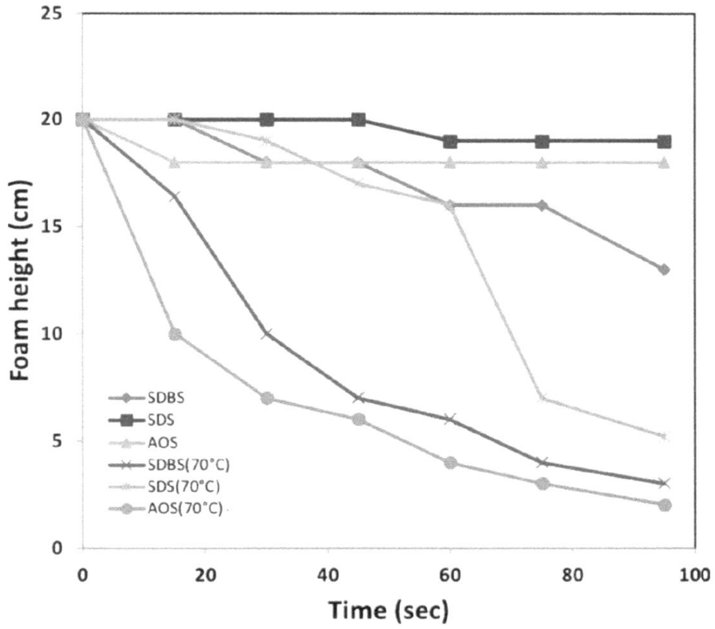


**Figure 6.3** Foam height for different surfactants in the absence of oil at 30°C and 70°C.

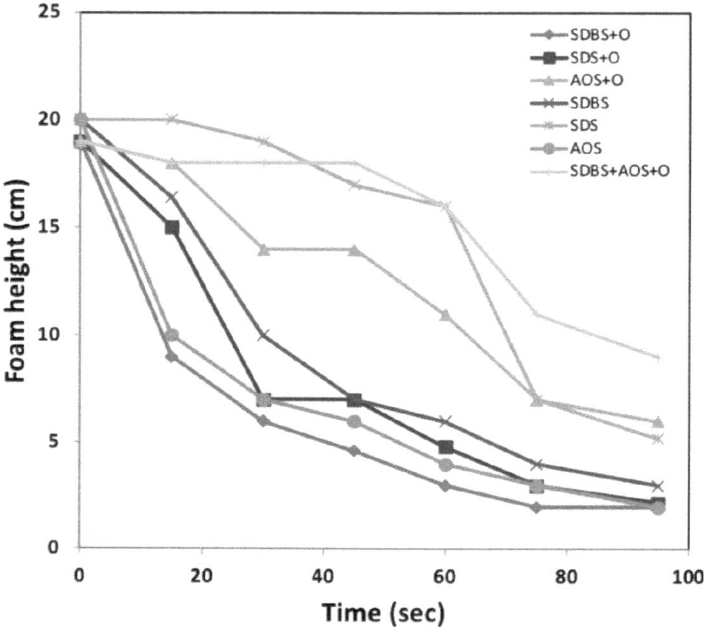


**Figure 6.4** Foam height for different surfactants in the presence of oil at 70°C.

the presence of crude oil this trend was changed and AOS showed better foam stability, followed by SDBS and SDS, which can be clearly seen in Fig. 6.4. When the solution of surfactant mixture containing SDBS and AOS was tested in the presence of oil, it showed better foam stability compared to surfactants alone. The rate of foam destruction increased with an increase in temperature. BL did not show any good foaming characteristics, and the foam disappearance trend was the same at high temperature.

## 6.3.4 Thermogravimetric Analysis

The thermal analysis of the formulated slug is illustrated in Fig. 6.5. Three mass loss steps were observed. The first region, with a mass loss of 22% from ambient to 100°C, corresponded to the loss of weakly bonded water molecules [89, 90]. Then, a 74.7% mass loss occurred sharply in the second region from 100°C to 150°C,

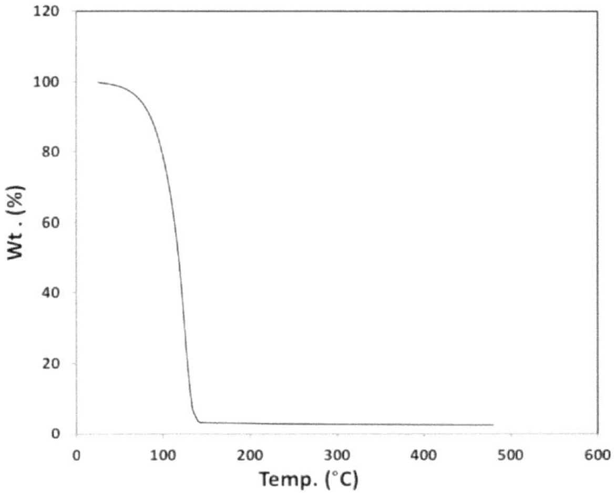


**Figure 6.5** Thermal stability curve for AS formulation.

revealing that AS formulation's molecules start to decompose at temperatures exceeding 100°C. The third degradation region from 150°C to 500°C represents a complex thermal decomposition with mass loss of 0.7%; the residual components of the AS formulation were thermally stable up to 500°C [91, 92]. But since the reported reservoir temperature of the Barail sand block is 70°C–90°C, the AS formulation would retain an average of 92.5% of its original structure and mass [93]. It can be concluded that the AS formulation is thermally stable under the desired reservoir temperature for application in EOR.

### 6.3.5 Core Flooding

Core flooding is a means to validate whether a promising chemical formulation that performed well in the previous experiments can recover residual oil in cores with low surfactant retention. The core-flooding experiment showed significant residual oil recovery of 69% of the water flood residual oil saturation. Table 6.5 lists dimensions, permeability and saturation core property data measured or calculated during core preparation and preflood (brine flood, oil flood, water flood) experiments.

**Table 6.5** Measured properties of the core sample during core flooding

| | | | |
|---|---|---|---|
| Mass, dimension | Mass (dry core only) | m (g) | 220.53 |
| | Total length | L (cm) | 8.8 |
| | Diameter | D (cm) | 3.8 |
| | Pore volume | PV (cc) | 14.93 |
| | porosity | $\phi$ | 17.19 |
| Permeability | Air (dry, bare core) | $k_{air}$ (md) | 11.5 |
| | Absolute (brine) | $k_{br}$ (md) | 8.8 |
| | Relative oil (at $S_{wi}$) | $k_{ro}$ (md) | 0.88 |
| | Relative water (at $S_{or}$) | $k_{rw}$ (md) | 0.13 |
| Saturation | Initial water | $S_{wi}$ (fraction) | 11.5 |
| | Initial oil | $S_{oi}$ (fraction) | 0.88 |
| | Residual oil | $S_{or}$ (fraction) | 0.642 |

## 6.4 Conclusion

This chapter concludes that on the basis of IFT criteria, a composition of 5% BL (w/w), 0.4% SDBS (w/w), 0.4% AOS (w/w) and 2% $Na_2CO_3$ (w/w) for the selected depleted field best suits as an AS slug for CEOR. This AS slug lowers the IFT to $10^{-3}$–$10^{-2}$, which will lead to an increase in $N_c$ favourable for EOR. The presence of ions other than $Na^+$ in FW enhances further reduction of the IFT between oleic and aqueous phases of the reservoir. The IFT of the crude oil–FW system increases slightly with temperature. However, the study indicated that temperature is not a pertinent parameter for lowering the IFT in the case of the fields under study. As the $Ca^{2+}$ and $Mg^{2+}$ concentration in the FW under consideration is less than 20 ppm, no scaling problem is expected for AS EOR. The formulated AS slug showed aqueous and thermal stability at reservoir temperature.

This novel formulation was tested in a core flood for the oilfield, and 69% of the water flood residual oil saturation was recovered. This is a very significant development with a high potential to decrease the cost of CEOR since the primary surfactant used here is BL, which is cheap and easily available as an effluent from paper mills.

## Acknowledgements

We gratefully acknowledge the following:

(a) Indo–US Project F. No. 194-1/2009(IC) dated February 2015, titled "Foam-Assisted $CO_2$ Flooding for the Depleted Reservoirs of Upper Assam Basin and in Candidate Reservoirs in Louisiana', of the 21st Century Knowledge Initiative programme under University Grants Commission

(b) Dibrugarh University for all laboratory facilities

## Chapter 7

# Characterisation of Reservoir Rock and Fluids for $CO_2$ Foam Enhanced Oil Recovery Application

**Ranjan Phukan,**[a] **Subrata Borgohain Gogoi**[b] **and Pankaj Tiwari**[a]
[a] *Chemical Engineering Department, Indian Institute of Technology, Guwahati, India*
[b] *Department of Petroleum Technology, Dibrugarh University, Assam, India*
rphukan29@gmail.com

The characterisation of reservoir rock and fluids is essential before embarking on any enhanced oil recovery (EOR) process. Moreover, the successful application of EOR processes requires a detailed reservoir description, including engineering and geological considerations. In this chapter, the reservoir rock, crude oil and formation water (FW) of an Upper Assam oilfield were characterised to evaluate the feasibility of implementing a $CO_2$ foam EOR process. First, routine core analysis was done to determine the important petrophysical properties of reservoir rocks, including porosity and permeability. Second, characterisation of the reservoir rock was done to evaluate its mineralogy and clay content through X-ray diffraction (XRD) and field emission scanning electron microscopy (FESEM) studies. Third, the important physicochemical properties of crude oil affecting the process of oil recovery, including American Petroleum Institute (API) gravity, viscosity, pour point, resin-

*Advances in Petroleum Technology*
Edited by Subrata Borgohain Gogoi

ISBN 978-981-4877-19-0 (Hardcover), 978-1-003-04993-7 (eBook)
www.jennystanford.com

asphaltene content and acid number, were determined. Fourth, FW analyses were done to evaluate its physciochemical properties. The porosity of the core plugs ranged from 11% to 28% and was found to bear an inverse correlation with the depth of the formation. The gas permeability of the core plugs was observed to be greater than the permeability value obtained by flowing liquid. The values of liquid and gas permeabilities of the core plugs ranged from 3 to 30 mD and 16 to 47 mD, respectively, which classified the oilfield formation as fair- to moderate-permeability reservoirs. Porosities and permeabilities of the core plugs were observed to be directly related as more porous rocks were also more permeable. Further, the XRD study of powered rock samples indicated the presence of silica (quartz) in the rock sample, which classified the reservoir rock as anionic type. Additionally, the FESEM micrograph of the rock sample showed the presence of smectite clays in the rock matrix. The experimental investigation confirmed that the crude oil was medium-gravity (API gravity 31°) with a dead oil viscosity of 4.46 cP at 70°C. The pour point of the crude oil was high (24°C ± 3°C), which indicated the paraffinic nature of the crude oil. The wax content was also fairly high at 12.47% ± 0.70%. The resin-to-asphaltene ratio of the crude oil was high enough for asphaltene stabilisation during $CO_2$ injection. The analysis also marked the crude as acidic and suitable for alkaline flooding. The FW analysis showed the presence of different types of dissolved cations and anions with a total salinity of 3458 ppm.

## 7.1 Introduction

The laboratory analyses of reservoir rock and fluids are necessary pre-requisites for characterising a particular oil reservoir so as to make a preliminary assessment of the effectiveness of any enhanced oil recovery (EOR) schemes. For the successful field-scale implementation of such EOR process, characterisation and understanding the properties of the reservoir rock, crude oil and formation water (FW) are very crucial. On the basis of a detailed oil reservoir description, including its engineering and geological aspects, a plausible EOR process could be selected on the basis of the technical screening criteria [94]. Screening criteria of different EOR methods are reported in the literature by various authors [95–97]. These criteria are used in many field applications, the laboratory and simulation studies to choose the best and most feasible EOR method for any candidate reservoir [98]. Foam has a wide range of applications in the oil industry. During drilling, foam can be used to carry the rock cuttings to the surface. In well stimulation treatments, foam can cause diversion of acids to improve the acid injection profile. In EOR, foams are popularly used for gas mobility control and improving oil recovery during gas/$CO_2$ applications. Foam generated in the reservoir rock by the co-injection or alternate injection of a chemical solution and gas exhibits certain favourable attributes, which makes it an attractive method for improving oil recovery. In gas-based EOR methods, foam application reduces the apparent viscosity of the gas and lowers the relative permeability of the liquid, making the mobility ratio (MR) favourable. Additionally, foam has the unique property of selective mobility reduction, which diverts fluid from higher- to lower-permeability zones of the reservoir, thus improving $E_{vo}$. The presence of surfactants in the foam-flooding process also helps to mobilise residual oil by lowering the oil–water interfacial tension (IFT). The foam-based EOR method of this chapter is a relatively new process termed 'alkaline-surfactant-alternated-gas/$CO_2$ (ASAG) flooding', which involves the alternate injection of an alkali-surfactant (AS) slug and $CO_2$ gas into the reservoir. The alternate injection creates a repeated imbibition–drainage cycle, which helps generate in situ foam in the reservoir rock. Although screening criteria have been suggested for chemical and $CO_2$ flooding on the basis of worldwide projects, no specific criteria have been reported for ASAG flooding due to limited number of field applications and laboratory studies. Aladasani [99] updated the EOR criteria published by Taber et al. [95], including new EOR categories considering projects reported from 1998 to 2010. Immiscible $CO_2$ flooding, water-alternating gas (WAG), hydrocarbon (HC)-WAG and surfactant-polymer/alkali were among the newly added EOR categories. Sheng [87] proposed screening criteria for AS flooding on the basis of the criteria for both alkaline flooding and surfactant flooding.

**Table 7.1** Screening criteria for AS immiscible $CO_2$ and ASAG $CO_2$ flooding

| S. no. | Method | °API gravity of oil | $\mu_o$ (cP) | Acid no. (mg/g) | K(mD) | $\phi$ (%) |
|---|---|---|---|---|---|---|
| 1 | AS | NC | ≥35 | Organic acid | ≤10 | NC |
| 2 | Immiscible $CO_2$ | ≥22 | ≤ 10 | NC | 1.5–4500 | 3–37 |
| 3 | Suggested for ASAG | ≥22 | ≤10 | Organic acid | ≥ 10 | 3–37 |

| S. no. | Rock type | Clay content | TDS in FW (ppm) | Divalent ions (ppm) | $S_o$ | Depth (ft.) | Temperature (°C) |
|---|---|---|---|---|---|---|---|
| 1 | Sandstone | Low | ≤50,000 | ≤100 | ≥35 | NC | ≥ 93.3 |
| 2 | Sandstone and carbonate | NC | NC | NC | ≥42 | ≥ 1800 | 28–125 |
| 3 | Sandstone | Low | ≤50,000 | ≤100 | ≥42 | ≥1800 | ≥93.3 |

Table 7.1 lists the technical criteria for separate AS flooding and immiscible $CO_2$ flooding. On the basis of this data, new criteria for ASAG flooding are also suggested in the table. Whereas $CO_2$ flooding is applicable for both sandstone and carbonate reservoirs, most of the chemical (AS and alkali-surfactant-polymer [ASP]) floods are conducted on sandstone reservoirs with only a few exceptions. AS and ASP floods are less effective in carbonate reservoirs because the high clay content in carbonate rocks results in surfactant adsorption. Moreover, the anhydrite, dolomite and gypsum occurring in carbonate reservoirs cause the consumption of alkali and form precipitates [100]. The clay content of sandstone reservoir rocks should also be low for surfactant-based EOR processes to prevent excessive surfactant adsorption [101]. Besides, reservoirs with FW containing high total dissolved salts and high hardness are not desirable for chemical floods as surfactants have a tendency to predicate and separate under these harsh conditions [102]. Another important parameter affecting the performance of chemical flooding is the presence of an excess amount of divalent ions in FW. The presence of these ions like $Ca^{2+}$ and $Mg^{2+}$ leads to overconsumption of alkali and scale formation due to their reaction with the alkali [103]. Over the years, it has been well-established that for light- to medium-gravity oil reservoirs, $CO_2$ flooding is one of the most effective EOR processes [104, 105].

Similarly, for chemical enhanced oil recovery (CEOR) methods like alkaline and AS flooding, crude oil containing light to intermediate components with some organic acids is preferred [95]. The injected alkali reacts with the acidic components of the crude oil to produce in situ surfactants, so the acidic crudes are favoured. Typically, the minimum acid number for the crude to be effective for alkaline and AS flooding is 0.3 mg KOH/g of oil [106]. Moreover surfactant systems are sensitive to high temperatures, so to prevent degradation of surfactants, the reservoir temperature needs to be typically less than 200°F [107]. Formation permeability and oil viscosity are less important parameters affecting the performance of surfactant-based EOR processes [87]. But, in $CO_2$ flooding, the possibility of viscous fingering of the injected gas increases with increase in oil viscosity, and consequently $CO_2$ flooding has been applied mainly in reservoirs having oil viscosity less than or equal to 3 [108]. As the ASAG technique is a combination of immiscible $CO_2$-WAG and chemical (AS) methods, the suggested screening criteria should suit both AS flooding and $CO_2$ flooding. However, the synergetic effect of ASAG flooding was not contemplated in the suggested criteria due to a lack of reported field projects.

The success of the ASAG process requires the selection of the appropriate candidate oil reservoir with favourable rock and fluid properties. Thus, a systematic investigation of these properties is required to be done to ascertain the applicability of the EOR process. The present study was done to determine the reservoir rock and fluid properties of an Upper Assam oilfield to match these properties with the screening criteria proposed for ASAG flooding. To fulfil this objective, the reservoir rock, crude oil and FW of an Upper Assam oilfield were characterised to evaluate the suitability of implementing the ASAG flooding process. For this purpose, routine core analysis was initially done to determine the important petrophysical properties of reservoir rocks, including porosity and permeability. Next, the reservoir rock was characterised to evaluate its mineralogy and clay content through X-ray diffraction (XRD) and field emission scanning electron microscopy (FESEM). Further, the important physicochemical properties of crude oil, including American Petroleum Institute (API) gravity, viscosity, pour point, resin–asphaltene content, wax content and acid number, were determined through standard tests. Additionally, FW was also analysed to estimate the important physicochemical properties.

## 7.2 Experimental

### 7.2.1 Materials

Core plugs (3.75 cm diameter and approximately 7 cm length), as shown in Fig. 7.1, used for core analysis and core flooding, were cut parallel to the bedding plane from conventional cores obtained from different depths of an Upper Assam oilfield ranging in 2480–2486 m, 2925–2943 m and 3940–3947 m. The oil used was the crude oil collected from a producing well of the oilfield. Crude oil was used to mimic the reservoir conditions and obtain

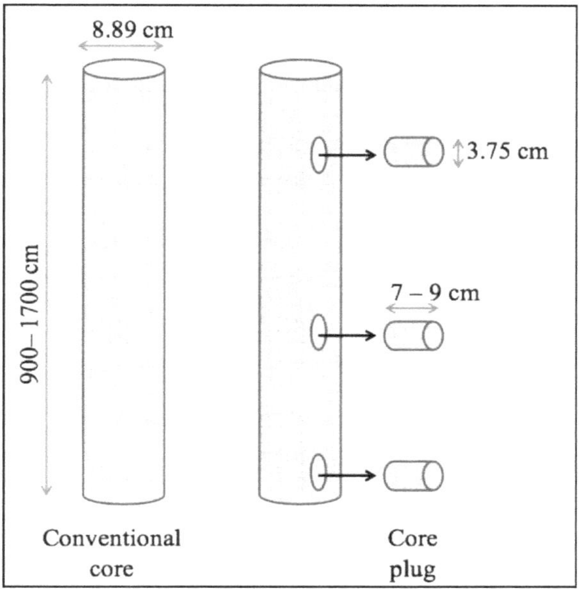


**Figure 7.1** Schematic representation of conventional cores and core plugs.

representative results. The crude oil was separated from water using a separating funnel at a temperature of 35°C before carrying out detailed analysis. FW of the oilfield was also collected and thoroughly analysed. The oilfield under study is one of the main crude oil producers of the Upper Assam Basin in India, covering an area of about 25 sq. km. The oilfield was discovered in the late 1960s, and commercial production initially commenced from the mid-1970s. The oil-producing zones have been identified as Tipam Sand, Barail Coal Sand and Barail Main Sand. Tipam Sand is subdivided into six zones from where the majority of oil production takes place. The sediments are mainly sandstone mixed with clays and shales. The driving mechanism of the oilfield is mostly depletion drive, and the reservoirs were developed through water injection. The reservoir selected for the study is the most salient and extensive one in the oilfield. The reservoir pressure initially was around 3400 psia at a depth of around 3000 m and the formation temperature about 70°C. However, the present pressure has declined to approximately

1800 psia and is lower than the saturation pressure (2500 psia) of the crude oil.

### 7.2.2 Core Analysis

The core plugs were cut from the conventional cores with the help of a core-cutting machine and a core-plugging machine. Thereafter, the plugs were end-faced using an end-facing machine to obtain plugs which were exactly uniform right-angled cylinders. The core plugs were required to be thoroughly cleaned of all interstitial fluids and dried before the determination of their petrophysical properties. The cleaning was done in the Soxhlet apparatus for about 40 working hours. The solvents used to dissolve and extract oil/brine from the core plugs were a mixture of 50% $CH_3OH$ and 50% $C_6H_42CH_3$. The cleanliness of the sample was determined from the colour of the solvent which was siphoned periodically from the extractor. After cleaning in the Soxhlet apparatus, the core plugs were again cleaned in an ultrasonic cleaner for 10 min by submerging the samples in acetone solution. The core plugs were then dried to remove any moisture present in the pore spaces. Drying times varied for core plugs but usually took from 30 to 50 working hours. The plugs were weighed after a definite interval of time (every 4 hours) till a constant weight was achieved, indicating that the plugs were completely dried.

The bulk volume (BV), grain volume and pore volume (PV) determination is a prerequisite for the laboratory determination of porosity. The BV can be measured with the help of calipers or the Archimedes (buoyancy) principle. The PV can be measured by the liquid saturation method or Boyle's law single-cell method. The grain volume can be measured from the dry weight of core plugs and sand grain density or Boyle's law double–cell method [109]. In this study, the BV of the core plugs was measured with the help of callipers and the grain volume was determined with a helium porosimeter (Core Test Systems, Inc., USA; model TPI-219), which was based on Boyle's law double-cell method. The liquid and gas permeabilities of core plugs were measured experimentally and the results were correlated. The liquid permeabilities were measured with a liquid permeameter (Vinci Technologies, France; model

LiquidPerm) and the gas permeabilities with a gas permeameter (Core Test Systems, Inc., USA; model TKA-209). The measured gas permeability values were used to determine the liquid permeability applying the Klinkenberg principle.

### 7.2.3 Crude Oil Analysis

The ASTM D1298-67 (IP 150/68) hydrometer method was followed for the determination of the density of crude oil samples. The dead oil viscosity of the oil samples was measured following the ASTM D445-65 method using a U-tube reverse flow viscometer (type BS/IP/RF). The standard test method IP 15/67 was followed to measure the pour point of the crude oil. The pour point is the lowest temperature, expressed as a multiple of 3, at which crude oil is observed to flow when cooled and examined under prescribed conditions. The asphaltene precipitation with n-heptane (ASTM D2007-80) was reported by Bon [129]. The experiments were performed under ambient conditions to measure the percentage by weight of asphaltenes in the crude oil. To estimate the resin content, the filtrate was contacted with silica gel. The filtrate, after removing asphaltene and resin, was used for measuring the wax content by mixing with methyl ethyl ketone (MEK). The acid number was measured by the standard titration method. Fourier transform infrared (FTIR) spectroscopic analysis of the crude oil was performed with an FTIR spectrophotometer (Shimadzu; model IR Affinity-1). The study was done to determine the functional groups present in crude oil, mainly the existence of carboxylic acids.

### 7.2.4 Formation Water Analysis

The pH and total dissolved solids (TDS) of FW were measured with a water analyser (Systronics; model 371). The density was measured with a pycnometer and viscosity using a U-tube viscometer. The positive ionic compositions like $Na^+$, $Ca^{2+}$ and $K^+$ were measured with a flame photometer (Systronics; model 128), whereas $Mg^{2+}$ was measured by titration with ethylenediaminetetraacetic acid (EDTA). For negative ionic compositions like $Cl^-$ and $HCO_3^-$, titrations were performed with $AgNO_3$ solution and HCl, respectively.

## 7.3 Results and Discussion

### 7.3.1 Core Analysis

Core analysis of reservoir rocks samples helps to provide a direct and quantitative evaluation of important reservoir properties. These properties assist in accurately describing reservoir rocks, total fluid content, fluid distribution and flow behaviour. Static and dynamic reservoir models can also be designed on the basis of formation evaluation done with core analysis [110]. For core analysis, a whole core is acquired from a particular depth of the producing formation through coring operation. As it is not possible to use whole/conventional cores in the laboratory for core analysis, short representative core plugs (2.54–3.81 cm in diameter) from the large conventional cores are selected. The core plugs are cut from the conventional cores parallel to the bedding plane, so that measurements of horizontal permeability may be made (Fig. 7.1). The core plugs used in this work were 3.75 cm in diameter and 6–8 cm in length. There are basically two main types of core analyses that are performed on core plugs. Conventional/routine core analysis includes measurements of petrophysical properties of core plugs like porosity, permeability and fluid saturation, together with the evaluation of grain density, rock texture and lithology. Special core analyses are done for measuring specific properties such as capillary pressure, relative permeability, wettability, electrical properties, formation damage and recovery factors. These special tests are also conducted to characterise reservoirs for EOR and to study the fluid–rock interactions.

### 7.3.2 Determination of Porosity

One of the most important reservoir rock properties is porosity, which is a measure of the space available for the storage of HCs. It is the effective porosity which is important in reservoir engineering calculations because it represents the interconnected pore space which contains the recoverable crude oil. The grain volume of the core plugs was measured with a helium porosimeter. The grain volume measurements were repeated, and the average

Table 7.2 Data of core plugs from different depths

| CP | $D$ (m) | Dia. (m) | $L$ (m) | BV ($cm^3$) | Wt. (g) | GV ($cm^3$) | $\phi$ (%) |
|---|---|---|---|---|---|---|---|
| CP1 | | 3.75 | 7.75 | 85.61 | 195.74 | 63.45 | 25.69 |
| CP2 | 2480–2486 | 3.75 | 7.02 | 77.54 | 161.15 | 56.16 | 27.57 |
| CP3 | | 3.75 | 6.81 | 75.02 | 156.11 | 54.65 | 27.15 |
| CP4 | | 3.75 | 8.11 | 89.58 | 199.34 | 71.28 | 20.44 |
| CP5 | | 3.75 | 7.43 | 82.07 | 186.28 | 65.52 | 20.14 |
| CP6 | | 3.75 | 6.90 | 76.22 | 159.61 | 61.03 | 19.96 |
| CP7 | | 3.75 | 6.95 | 76.77 | 160.27 | 60.81 | 20.76 |
| CP8 | | 3.75 | 7.50 | 82.85 | 188.82 | 65.91 | 20.49 |
| CP9 | | 3.75 | 7.61 | 84.06 | 190.36 | 66.37 | 21.04 |
| CP10 | 2925–2943 | 3.75 | 6.75 | 74.56 | 154.72 | 59.95 | 19.57 |
| CP11 | | 3.75 | 7.55 | 83.40 | 189.83 | 65.50 | 21.46 |
| CP12 | | 3.75 | 7.00 | 77.32 | 158.54 | 61.81 | 20.09 |
| CP13 | | 3.75 | 6.89 | 76.11 | 159.34 | 60.44 | 20.52 |
| CP14 | | 3.75 | 7.48 | 82.63 | 187.86 | 65.76 | 20.39 |
| CP15 | | 3.75 | 7.53 | 83.18 | 188.65 | 65.41 | 21.35 |
| CP16 | | 3.75 | 6.97 | 76.99 | 162.32 | 61.04 | 20.73 |
| CP17 | | 3.75 | 7.60 | 83.95 | 189.69 | 66.09 | 21.27 |
| CP18 | | 3.75 | 7.58 | 83.73 | 188.16 | 73.49 | 12.23 |
| CP19 | 3940–3947 | 3.75 | 6.88 | 76.00 | 156.97 | 67.15 | 11.64 |
| CP20 | | 3.75 | 7.50 | 82.85 | 188.56 | 73.36 | 11.45 |

Table 7.3 Porosity-permeability of core samples as a function of depth

| **$D$ (m)** | 2480–2486 | 2925–2943 | 3940–3947 |
|---|---|---|---|
| **$\phi$ range (%)** | 25–28 | 19–21 | 11–13 |
| **Avg. liquid $K$ (mD)** | 27.21 | 6.32 | 3.25 |
| **Estimated liquid $K$ (mD)** | 31–39 | 7–10 | 4–7 |

of three values was taken to calculate the porosity of each core plug. The results of the porosity measurements are presented in Table 7.2. As can be seen, the values of porosities of the core plugs varied from 25%–28% at a shallower depth, 19%–21% at intermediate depths and 11%–13% at deeper depths. Similar values of porosity of the Upper Assam oilfield's core plugs were obtained by previous workers [111]. The porosity/depth data of the core plugs is summarised in Table 7.3.

The low porosity values were observed with core plugs obtained from a deeper depth (3940–3947 m), while relatively high porosity was observed for core plugs from depths of 2480–2486 m. Thus, the porosities of the rock plugs were found to decrease with the depth of the formation from where the conventional cores were collected. This is consistent with the known fact that porosity bears an inverse correlation with depth. The porosity–depth correlation is influenced by many factors, but the most obvious reason for this inverse relation is because of the combined effects of compaction and cementation [112]. An increase in the depth of progressive burial will increase the overburden pressure or compaction, resulting in the reduction of porosity. As the effective stress and temperature increase with the burial depth, the porosity–depth correlations are also dependent on effective stress. The other factors which influence porosity profiles are subsidence history, oil/gas saturation and mineralogy. However, it is difficult to isolate the effect of any single factor or processes responsible for the observed relation [113].

#### 7.3.2.1 Determination of permeability ($K$)

The liquid and gas permeabilities were measured at 31°C and gas/$N_2$ viscosity ($\mu$) of 0.02 cP. The liquid permeability experimental results of all the core plugs are shown in Table 7.4 where, $K$ is measured on the basis of Eq. 7.1, while Table 7.5 presents the gas permeability measurement results of three core plugs of different depths, where $K$ is measured on the basis of Eq. 7.2. The values of liquid permeability of the core plugs ranged from 3 to 30 mD, while the gas permeability varied from 16 to 47 mD.

$$K_{\mathrm{L}} = \frac{\mu Q L}{\Delta P A} \tag{7.1}$$

$$K = \frac{2000\mu_g L P_b q_{sc}}{A(P_1^2 - P_2^2)}, \tag{7.2}$$

where $P_{\mathrm{m}}$ is $\frac{P_1+P_2}{2}$, $\mu$ is viscosity of fluids (oil or water) in cP, $Q$ is the volumetric flow rate of any fluid in $\mathrm{cm}^3\ s^{-1}$, $\Delta P$ is pressure difference in atm (inlet pressure – outlet pressure), $A$ is the area of the core plug in $\mathrm{cm}^2$ and $L$ is the length of the core plug in cm.

**Table 7.4** Results of liquid permeability measurement (Confining Pr.: 380 psi; Temp.: 31°C; $\mu_o$: 1 cP)

| CP | Dia. (cm) | $L$ (m) | VC ($cm^3$) | $t$ (s) | $\Delta P$ (atm) | $\mu_w$ (cP) | $A$ ($cm^2$) | $Q$ ($cm^3s^{-1}$) | $K_L$ (mD) | $K_{avg}$ (mD) |
|---|---|---|---|---|---|---|---|---|---|---|
| CP1 | | 7.75 | 10.00 | 204.00 | 1.35 | 1.00 | 11.05 | 0.05 | 25.50 | |
| CP2 | 3.75 | 7.02 | 10.00 | 188.00 | 1.14 | 1.00 | 11.05 | 0.05 | 29.61 | 27.21 |
| CP3 | | 6.81 | 10.00 | 195.00 | 1.19 | 1.00 | 11.05 | 0.05 | 26.53 | |
| CP4 | | 8.11 | 10.00 | 420.00 | 2.47 | 1.00 | 11.05 | 0.02 | 7.09 | |
| CP5 | | 7.43 | 10.00 | 480.00 | 2.62 | 1.00 | 11.05 | 0.02 | 5.36 | |
| CP6 | | 6.90 | 10.00 | 489.00 | 1.99 | 1.00 | 11.05 | 0.02 | 6.40 | |
| CP7 | | 6.95 | 10.00 | 514.00 | 2.33 | 1.00 | 11.05 | 0.02 | 5.26 | |
| CP8 | | 7.50 | 10.00 | 476.00 | 2.01 | 1.00 | 11.05 | 0.02 | 7.11 | |
| CP9 | | 7.61 | 10.00 | 415.00 | 2.07 | 1.00 | 11.05 | 0.02 | 8.02 | |
| CP10 | 3.75 | 6.75 | 10.00 | 585.00 | 2.30 | 1.00 | 11.05 | 0.02 | 4.54 | 6.32 |
| CP11 | | 7.55 | 10.00 | 482.00 | 2.39 | 1.00 | 11.05 | 0.02 | 5.92 | |
| CP12 | | 7.00 | 10.00 | 465.00 | 2.07 | 1.00 | 11.05 | 0.02 | 6.58 | |
| CP13 | | 6.89 | 10.00 | 497.00 | 2.27 | 1.00 | 11.05 | 0.02 | 5.52 | |
| CP14 | | 7.48 | 10.00 | 465.00 | 2.23 | 1.00 | 11.05 | 0.02 | 6.53 | |
| CP15 | | 7.53 | 10.00 | 480.00 | 2.33 | 1.00 | 11.05 | 0.02 | 6.08 | |
| CP16 | | 6.97 | 10.00 | 500.00 | 2.47 | 1.00 | 11.05 | 0.02 | 5.12 | |
| CP17 | | 7.60 | 10.00 | 385.00 | 2.00 | 1.00 | 11.05 | 0.03 | 8.93 | |
| CP18 | | 7.58 | 10.00 | 600.00 | 3.01 | 1.00 | 11.05 | 0.02 | 3.79 | |
| CP19 | 3.75 | 6.88 | 10.00 | 650.00 | 3.15 | 1.00 | 11.05 | 0.02 | 3.04 | 3.25 |
| CP20 | | 7.50 | 10.00 | 689.00 | 3.38 | 1.00 | 11.05 | 0.01 | 2.91 | |

Thus, the permeability measured by flowing gas was observed to be greater than the permeability value obtained by flowing liquid. The observed difference between liquid and gas permeability of the core plugs could be explained on the basis of the Klinkenberg effect. The measured range of permeability of the core plugs classifies the Upper Assam formation as fair- to moderate-permeability reservoirs [113] The porosity and permeability of the core plugs were directly related as more porous rocks were more permeable and vice versa (Table 7.3). However, the effects of a number of parameters like grain size, compaction, packing and diagenetic processes related to development influence the porosity-permeability of a particular formation [114]. Nind [115] reported that there is no direct correlation for porosity and permeability; however, permeability

**Table 7.5** Results of gas permeability measurement

| CP | Dia. (cm) | $L$ (m) | $P_1$ (atm) | $P_2$ (atm) | $P_m$ (atm) | $1/P_m$ ($atm^{-1}$) | $Q$ ($cm^3s^{-1}$) | $K_G$ (mD) |
|---|---|---|---|---|---|---|---|---|
| | | | 2.02 | 1.00 | 1.51 | 0.66 | 5.01 | 46.32 |
| | | | 2.36 | 1.00 | 1.68 | 0.6 | 7.23 | 45.02 |
| CP1 | 3.75 | 7.75 | 2.70 | 1.00 | 1.85 | 0.54 | 9.70 | 43.86 |
| | | | 3.04 | 1.00 | 2.02 | 0.49 | 12.23 | 42.22 |
| | | | 3.38 | 1.00 | 2.19 | 0.46 | 15.33 | 41.83 |
| | | | 1.34 | 1.00 | 1.17 | 0.85 | 0.98 | 31.30 |
| | | | 1.60 | 1.00 | 1.34 | 0.75 | 2.00 | 27.79 |
| CP6 | 3.75 | 6.90 | 2.02 | 1.00 | 1.51 | 0.66 | 3.23 | 26.58 |
| | | | 2.36 | 1.00 | 1.68 | 0.6 | 4.25 | 23.55 |
| | | | 2.70 | 1.00 | 1.85 | 0.54 | 5.72 | 23.01 |
| | | | 1.34 | 1.00 | 1.17 | 0.88 | 0.68 | 21.82 |
| | | | 1.55 | 1.00 | 1.27 | 0.79 | 1.12 | 20.31 |
| CP19 | 3.75 | 6.88 | 1.82 | 1.00 | 1.41 | 0.71 | 1.67 | 18.35 |
| | | | 2.09 | 1.00 | 1.54 | 0.65 | 2.40 | 18.08 |
| | | | 2.36 | 1.00 | 1.68 | 0.6 | 2.88 | 15.92 |

generally increases with porosity for sandstones reservoirs formed under a similar depositional environment.

Figures 7.2, 7.3 and 7.4 show the permeabilities to gas ($N_2$) of three core plugs of different depths plotted against the inverse of the mean pressure ($P_m$) based on the Klinkenberg principle [116].

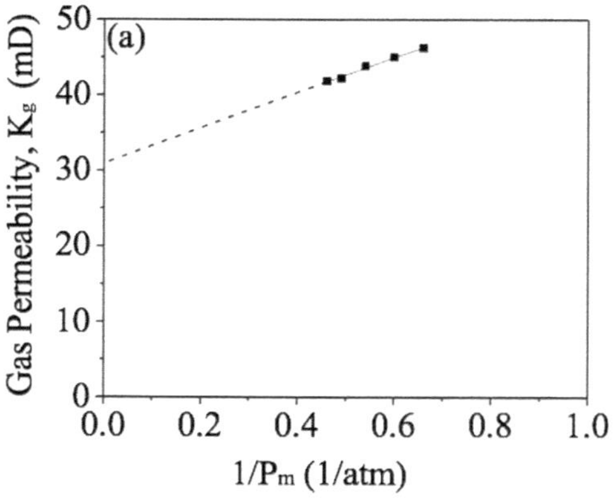


**Figure 7.2** Gas and estimated liquid permeabilities of core plugs (2480–2486 m).

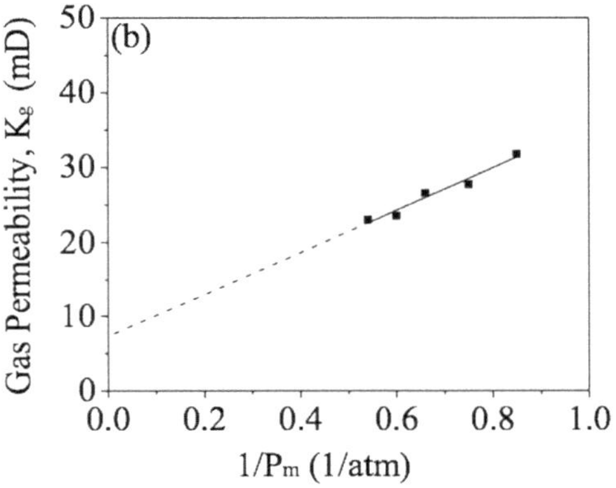


**Figure 7.3** Gas and estimated liquid permeabilities of core plugs (2925–2943 m).

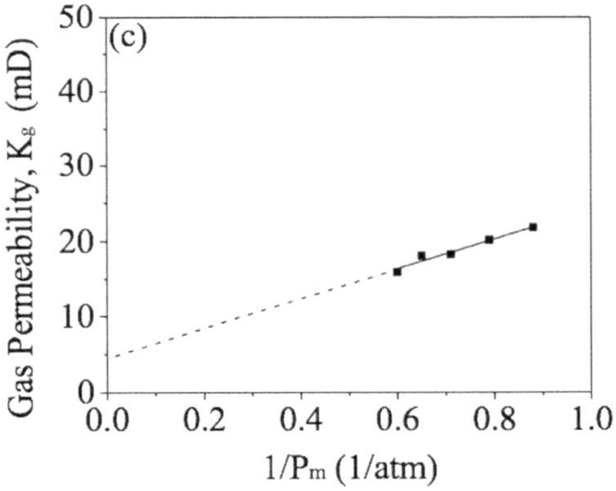


**Figure 7.4** Gas and estimated liquid permeabilities of core plugs (3940–3947 m).

The gas permeability linearly increased with an increase in $\frac{1}{P_m}$ and extrapolating to the point where $\frac{1}{P_m} = 0$ the liquid permeability was estimated. From Table 7.3, it can be seen that the estimated liquid permeability and measured liquid permeability to brine are in agreement. Although the estimated permeability was found to be slightly greater than actual liquid permeability, this difference is much smaller compared to the difference between the permeability of the core plugs to liquid and gas.

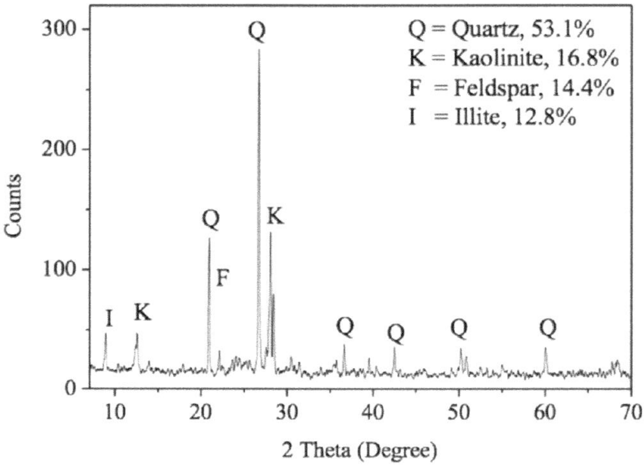


**Figure 7.5** XRD patterns of the core sample.

### 7.3.3 Reservoir Rocks

The characterisation of the reservoir rock was done to evaluate its mineralogy and clay content through XRD and FESEM studies, respectively. XRD is known to be an important tool to identify the mineralogy of reservoir rock [116]. X-ray diffractograms were recorded by a Bucker X-ray D8 advance diffractometer. Clean core samples were grounded to a fine powder sample, and X-ray diffractograms were collected for a range of Bragg angle $2\theta$ ($0°$ to $90°$). Figure 7.5 presents the XRD pattern of the powdered core sample with characteristic peaks at 6.05°, 8.96°, 12.59°, 20.96°, 26.72° and 28.07°. This peak information and analysis with MATCH! 3 indicated the presence of silica (quartz) as the dominant mineral. The large and abundant quartzpeaks in the XRD diffractograms support the above observation. The occurrence of other minerals like feldspar and clays (kaolinite and illite) was also indicated by the XRD study. Since silica carries a negative charge at normal FW pH [117], the reservoir rock was of anionic type. A FESEM study of the rock samples was done for the identification of clay minerals. The sample was prepared by gently breaking the core sample and attached to a FESEM specimen plug after coating with

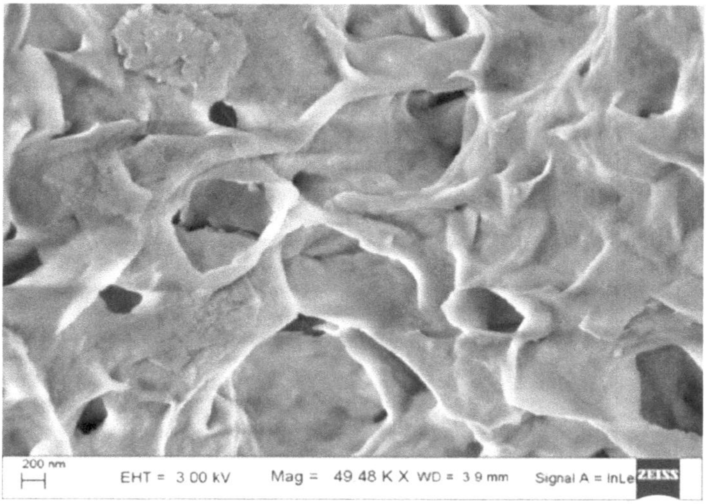


**Figure 7.6** FESEM micrographs of the core sample.

a conductive metal (gold). This coating is required to obtain a clear image of the rock sample, but should also be so thin (200 Å) that it does not hinder the identification of specific minerals. The FESEM images were obtained and compared with standard images to identify minerals. A FESEM micrograph of the rock sample is shown in Fig. 7.6. On the basis of the analysis of the image using established literature [118], the presence of the smectite group of clays was indicated as a thin, webby crust in the matrix of the core sample. The minerals normally present in sandstone reservoirs are quartz, feldspar and clay minerals. The common types of clay minerals present are montmorillonite, smectite , illite, chlorite and kaolinite, depending upon the depositional environment [119, 120]. The presence of these clay minerals makes the surfactants injected for the purpose of EOR susceptible to be adsorbed onto the reservoir rocks [121].

### 7.3.4 Crude Oil Analyses

The important physicochemical properties of crude oil affecting the process of recovery include API gravity, viscosity, pour point, resin–

**Table 7.6** Physicochemical properties of crude oil

| S. no. | Properties | Value | Remarks |
|---|---|---|---|
| 1 | Density | 871 kg/m$^3$ at 15.6°C | Medium-gravity oil |
| 2 | API gravity | 31° API | |
| 3 | Dynamic viscosity | 4.46 ± 0.5 cP at reservoir temperature of 70°C | Decreases with temperature |
| 4 | Pour point | 24°C ± 3°C | pour point was high indicating high paraffin content |
| 5 | Asphaltene content | 1.4 ± 0.2 wt% | High R/A indicating asphaltene stabilisation |
| 6 | Resin content | 16.73 ± 0.5 wt% | |
| 7 | Wax content | 12.47 ± 0.70 wt% | Waxy crude |
| 8 | Acid number | 1.2 ± 0.2 (mg of KOH)/(g of crude oil) | Acidic crude |

asphaltene (R/A) content and acid number. The knowledge of these properties is essential before embarking on any EOR process and for designing the components for its successful application. The results of the crude oil analyses are discussed below and summarised in Table 7.6.

### 7.3.4.1 API gravity

In the oil industry, it is customary to express the density of crude oil in °API. With water having an API gravity of 10, liquids lighter than water will have API gravity greater than 10. On this scale, gasoline has a gravity of 60°API, while crude oils may have a typical value of around 35°API [35]. The density of crude oil measured with hydrometers was found to be 863 ± 3 kg/m$^3$ at an ambient temperature of 28°C. The corresponding density at 15.56°C or 60°F was 871 kg/cm$^3$ after applying the necessary correction factor (0.9908) obtained from Table 7.3a of the Petroleum Measurements Tables [122]. The API gravity of the crude calculated using Eq. 7.3 was 31°API. Crude oils are generally classified on the basis of their measured API gravity as light, medium or heavy [123]:

$$^{\circ}\text{API} = \frac{141.5}{\text{Sp. gr. at } \frac{15.6^{\circ}\text{C}}{15.6^{\circ}\text{C}}} - 131.5 \tag{7.3}$$

**Table 7.7** Classification of crude oil

| Type of crude oil | °API gravity | Density of crude oil (g/cc) |
|---|---|---|
| Light | $\geq 31.1$ | $\leq 0.87$ |
| Medium | $22.3 - 3.1$ | $0.87 - 0.92$ |
| Heavy | $\leq 22.3$ | $0.92 - 1.00$ |
| Extra-heavy oil or bitumen | $\leq 10.0$ | $\geq 1.00$ |

As per this classification (Table 7.7), the crude oil of the Upper Assam oilfield used in this study falls under the category of medium oil. $CO_2$ flooding being the leading EOR technique for light and medium oil reservoirs, this preliminary information suggested that the crude oil under study was suitable for $CO_2$-EOR application.

### 7.3.4.2 Crude oil viscosity

Viscosity ($\mu_o$) is an important physical property of crude oil that controls the rate of flow and recovery of crude oil from reservoirs. Crude oil viscosity is a strong function of the temperature, pressure, oil gravity and gas solubility. An increase in temperature causes a decrease in crude oil viscosity, while an increase in pressure increases oil viscosity, provided the only effect of pressure is to compress the liquid. Additionally, a decrease in the solution gas also increases crude oil viscosity. The viscosity of crude oils based on pressure can be classified as [124]:

(1) Dead oil $\mu_o$, which is the $\mu_o$ of crude oil at atmospHeric pressure (no gas in solution) and system temperature
(2) Saturated oil $\mu_o$, which the viscosity of the crude oil at saturation pressure and reservoir temperature
(3) Undersaturated oil $\mu_o$, which is the $\mu_o$ of the crude oil at a pressure above the bubble-point and reservoir temperature

Figure 7.7 shows the plot of the dynamic $\mu_o$ of the crude oil sample as a function of temperature. As expected the dynamic viscosity decreased with temperature increase from a value of $10.21 \pm 0.3$ cP at room temperature (28°C) to $4.46 \pm 0.5$ cP at reservoir temperature (70°C). Viscosities of crude oil play an

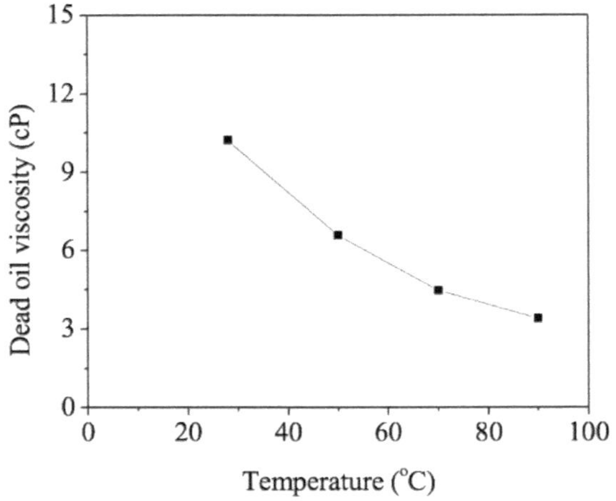


**Figure 7.7** Dynamic viscosity of oil versus temperature.

important role during its recovery. Experimental investigations, as well as field application, have shown that high viscosity leads to high remaining oil saturation after water flooding and conformance problems during chemical and gas EOR processes.

### 7.3.4.3 Pour point

The pour point of a liquid is the temperature at which it becomes a semi-solid and loses its flow characteristics. The pour point of the crude oil was high (24°C ± 3°C). This was based on the rule of thumb that high-pour-point crude oils generally have pour points in the range of 15°C to 52°C and may be solid at room temperature. It has been observed that a high pour point is generally associated with crude oils with high paraffin content and a low pour point with crudes of more aromatic nature [125]. Assam crude oil, in general, has been reported to have higher pour points (30°C) and higher wax content (11 wt%) [126].

### 7.3.4.4 Resin, asphaltene and wax content

Asphaltenes are polar, polyaromatic, high-molecular-weight HCs and the wax-free fraction of crude oil which are soluble in toluene but insoluble in alkanes (n-heptane to n-pentane) [127]. Asphaltene tends to remain in solution under reservoir temperature and pressure conditions stabilised by resins adsorbed on its surface. The asphaltene–resin ratio and high-/low-molecular-weight component ratio determines which crude oil can precipitate asphaltenes. Asphaltenes may start to precipitate if the stability with crude oil is destabilised because of changes in temperature and/or pressure during primary depletion. Asphaltene may also become unstable as a result of the co-mingling of fluid streams as well as by gas injection during EOR operations [128]. During $CO_2$ flooding, the asphaltene–resin ratio of crude oil is altered, causing asphaltene precipitation and thereby its deposition [129]. This is one of the major problems which confront petroleum engineers during $CO_2$-flooding applications. Precipitation of asphaltene could occur in reservoir, wellbore and production facilities and could adversely affect the oil recovery, operational facilities and rock properties. This often leads to expensive and disruptive remedial measures to reduce asphaltene problems. Even in reservoirs without asphaltene precipitation during primary and secondary recovery, injections of $CO_2$ often cause the asphaltene to precipitate within a very short period. Resins have the effect of keeping asphaltenes in solution. A high R/A ratio indicates that asphaltenes are less likely to come out of solution [130]. Leontaritis and Mansoori [131] presented a condition for asphaltene stability as follows:

(a) Steady state: $R/A \geq 3.0$
(b) Meta steady state: $2.0 \leq R/A \leq 3.0$
(c) Unsteady state: $R/A \leq 3.0$

The asphaltene content of the crude oil was found to be 1.4 $\pm$ 0.2 wt% and resin content 16.73 $\pm$ 0.5 wt%. So, the R/A ratio for the crude oil was greater than 3, so on the basis of the Leontaritis criterion, there were enough amounts of resins for asphaltene stabilisation. The wax content of the crude oil was fairly high at 12.47 $\pm$ 0.70%. A similar observation on the wax content of Assam crude oil was made by Kandwal et al. [126].

### 7.3.4.5 Acid number

Acid number is a measure of the amount of acidic components present in crude oil. It is expressed as the milligrams of KOH which is required to neutralise 1 g of oil sample. For a reservoir to be a good candidate for alkaline flooding, the acid number of crude oil should be at least 0.5 ml/g. Crude oils with an acid number greater than 0.5 mg KOH/g of crude oil are known as acidic crudes [132]. The acid number of the present crude oil was 1.2 ± 0.2 mg KOH/g of crude oil. Thus acid number was more than 0.5 mg KOH/g implying that the crude oil was acidic and is suitable for alkaline flooding.

The FTIR of the crude oil was recorded between 400 and 4000 $cm^{-1}$ to study the presence of major functional groups present. The dominant spectra of crude oils normally include the absorption bands of aliphatic C–H bonds along with other additional bands of groups containing oxygen, aromatics, nitrogen and sulphur. The IR spectrum of the crude oil is shown in Fig. 7.8. The peak at 2932 $cm^{-1}$ indicated the presence of single bond alkanes, while at 1724, 1603, 1454 and 1377 $cm^{-1}$ account for carboxylic acids, aromatics, $-CH_2$ and $-CH_3$ functional groups, respectively. The other peak

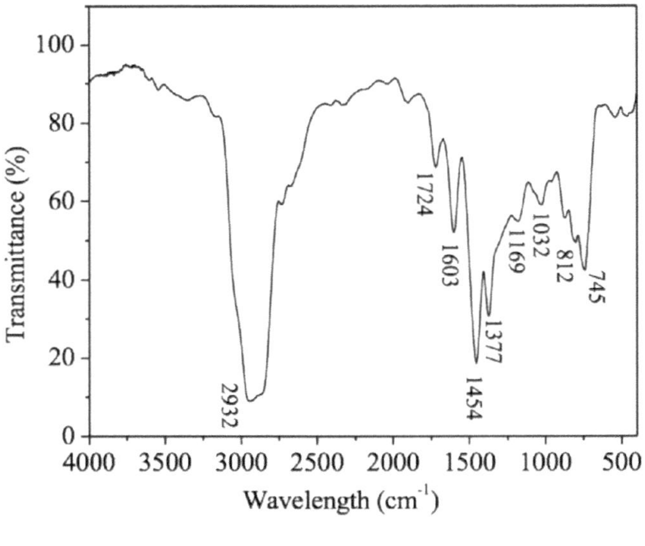


**Figure 7.8** FTIR spectrum of crude oil.

at 745 $cm^{-1}$ indicates the existence of long-chain alkyl groups [133, 134]. Thus the presence of acidic groups in the crude oil was identified by FTIR analysis and supported by the acid number measurement.

### 7.3.4.6 Formation water analyses

The water which naturally exists along with oil and gas in the reservoir is called reservoir/FW. Its most important physical properties are density and viscosity. The pH of FW is typically between 4 and 9. The pH value below and above this range is not desirable because of the tendency of corrosion and scaling, respectively [134]. FW almost invariably contains dissolved salts and the chemical analyses have indicated a wide variation in the type and amount of the dissolved ions. The commonly present cations and anions are $Na^+$, $K^+$, $Ca^{2+}$, $Mg^{2+}$, $Ba^{2+}$, $Cl^+$, $SO_4^{2-}$, $HCO_3^-$ and $CO_3^{2-}$ [135]. The total salinity of FW is calculated by summing up the cationic and anionic concentrations [135]. Table 7.8 lists the physical and chemical properties of the FW used in the study. The presence of divalent ions ($Ca^{2+}$ and $Mg^{2+}$) in FW was observed, although in small amounts. The total salinity of the water sample was 3458 mg/l. Synthetic formation brine (SFB) at this salinity was prepared to mimic the FW and was used for the preparation of solution. SFB

**Table 7.8** Physical and chemical properties of formation water (FW)

| Property | Measured value |
|---|---|
| Density, kg/$m^3$ (at 28°C) | 1015 |
| Viscosity, cP (at 28°C) | 0.95 |
| pH (at 28°C) | 8.35 |
| Total dissolved solids (ppm) | 3524 |
| Major components (ppm) | |
| Sodium ($Na^+$) | 1550 |
| Calcium ($Ca^{2+}$) | 85 |
| Potassium ($K^+$) | 14 |
| Magnesium ($Mg^{2+}$) | 17 |
| Chloride ($Cl^-$) | 730 |
| Bicarbonate ($HCO_3^-$) | 1065 |
| Salinity (mg/l) | 3461 |

was diluted with distilled water to varying degrees to obtain SFB of different salinities for phase behaviour studies and core-flooding experiments.

## 7.4 Conclusion

In this work, the characterisation of crude oil, core samples and FW of the Upper Assam oilfield was done to evaluate the applicability of ASAG flooding, which is a foam-based EOR process. The routine core analyses were done for the measurement of porosity and permeability of core samples. But prior to core analysis, core plugs were prepared from conventional cores obtained from particular depths of the oil reservoir and further subjected to cleaning followed by drying. The porosities of the core plugs ranged from 11% to 28% and were found to bear an inverse correlation with the depth of the formation. The gas permeability measured by flowing gas was observed to be greater than the permeability value obtained by flowing liquid. The values of the liquid and gas permeability of the core plugs ranged from 3 to 30 mD and 16 to 47 mD, respectively, which classifies the oilfield formation as fair- to moderate-permeability reservoirs. The porosities and permeabilities of the core plugs were observed to be directly related as the more porous cores were also more permeable. Further, XRD study of powered core samples indicated the presence of silica (quartz), which classified the reservoir rock as anionic type. So, using anionic surfactants as foaming agents will result in lower adsorption of surfactants on the reservoir rocks. Additionally, the FESEM micrograph of the rock sample showed the presence of smectite clays in the core matrix. Due to the presence of these clay minerals, surfactants injected are susceptible to be adsorbed onto the reservoir rocks. The use of alkali and other secondary surfactants as sacrificial adsorbates should be implemented to reduce the loss of surfactants. The experimental investigation of the crude oil confirmed that the crude oil was medium-gravity (API gravity 31°), and dead-oil viscosity of 4.46 cP (at 70°C). The pour point of the crude oil was high (24°C ± 3°C) which indicated the paraffinic nature of the crude oil. The wax content was also fairly high at 12.47% ± 0.70%. The R/A ratio of the crude oil was high enough for asphaltene stabilisation. The analysis also marked the crude as acidic and suitable for alkaline flooding. FW analysis showed the presence of different types of dissolved cations and anions with total salinity of 3458 ppm.

## Acknowledgements

The authors gratefully acknowledge the experimental facilities extended by the Department of Petroleum Technology, Dibrugarh University, and the Chemical Engineering Department, Indian Institute of Technology, Guwahati.

## Chapter 8

# Formulation of an Optimum Surfactant Combination for EOR Applications

**Sekhar Gogoi,[a] Shilpi Sarmah[b] and Subrata Borgohain Gogoi[b]**

[a] *Assam Energy Institute, Sivasagar, a centre of Rajiv Gandhi Institute of Petroleum Technology, India*

[b] *Department of Petroleum Technology, Dibrugarh University, Dibrugarh, Assam, India*

sgogoi@rgipt.ac.in

A variety of surfactants can be utilised for application in enhanced oil recovery (EOR) conducted at the microfluidic level, that is, a lab-on-a-chip scale. The surfactant can be in powdered form or in liquid form. Surfactants taken for this study were chosen from anionic and non-ionic surfactants. A total of five surfactants were considered for the experimental procedure of surface tension (ST) and interfacial tension (IFT). For ST, the solid surfactants were first dissolved in distilled water to prepare a standard solution. This surfactant standard solution was then incrementally added, thereby increasing in its concentration, in fresh distilled water to obtain 30 samples having varying concentrations, in a stepwise manner. The liquid surfactants were directly added incrementally in distilled water. For IFT, the same procedure was followed. Since IFT requires two liquid phases, the second liquid phase considered was readily available liquid paraffin (light). It was added to each sample of surfactant solution before experimenting. A Kruss Easy

*Advances in Petroleum Technology*
Edited by Subrata Borgohain Gogoi

ISBN 978-981-4877-19-0 (Hardcover), 978-1-003-04993-7 (eBook)
www.jennystanford.com

Dyne tensiometer was used for measuring the ST and IFT for all the samples. The results were utilised in creation of IFT and ST variation graphical plots. A comparison was made among the participating surfactants so as to ascertain their critical micelle concentration (CMC) at appropriate tension values. Although test samples having the lowest CMC have to be considered, a factor of costing comes into play. Therefore, a cost analysis was done using Indian pricing (INR) per 500 g/500 ml of the test samples. Taking both technical and economic considerations, the two most favourable and optimistic samples were selected. A better combination can also be made of one anionic and one non-ionic sample to complement each other's properties. This surfactant combination can then be further blended with alkali to create a better-performing slug for recovery process.

## 8.1 Introduction

The flow of fluids is a vast subject. It finds an important place in the oil and gas industry, both upstream and downstream. This chapter will deal with the upstream side, on the recovery of hydrocarbons (HCs) from a reservoir which are depleted and have need of external resources to recover the remaining oil. For this the study of flow patterns and behaviours becomes necessary inside the reservoir rocks' subsurface. As rocks are opaque, only an estimation can be made about flow rate conditions through core-flooding experiments. Thus, the limitations of laboratory core-flooding experiments in impeding a fundamental investigation of oil reservoirs, to get visibility, microfluidics is a necessary tool which utilises synthetic micromodels using microfabrication techniques [136].

Lab on a chip, as microfluidics is known in the field of research, employs the use of transparent microchips having pores and connectivity. Through these pores, fluids are injected to understand their patterns and behaviour. As supported by Lifton 2016 [137], microfluidics studies the movement, handling and manipulation of fluids in very smaller volumes in the range of nano- and microscales. Due to the microlevel range it can help in understanding of the process which governs and controls the fluid volume, fluid flow rate and fluid extraction through microchips. A slug will help in pushing the required oleic phase trapped inside the chips. The injection of a slug inside the chips will lead to a multiphase flow. Oil companies give extraordinary emphasis on extracting the residual reserve of crude oil though an enhanced oil recovery (EOR) process by injecting formulated fluids into reservoirs [138]. Zhong and Mayer (2001) [139] stated that viscous forces exerted by an aqueous phase are not large enough to exceed capillary forces which are trapping the denser-than-water non-aqueous-phase liquids (DNAPL). These capillary forces are a result of the high interfacial tension (IFT) between the two phases. For this a surfactant removal mechanism is employed. The macroscale process involves immiscible displacement between phases, leading to mobilisation. This macroscale process is controlled by microscale physics. Hence a microscale study becomes important for understanding the entrapment, solubilisation and mobilisation processes. Negin et al. (2016) [140] reported that surfactants play a role in the recovery process by decreasing the IFT and shifting reservoir wettabilities towards strongly water-wet. Sodium dodecyl sulphate (SDS) is an anionic surfactant belonging to the alkyl sulphate group. If introduced with nanoparticles, their surface activity can be enhanced below the critical micelle concentration (CMC).

Characterisation of the physiochemical properties of nanoemulsions is done by deploying Tergitol surfactants [140]. They can be Tergitol 15-S-5, Tergitol 15-S-7 (T-7), Tergitol 15-S-9 (T-9) and Tergitol 15-S-12 (T-12). T-9 was utilised because of its availability. Experiments have shown that 30% of original oil in place (OOIP) was recovered by injection of nanoemulsions. Also, Sarmah et al. (2019) [67] studied the above non-ionic surfactants and observed their particle size to be less than 10 nm, thus assisting mobilisation of oil droplets even from smaller pores. The use of a polymer solution is not attempted in microchips due to the higher viscous nature of polymers, which may find difficulty in flow through micropores of these chips. Also, the alkali gives an added advantage to surfactant performance.

# 8.2 Experimental

## 8.2.1 Material

The chemicals used for the experimental work are shown in Table 8.1, while the instrument and glassware are in Tables 8.2 and 8.3, respectively.

**Table 8.1** List of chemicals

| S. no. | Material | Source |
|---|---|---|
| 1 | Sodium dodecyl sulphate (SDS) | Merck Life Science, Mumbai |
| 2 | Sodium dodecylbenzene sulphonate | Loba Chemie, Mumbai |
| 3 | Tergitol 15-S-9 | Sigma-Aldrich |
| 4 | Triton X-405 solution | |
| 5 | Triton X-100 | HiMedia Laboratories |
| 6 | Acetone | Merck Life Science, Mumbai |
| 7 | Liquid paraffin light | Fisher Scientific |

**Table 8.2** List of instruments

| S.no. | Instruments | Make/Model |
|---|---|---|
| 1 | Kruss, Easy Dyne tensiometer | Kruss GmBH |
| 2 | Electronic weight balance | Scale-Tec |

**Table 8.3** List of glassware

| S. no. | Glassware | Source |
|---|---|---|
| 1 | Tarson | |
| 2 | Micropipette | |
| 3 | Pipette | AB Associates, Assam |
| 4 | Beaker | |
| 5 | Glass stirrer | |
| 6 | Syringe | |

## 8.2.2 Methodology

### 8.2.2.1 Measuring the surface tension

A standard solution was prepared in a beaker of 100 ml distilled water, and 0.1 ml of standard solution/surfactant solution was added incrementally in a stepwise manner to obtain 30 samples having different concentrations.

### 8.2.2.2 Measuring the interfacial tension

In a beaker containing 15 ml of distilled water, an increment of 0.015 ml of standard solution was added using a pipette and a micropipette to create 30 samples. For each sample, 10 ml of liquid paraffin light was added to create two layers, thus generating an interface between the sample and liquid paraffin light.

### 8.2.2.3 Measuring the ST and IFT

The ST/IFT for each sample was found using the Du Nouy ring method of the tensiometer. Graphical plots of concentration versus ST/IFT were created. The plots were compared to find the least CMC values among the surfactants.The selection of surfactants was done on the basis of the CMC value as well as availability and cost.

## 8.3 Results and Discussion

The combined ST and IFT of different surfactants at different concentrations of 30 samples are given in Tables 8.4 and 8.5, respectively. The corresponding plot variation for concentration is in Figs. 8.1 and 8.2, respectively. As concentration increased, the ST showed a decrease in the trend up to a certain concentration, known as the CMC. After it had crossed the CMC value, with more increase in concentration, the ST showed a gradual increasing trend with few drops in value while travelling upwards. The SDS plot was wavy in nature, while the TX-405 plot showed a gradual constant trend after reaching CMC values. The largest range of ST had been shown by Triton X-405 (TX-405), while the smallest range was shown

**Table 8.4** ST (mN/m) of surfactants

| S. no. | Conc. (%) | SDS | TX-405 | T-9 | SDBS | TX-100 |
|---|---|---|---|---|---|---|
| 1 | 0.0999 | 63.1 | 60.0 | 29.1 | 63.8 | 30.8 |
| 2 | 0.199600798 | 63.0 | 50.0 | 28.3 | 56.0 | 31.6 |
| 3 | 0.299102692 | 59.7 | 48.7 | 25.7 | 54.5 | 31.0 |
| 4 | 0.398406375 | 57.6 | 47.6 | 28.9 | 57.2 | 31.1 |
| 5 | 0.497512438 | 61.6 | 44.5 | 29.6 | 54.5 | 30.2 |
| 6 | 0.596421471 | 59.2 | 41.8 | 28.6 | 53.2 | 30.9 |
| 7 | 0.695134062 | 60.3 | 40.9 | 27.9 | 51.6 | 30.9 |
| 8 | 0.793650794 | 57.9 | 38.1 | 27.8 | 51.0 | 31.6 |
| 9 | 0.89197225 | 56.8 | 38 | 27.7 | 48.7 | 31.2 |
| 10 | 0.99009901 | 55.8 | 37.2 | 27.1 | 50.5 | 31.1 |
| 11 | 1.088031652 | 55.2 | 38.9 | 27.3 | 47.7 | 30 |
| 12 | 1.185770751 | 52.6 | 39.3 | 27.6 | 45 | 31.4 |
| 13 | 1.283316881 | 52.7 | 40.0 | 27.5 | 44.2 | 31.6 |
| 14 | 1.380670611 | 54.6 | 40.0 | 27.5 | 44.3 | 31 |
| 15 | 1.477832512 | 42.2 | 40.6 | 28.0 | 42.4 | 30.4 |
| 16 | 1.57480315 | 52.4 | 40.3 | 28.8 | 43.0 | 31.0 |
| 17 | 1.671583088 | 53.2 | 40.5 | 29.5 | 43.6 | 30.0 |
| 18 | 1.768172888 | 51.5 | 40.4 | 28.4 | 42.6 | 28.0 |
| 19 | 1.864573111 | 52.5 | 39.7 | 29.9 | 42.7 | 28.9 |
| 20 | 1.960784314 | 50.9 | 41.0 | 29.6 | 41.4 | 27.9 |
| 21 | 2.056807052 | 50.6 | 41.1 | 29.3 | 40.2 | 28.4 |
| 22 | 2.152641879 | 51.9 | 39.7 | 28.9 | 40.0 | 27.8 |
| 23 | 2.248289345 | 48.9 | 41.2 | 29.6 | 39.1 | 27.8 |
| 24 | 2.34375000 | 50.5 | 41.7 | 29.2 | 38.0 | 27.6 |
| 25 | 2.43902439 | 49.8 | 40.7 | 29.2 | 38.3 | 28.3 |
| 26 | 2.53411306 | 50.2 | 41.0 | 29.2 | 39.2 | 28.1 |
| 27 | 2.629016553 | 49.4 | 40.0 | 29.8 | 35.8 | 28.4 |
| 28 | 2.723735409 | 49.7 | 40.4 | 29.1 | 36.0 | 27.7 |
| 29 | 2.818270165 | 48.6 | 41.5 | 28.9 | 37.2 | 29.8 |
| 30 | 2.912621359 | 47.1 | 41.0 | 28.5 | 36.6 | 29.4 |

by T-9 and Triton X-100 (TX-100). The sodium dodecyl benzene sulphonate (SDBS) ST plot decreased gradually with an increase in concentration. The CMC values of each of the surfactants had been observed. The IFT plots had shown a similar trend as ST; for SDBS, there has been a sharp decrease in the ST value till the CMC. The plot for SDS was wavy in nature, while the plot for TX-405 showed a gradual constant trend after reaching the CMC. The range of ST/IFT values for T-9 and TX-100 were narrower compared to other

**Table 8.5** IFT (mN/m) of surfactants

| S. no. | Conc. (%) | SDS | TX-405 | T-9 | SDBS | TX-100 |
|---|---|---|---|---|---|---|
| 1 | 0.0.099900100 | 29.8 | 20.4 | 4.2 | 29.4 | 3.8 |
| 2 | 0.199600798 | 31.0 | 15.6 | 3.9 | 29.6 | 3.0 |
| 3 | 0.299102692 | 30.5 | 14.5 | 3.6 | 29.8 | 3.0 |
| 4 | 0.398406375 | 31.6 | 15.3 | 3.2 | 25.8 | 3.3 |
| 5 | 0.497512438 | 30.2 | 14.4 | 3.3 | 24.6 | 3.0 |
| 6 | 0.596421471 | 31.2 | 15.1 | 3.2 | 24.0 | 3.0 |
| 7 | 0.695134062 | 30.5 | 15.1 | 3.6 | 24 | 3.1 |
| 8 | 0.793650794 | 30.4 | 14.8 | 3.3 | 21.6 | 2.7 |
| 9 | 0.89197225 | 26.2 | 14.6 | 3.4 | 20.5 | 2.7 |
| 10 | 0.99009901 | 29.8 | 14.9 | 3.2 | 19.2 | 2.6 |
| 11 | 1.088031652 | 24.8 | 15.2 | 3.1 | 18.4 | 2.4 |
| 12 | 1.185770751 | 24.4 | 14.5 | 3.2 | 19.1 | 2.3 |
| 13 | 1.283316881 | 25.9 | 15.2 | 3.0 | 18.1 | 2.8 |
| 14 | 1.380670611 | 23.8 | 15.3 | 3.0 | 17.9 | 2.7 |
| 15 | 1.477832512 | 25.8 | 14.4 | 2.7 | 15.9 | 2.5 |
| 16 | 1.57480315 | 25.6 | 15.6 | 2.7 | 14.7 | 2.9 |
| 17 | 1.671583088 | 22.6 | 15.3 | 2.8 | 14.7 | 2.5 |
| 18 | 1.768172888 | 25.4 | 15.3 | 2.8 | 14.4 | 2.3 |
| 19 | 1.864573111 | 26.5 | 15.3 | 2.7 | 14.4 | 2.5 |
| 20 | 1.960784314 | 25.2 | 15.3 | 2.7 | 13.8 | 2.6 |
| 21 | 2.056807052 | 25.5 | 15.9 | 2.7 | 11.3 | 2.5 |
| 22 | 2.152641879 | 24.0 | 15.3 | 2.8 | 11.6 | 2.5 |
| 23 | 2.248289345 | 21.2 | 15.3 | 2.7 | 11.9 | 2.8 |
| 24 | 2.343750000 | 25.1 | 15.6 | 2.8 | 12.8 | 2.6 |
| 25 | 2.439024390 | 24.6 | 16.2 | 2.6 | 11.3 | 2.1 |
| 26 | 2.534113060 | 25.7 | 15.3 | 2.8 | 12.2 | 2.6 |
| 27 | 2.629016553 | 23.1 | 15.7 | 2.7 | 10.7 | 2.5 |
| 28 | 2.723735409 | 20.8 | 15.1 | 2.5 | 9.7 | 2.6 |
| 29 | 2.818270165 | 23.7 | 15.5 | 2.5 | 11.5 | 2.9 |
| 30 | 2.912621359 | 23.8 | 15.4 | 2.3 | 10.9 | 2.7 |

surfactants. The CMC values for each surfactant had been noted. From both ST and IFT plots, it was seen that each of the surfactant had given a similar trend for both ST and IFT. Table 8.6 shows the cost of surfactants in Indian currency. The IFT values were lower than the ST values for the same surfactants. Liquid paraffin had been considered as the oleic phase in IFT measurements. Comparing the technical factor, the T-9 and TX-100 surfactant combination was found to be the most desired slug, as shown in Table 8.7. But

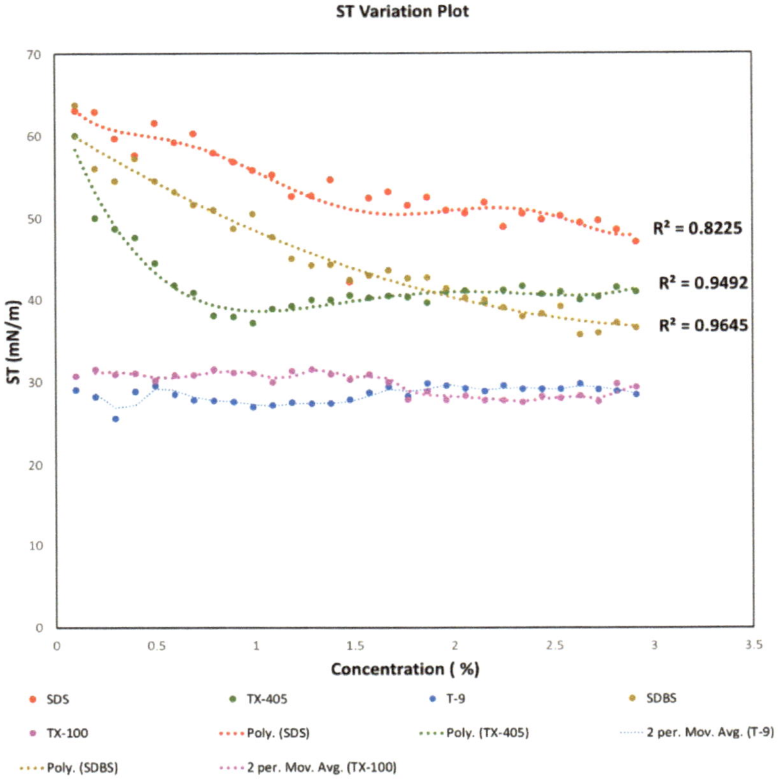


**Figure 8.1** ST (mN/m) variation with conc. (%).

considering the cost factor, it was found that T-9 is expensive as compared to the other surfactants, only slightly lesser than TX-405, as shown in Table 8.6. Hence, the optimum surfactant combination considering both technical and cost factors for formulating a slug is SDS and TX-100.

## 8.4 Conclusion

For conducting EOR flooding on microfluidics, surfactant selection needs to be done. The slug contains a surfactant and an alkali.

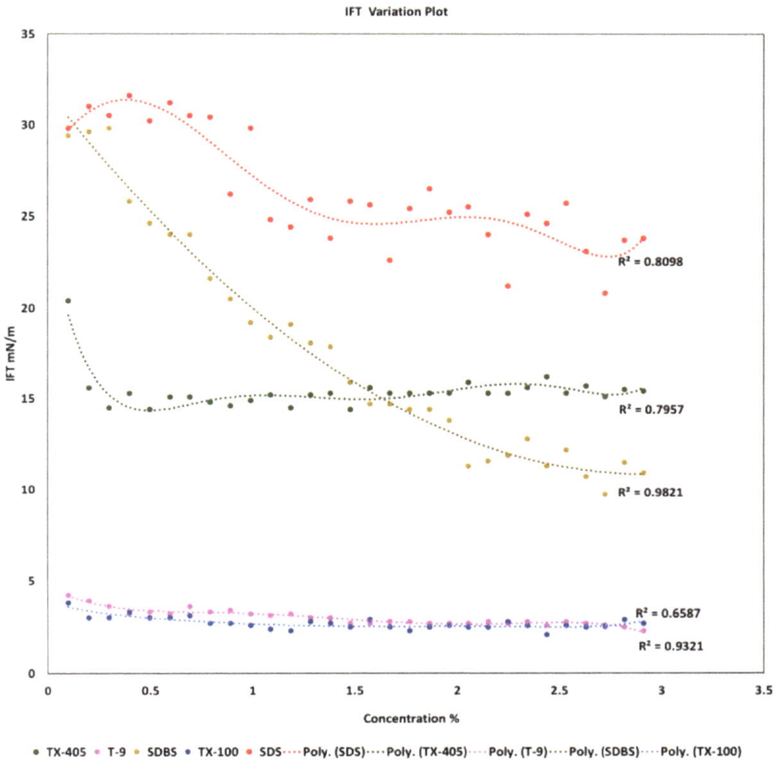


**Figure 8.2** IFT (mN/m) variation with conc. (%).

**Table 8.6** Cost of surfactants

| S. no. | Surfactants | Quantity | Unit | Price (INR) |
|---|---|---|---|---|
| 1 | SDS | 500 | g | 1056.00 |
| 2 | SDBS | 500 | g | 2324.00 |
| 3 | T-9 | 500 | ml | 4478.31 |
| 4 | TX-405 | 500 | ml | 5339.09 |
| 5 | TX-100 | 100 | ml | 1102.00 |

**Table 8.7** Surfactant combination for slug formulation

| S. no. | Surfactant | CMC for ST (%) | ST (mN/m) | CMC for IFT (%) | IFT (mN/m) |
|---|---|---|---|---|---|
| 1 | T-9 | 1 | 27 | 2.4 | 2.6 |
| 2 | TX-100 | 1.2 | 31 | 1.2 | 2.3 |
| 3 | SDS | 1.7 | 50 | 1.7 | 24.5 |
| 4 | SDBS | 2.6 | 37 | 2.6 | 11 |
| 5 | TX-405 | 1 | 38 | 0.5 | 14.5 |

Surfactants help in reducing the ST/IFT in fluids, allowing for a better flow in porous media. Solid powdered surfactants as well as surfactant solutions were experimented upon to understand the ST and IFT. After analysing the samples for each surfactant and studying the plots of ST/IFT against concentration and also taking the cost factor, it was observed that SDS and TX-100 were the best-fit surfactant combination for slug formulation. Further, this combined pair has one solid and one liquid solution.

## Acknowledgements

We would like to acknowledge the following for their valuable contribution:

(1) 21st Century Knowledge initiative programme under University Grants Commission Indo-US Project F.no. 194-1/2009 (IC) dated February 2015, entitled 'Foam-Assisted $CO_2$ Flooding for the Depleted Reservoirs of Upper Assam Basin and in Candidate Reservoirs in Louisiana'

(2) University Grants Commission and UK–India Education and Research Initiative UGC-UKIERI Joint Research Programme (UKIERI-III) Project No. F.No. 184-1/2018(IC) entitled 'Novel EOR Technique for the Depleted Oil Reservoirs of Upper Assam Basin with Simultaneous $CO_2$ Capture and Sequestration'

# Chapter 9

# Microbial Enhanced Oil Recovery: A Technological Perspective

**Saurav Haloi and Tapas Medhi**
*Department of Molecular biology and Biotechnology, Tezpur University, Assam, India*
saurav.abl.tu@gmail.com, tmedhi@tezu.ernet.in

Microbial enhanced oil recovery (MEOR) is an alternative tertiary method of tertiary enhanced oil recovery (EOR) with apparent advantages of low toxicity, bioequivalence, biodegradability and economic feasibility for field implementation. The in situ and ex situ production of microbial metabolites and their application in a laboratory-scale study as well as in actual oilfield condition have their own merits and limitations and are an outcome of a multidisciplinary approach entailing tools from science and engineering. Common MEOR practices employ either an indigenous bacterial population or cultured ones or their metabolic products for this purpose. Biosurfactants (BSs) are amphipathic molecules produced by bacterial genera *Bacillus*, *Pseudomonas*, *Alcaligenes*, *Burkholderia*, *Clostridium*, *Rhodococcus*, etc., which can be used for efficient EOR through surface and interfacial tension (IFT) mitigation, wettability alteration and viscosity reduction to increase permeability, etc. But the main drawbacks to using BSs of microbial origin in MEOR are their higher production cost and low production

*Advances in Petroleum Technology*
Edited by Subrata Borgohain Gogoi

ISBN 978-981-4877-19-0 (Hardcover), 978-1-003-04993-7 (eBook)
www.jennystanford.com

rates. Therefore, in this chapter, we primarily focussed on different biotechnological approaches for incrementing BS production. Also, we summarised the success story of BS-based MEOR technology at the laboratory scale as well as the field scale, highlighting the economical crude oil recovery.

## 9.1 Introduction

Recovery of crude oil from a depleted reservoir is one of the most promising areas and stands as a major challenge in petroleum industry research due to the rising demand for crude oil in daily life. According to Al-Sulaimani et al., only 30%–40% of crude oil can be recovered through conventional recovery methods such as primary and secondary recovery methods [141]. An anticipated amount of nearly 7 trillion barrels of conventional and heavy oil remain in the reservoir porous media worldwide after using these conventional technologies [142]. Enhanced oil recovery (EOR) is a tertiary oil recovery method, where synthetic chemicals (surfactants and polymers), heat (hot water) and gases ($CO_2$, $N_2$) are used to recover the trapped oil from the depleted reservoir. But these methods or technologies have come up with various difficulties such as high capital investment, system complexity, secondary pollution and hydrogen sulphide production. In such circumstances, an economically attractive and environment-friendly tertiary intervention could be the use of microbes and their by-products such as biosurfactants (BS), biopolymers, bioacids, biosolvents and biogas in microbial enhanced oil recovery (MEOR) [143, 144].

To this end, microbes, especially bacteria, have drawn substantial attention because of their size, shorter generation time and capability to survive in reservoir conditions. On the basis of the recovery strategies by using microbes and their by-products in actual operations, MEOR can be divided into in situ and ex situ, as shown in Fig. 9.1. In situ and ex situ MEOR, both have their own merits and demerits in actual oilfield applications. In either case, MEOR borrows technical knowledge from disciplines like microbiology, biochemistry, geophysics, petroleum engineering

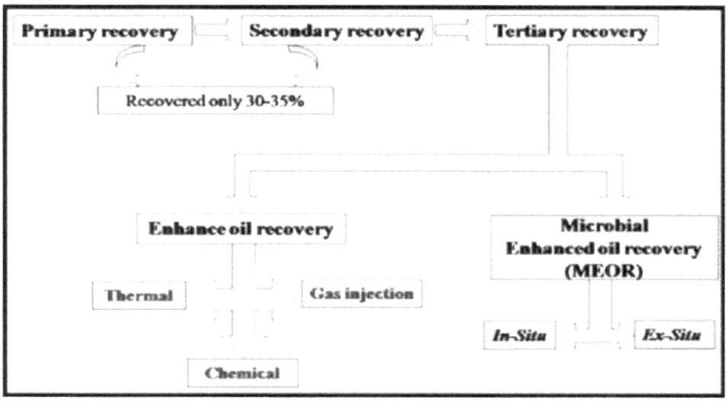


**Figure 9.1** Flow diagram of oil recovery methods.

and mechanical engineering for effective recovery of oil in an environment-friendly manner.

In this case, bulk and economic production of BSs having appropriate physicochemical properties poses a major challenge to attain industrial readiness, which therefore calls for innovations in biotechnological research for sufficient production of tailor-made organisms capable of performing in reservoir conditions. The low inherent toxicity coupled with functional ability to withstand harsh environmental conditions, including a wide range of temperatures, pH and salinity, lower critical micelle concentration (CMC) and superior selectivity, provides added mileage to research and development (R&D) efforts for their prospective use in different industries. Production of BSs using renewable sources or industrial waste also has an advantage over chemical counterparts.

Being amphipathic in nature, BSs can act as an interphase between two immiscible substrates like oil and water or oil and rock plugs or oil and gas to reduce the interfacial tension (IFT), which results in reduction of capillary force and increases the capillary numbers, facilitating recovery of oil from depleted reservoirs [145]. A BS-based aqueous solution also forms oil-in-water emulsions to block the highly permeable zones. Otherwise, the displacing BS containing fluids only flows through the highly permeable zones and the oil trapped in the smaller pores will not be recovered. There are three strategies for application of BS-based EOR technology, as reported by Banat et al. (2000) [146]:

(1) Production and extraction of BSs at the laboratory scale using fermentation technology, followed by injection to the depleted reservoir along with saline water (ex situ MEOR)
(2) Stimulation of indigenous microbes by flooding selective nutrients to the reservoir (in situ MEOR)
(3) Isolation of anaerobic or facultative BS-producing bacteria at the lab scale and optimisation of their conditions for maximum BS production followed by the injection of the isolates along with nutrients into the reservoir (in situ MEOR)

Microorganism-based EOR offers an inexpensive alternative to other EOR processes which require long-term goals to design robust processes functional in different reservoir conditions. The multifarious mechanisms adopted by microbes at the same time, BSs production rates and cost and the effectiveness of which may alter in response to petroleum reservoir environments are probably the main reasons working against such mechanistic progress. Here, we have highlighted different EOR methods in comparison to the success or usefulness of microbial BSs for efficient oil recovery from a depleted reservoir. The bottlenecks in bulk and economic BS production, including biotechnological and bioprocess engineering means of higher production, are also discussed to eliminate hurdles for their commercialisation.

## 9.2 Crude Oil and Its Importance in India

Naturally occurring crude oil is a flammable liquid made up of a complex mixture of alkenes, cycloalkanes and aromatic hydrocarbons (HCs), and its molecular composition always varies with the reservoir conditions and geological makeup. Presently, the majority of the world's energy is coming from crude oil, and 40% of the world's total energy demand has been fulfilled by crude oil. According to the International Energy Agency (IEA), the world consumed about 99.558 million bbl of oil per day (bbl/day) on average in 2018, where India only ranked behind the United States and China as the third-highest oil-consuming country of the world as per the Energy Information Administration (EIA, 2017; https://www.iea.org/). The EIA also predicted that India would be only behind the United States at the end of the year 2019. According to World Oil Outlook (2017), the demand for primary energy of electronic society will be increased by 35% at the year 2040, where India and China are expecting to contribute the highest in the growing requirement. In 2018, the Ministry of Petroleum and Natural Gas, Government of India, also reported that the demand for crude oil is constantly increasing day by day. But the production rate remains constant, and more than 75% of our demand has been met by importing crude oil. As per Indian petroleum and natural gas (PNG) statistics (2017–2018), India has estimated oil reserves of 594.49 million tonnes as of March 2018, and the largest oil reserves are present in the western offshore (40%), followed by Assam (27%), as reported in 'Energy Statistics 2019', Ministry of Statistics and Programme Implementation [147]. Therefore, there is an undisputable urgency to improve tertiary oil recovery from the reservoirs, and biosurfactant-based EOR could be one of the best alternatives to serve that purpose.

## 9.3 Conventional Oil Recovery Methods

Primary recovery is the first conventional oil recovery method where the natural reservoir pressure is sufficient enough to flow the oil from reservoir to surface. This method can recover only 5%–10% of oil from the original oil in place (OOIP) [144]. Therefore, external energies in the form of water and natural gases (air, $CO_2$, etc.) are used by petroleum industries to displace the trapped oil from reservoir to surface in secondary recovery methods. On the basis of the crude oil and reservoir rock characteristics, the secondary method recovers only 25%–35% of reservoir oil [148, 149]. That means, 50%–65% of oil is trapped in the rocks after using these conventional methods, and this remaining trapped oil blobs are the potential target of EOR methods [150]. According to Al-Sulaimani et al. [141], nearly 2 trillion barrels of conventional and 5 trillion barrels of heavy oils remain in reservoirs worldwide after the operation of conventional recovery methods. The IFT between water and oil or oil and rock plugs and heterogeneity of the rock plugs are considered as the main causes of incompetent recovery through these methods.

## 9.4 EOR

EOR is a tertiary oil recovery method, as shown in Fig. 9.1, which is divided into thermal, chemical and miscible flooding, depending on the mechanism of operation. The main target of this method is to recover the trapped oil existing in the reservoir after conventional recovery methods. The trapped oil is tough to recover from the reservoir due to wettability, oil viscosity and capillary pressure. The amount of residual oil is mainly dependent on the capillary number ($N_{\text{Cap}}$), which describes the ratio of the fluid's viscosity and IFT [151], as defined in Eq. 9.1

$$C_{\text{a}} = \frac{\mu V}{\sigma}, \tag{9.1}$$

where $\mu$ is viscosity of oil (Pa-s), $V$ is Darcy's velocity (m/s) and $\sigma$ is ST or IFT between two immiscible liquids (N/m).

### 9.4.1 Thermal EOR

Stream injection, hot water and in situ combustion are applied to the heavy oil reservoirs in thermal EOR [152, 153]. Stream and hot water reduce the viscosity of the oil and allow the oil to flow more easily from the reservoir to production well. In stream flooding, stream is continuously flooded through vertical injection wells to the reservoir, which reduces the viscosity of the oil by increasing temperature [152]. In in situ combustion, fair flooding is performed throughout the reservoir, which improves the mobility of the unburned oil to bring it to the production wells [154]. Globally, these methods are the most commonly used tertiary oil recovery methods for significant recovery of residual oil from depleted reservoirs [155, 156].

## 9.4.2 CEOR

Synthetic chemicals such as surfactants, polymers, alkaline solutions and their mixtures are injected with saline water into the reservoir. The IFT between two immiscible fluids is the dominant cause of poor recovery during conventional recovery methods. Therefore, chemical surfactants such as sodium dodecyl sulphate (SDS), cetyl trimethyl ammonium bromide (CTAB) and cetyl pyridinium chloride (CPC) are used to reduce the IFT between oil and water or oil and rock plugs to lead to a reduction in capillary pressure [157]. Surfactant-based slugs improve the recovery efficiency of 10%–20% from the depleted reservoir [158].

The capillary pressure ($P_c$) is defined in Eq. 9.2:

$$P_c = P_{nw} - P_w, \tag{9.2}$$

where $P_{nw}$ is the pressure of nonwetting fluid (Pa-s) and $P_w$ is the pressure of wetting fluid (Pa-s).

The Young–Laplace equation also describes the capillary pressure, as given in Eq. 9.3:

$$P_c = \frac{2\sigma \cos\theta}{r}, \tag{9.3}$$

where $\sigma$ is the IFT between oil and water or brine (mN/m), $\theta$ is the contact angle between the oil and rock (rad) and $r$ is the pore radius (cm).

According to Glatz (2013), an unfavourable mobility ratio (MR), as in Eq. 9.4, between the displacing and displaced fluids is also an important factor in not recovering the maximum oil with conventional recovery methods [159]. Therefore, partially hydrolysed polyacrylamide (HPAM) is added most popularly to the displacing fluids to control the MR, which is defined below [160]:

$$\gamma = \frac{\lambda_w}{\lambda_o} = \frac{\frac{K_w}{\mu_w}}{\frac{K_o}{\mu_o}} = \frac{\frac{k_w}{\mu_w}}{\frac{k_o}{\mu_o}}, \tag{9.4}$$

where $\lambda_w$ is the mobility of water or brine $(cm^3s)/g$, $\lambda_o$ is the mobility of oil $(cm^3s)/g$, $K_w$ is the relative permeability of water with respect to oil, $K_o$ is the relative permeability of oil with respect to water, $\mu_w$ is the viscosity of water (cP), $\mu_o$ is the viscosity of oil (cP), $k_w$ is the effective permeability of water (mD) and $k_o$ is the effective permeability of oil (mD).

### 9.4.3 Miscible Flooding

This method includes the flooding of $CO_2$, $N_2$, liquefied petroleum gas (LPG) and other miscible solvents like alcohol. $CO_2$ is generally used for recovery of light crude oil, which causes the swelling and viscosity reduction for extraction of trapped oil [161]. LPG is the most reliable miscible flooding agent, where natural gases (ethane, propane and butane) are injected in liquid form to reduce the viscosity as well as surface tension (ST) for enhancing the oil recovery efficiency [162]. But these EOR methods have some limitations to use in oil industries, such as high technical cost, pollution or toxicity to the environment or secondary pollution, system complexity and hydrogen sulphide production [143, 144, 163]. In such circumstances, MEOR is the suitable alternative of EOR methods, where microbes and their by-products are used for increasing the oil recovery efficiency [143, 163–165].

## 9.5 MEOR

MEOR is a multidisciplinary field that not only includes microbiology but also is associated with geology, chemistry, petroleum engineering and chemical engineering. Because of the size and ability to survive in reservoir conditions, bacteria have attracted great attention in this technology and make this method economically advantageous [141, 143, 163]. Bacteria and their by-products such as biosurfactants, biopolymers, biosolvents, bioacids and biogas are used in this method [164–166], where their microbial activity and by-products affect the interfacial properties of the oil-water-mineral system for enhancing the oil recovery through viscosity reduction, IFT reduction and permeability increase [166]. Moreover, microbes also promote the degradation of heavy oils to light oils for enhanced mobility to increase recovery efficiency.

### 9.5.1 History of MEOR

In 1926, Beckman employed microbes for the first time as a MEOR agent. Since 1947, numerous investigations have been done in MEOR technology [141]. But the first breakthrough came in 1946, when Zobell and his colleagues were given the first patent using HC-degrading and sulphate-reducing anaerobic bacteria *Desulfovibrio hydrocarbonoclastices* [167]. They tried to explain the involvement of bacteria and their by-products in the process of releasing trapped oil from the reservoir in these methods [167]. A slug consisting of the bacterial strains *Clostridium spp., Aerobacter aerogenes, Escherichia coli* and *Bacillus polymyxa* was patented by Updegraff in 1957, and flooded to the reservoir with carbohydrate medium [168]. But the first MEOR field trial was performed in the Lisbon oilfield, Union County, AR, in 1954, and the periods of 1970–2000 in the MEOR research had been devoted to exploring reservoir characterisation and microbiota exploration for successful MEOR implementation [143].

According to Banat et al. and Evans and Furlong, MEOR is dependent on ubiquitous oil reservoir conditions, including temperature, pressure, pH, porosity, salinity, geologic make-up of the reservoir, available nutrients and the presence of diverse indigenous microorganisms [146, 169]. Due to multiple effects of the microorganisms on the reservoir environment and oil, MEOR's action is most probably highly complex. Mechanisms for MEOR include gas formation to increase reservoir pressure, acid production to degrade the carbonate matrices and reduction of oil viscosity, BS production to reduce IFT and ST, solvent production to reduce oil viscosity, biomass accumulation for plugging and degradation of high-molecular-weight HCs present in the oil [165]. MEOR is also considered as a cost-effective process for implantation in the field levels over EOR [170]. It is unfortunate that MEOR has been unable to gain much attention in oil industry research, the causes for which have explained by researchers on various occasions as well. Among them, some of the reasons are: (i) the negative perception about the handling of microbes during MEOR, although the bacterial consortiums are verified as safe for handling and environments as proved in lab-scale experiments [171, 172]; (ii) the absence of standardised fields trial results and lack of posttrial analysis [163]; and (iii) investment in terms of sophisticated and expensive laboratory facilities to explore the reservoir microbiota and to isolate effective bacterial strains [141].

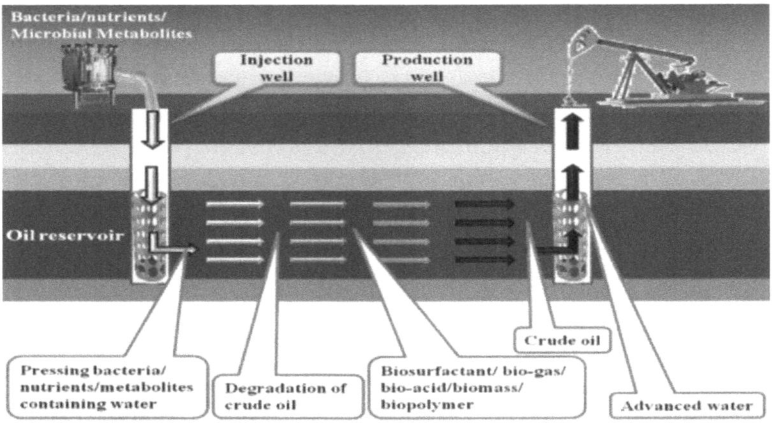


**Figure 9.2** Schematic of MEOR operation in the field.

Figure 9.2 describes schematically the MEOR operations in field applications.

## 9.5.2 Types of MEOR

On the basis of the modes of operation, MEOR is divided into ex situ and in situ MEOR technology. In in situ MEOR methods, indigenous microbes present in the reservoirs are stimulated by providing optimum conditions for production of favourable metabolites, thereby facilitating increased oil recovery. Selective microbial consortia are also injected into the reservoir with the optimised growth environment. For in situ operation, anaerobic or facultative bacteria are required due to the anaerobic reservoir conditions [173–175], and the success of this technology is mainly dependent on the selection of reservoir types and bacterial strains with their viability and effectiveness for the production of required amounts of by-products for residual oil recovery. Sulphate-reducing bacteria are the drawback of in situ technology. Therefore, special care has to be taken during the injection of nutrients or water because of the fact that the presence of sulphate in the injected water may activate the indigenous sulphate-reducing bacteria, leading to corrosion by $H_2S$ production and plugging by iron sulphide

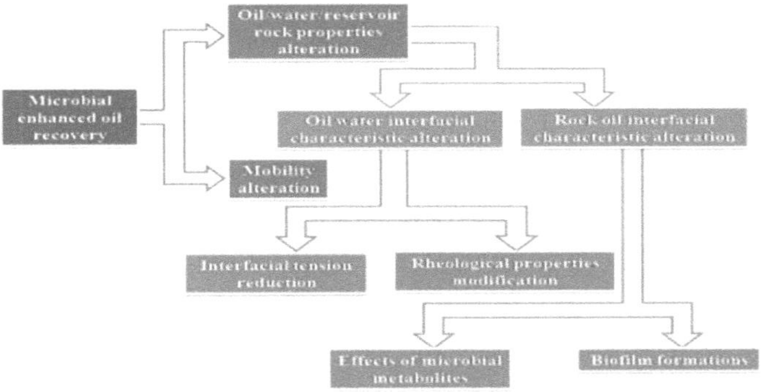


**Figure 9.3** Flow diagram of different mechanisms of MEOR.

production, which are also major concerns of the oil industries [164]. In ex situ MEOR, bacterial by-products are produced at the laboratory scale, followed by scale-up and injection to the depleted reservoir with or without the microbial cell for enhancement of recovery [141, 176, 177]. The main advantage of this method over the in situ method is the suitability of aerobic bacteria for MEOR operations. Figure 9.3 depicts the flow diagram of different MEOR mechanisms.

## 9.5.3 Microbes and Their Metabolites in MEOR

Microorganisms and their metabolites like BSs, biopolymers, bio-solvents, bioacids, biogas, etc., in MEOR applications have attracted attention recently owing to their bioavailability, low toxicity and biodegradation properties [146, 166, 176, 178, 179]. One of the major advantages of MEOR over EOR methods is the economic viability, because minimum modification in the existing oil recovery facilities is sufficient for MEOR field applications [149, 180].

The main causes of not being able to recover existing oil present in the reservoir are poor permeability, high viscosity, high IFT between oil and water and wettability of the rocks. Therefore, HC-degrading microbes break the heavy-molecular-weight HCs to low-molecular-weight HCs and reduce viscosity to increase permeability.

**Table 9.1** Bacterial strains used in MEOR technology and their by-products with the mode of mechanism

| S. no. | Bacteria | Source | Microbial by-products | Mechanisms | Ref. |
|---|---|---|---|---|---|
| 1 | *Chelatococcus daeguensis* HB–4 | Baolige Oilfield, China | Bioemulsifier | $\rho_o$ reduction and degradation of heavy oil | [266] |
| 2 | *Rhizobium viscosum CECT* 908 | Soil of Guatemala | Biopolymer | MR reduction and increase $E_s$ | [267] |
| 3 | *Ochrobactrum pseudintermedium* C1 and *B. cereus* K1 | IISCO Steel plant, Burnpur, India | Exopolysaccharides (bioemulsifier) | emulsification and IFT reduction | [268] |
| 4 | *Luteimonas huabeiensis* sp. nov. HB-2 | Baolige Oilfield, China | Biosurfactant (lipopeptide) | Reduction in IFT and $\rho_o$ | [269] |
| 5 | *P. aeruginosa, B. atrophaeus* and *B. cereus* | Yanchang Oilfield, China | Biosurfactant and enzyme solution | IFT reduction and increased $E_d$ | [270] |
| 6 | *Pseudomonas putida* MTCC 2467 | Mehrauli, Delhi | Biosurfactant | Increased $E_d$ and IFT reduction | [271] |
| 7 | *Candida tropicalis* MTCC230 | – | Biosurfactant (surfactin) | ST and IFT reduction | [272] |
| 8 | *Bacillus megaterium* | Andaman and Nicobar Islands, India | Biosurfactant | ST and IFT reduction | [273] |
| 9 | *Bacillus licheniformis* | Agni kund, West Bengal, India | Biopolymer | Increase of $\rho_w$, pore plugging and biocementing | [273] |
| 10 | *Clostridium acetobutylicum* ATCC824 | Garden soil in Connecticut, USA | Biosolvent (acetone and butanol) | $\rho_o$ reduction and emulsification | [274] |
| 11 | *Thermoanaerobacter* sp. | Mexican carbonate reservoirs | Acid (acetate and lactate), solvent (ethanol), and gas ($H_2$ and $CO_2$) production, | Increased reservoir pressure, $\rho_o$ reduction, changed in $K_{abs}$ of porous media and facilitates oil mobility | [173] |
| 12 | *Methanobacteriales* and *Methanomicrobiales* | Baolige oilfield, China | Biogas ($CH_4$) | Increased reservoir pressure | [275] |
| 13 | *Oceanobacillus* sp. and *Staphylococcus* sp. | Qinghai oilfield, China | Biogas production | Increased reservoir pressure | [276] |
| 14 | *Bacillus subtilis* RI4914 | Production water of Rio-Itaúnas Brazil | Biosurfactant, biosolvent, and biopolymer | Reduction in FT and $\rho_o$ and increased $\rho_w$ | [277] |

Microbes also produce their metabolites by utilising oil as a carbon source and increase recovery from depleted reservoirs. BSs reduce the ST as well as the IFT between oil and water, whereas biopolymers increase the viscosity of the displacing fluids and reduce the MR between displacing and displaced fluids to increase the oil recovery efficiency. Moreover, biogases increase the reservoir pressure and biosolvents reduce the viscosity of reservoir fluids [144, 181]. Table 9.1 summarises some of the reported microorganisms and their by-products with the mode of action for MEOR.

## 9.6 Biosurfactants

BIOlogically produced SURface AgeNTS are called BIOSURFACTANTS, which exhibit ST reduction, IFT reduction and emulsifying activity (E24%) [182]. Moreover, microbial-origin BSs also possess various advantages over chemically synthesised surfactants, such as SDS, CTAB and Tween-80, in terms of lower toxicity, bioavailability, biocompatibility and biodegradation properties [145, 183–187]. BSs can be anionic, non-ionic or cationic in nature. On the basis of their nature and structure, BSs are classified into glycolipids, lipopeptides, polysaccharide–protein complexes, phospholipids, fatty acids and neutral lipids, which are appraised to be secondary metabolites. Due to the presence of a hydrophilic head and a hydrophobic tail, BSs act as an intermediate between two substrates of different polarities and reduce the IFT.

Moreover, on the basis of molecular weight, BSs are also divided into low-molecular-weight and high-molecular-weight BSs. Low-molecular-weight BSs include glycolipids and lipopeptides, which reduce the ST and IFT. High-molecular-weight BSs include polymers, which are a combination of polysaccharides, proteins, lipopolysaccharides, lipoproteins or complex mixtures of these biopolymers. Generally, high-molecular-weight BSs have less efficient ST or IFT reduction activity, whereas the E24% activities of BSs vary with the substrates. *Acinetobacter* is the most reported genus for the production of high-molecular-weight BSs [188, 189].

## 9.6.1 Glycolipids

In glycolipids, carbohydrates link to a long hydrophobic chain by an ester group. Rhamnolipids, sophorolipids and trehalolipids are the most studied glycolipids. But rhamnolipids are widely known BSs, which are produced extracellularly as a mixture of mono- and di-rhamnolipids by *Pseudomonas* sp. [190–192].

### 9.6.1.1 Rhamnolipids

The first rhamnolipid was reported from *Pseudomonas aeruginosa* by Bergstrom in 1946 with the structural components of L-rhamnose and *β-hydroxydecanoic* acid [193]. But the exact chemical structure of rhamnolipid produced by *P. aeruginosa* was first revealed by Jarvis and Johnson in 1949 [194]. The carbohydrate part of rhamnolipids is made up of a rhamnose group ($C_6H_{12}O_5$), and on the basis of that number in the hydrophilic head, rhamnolipids are divided into mono-rhamnolipids (presence of one rhamnose group) and di-rhamnolipids (presence of two rhamnose groups). In di-rhamnolipids, rhamnose moieties are linked with each other through an *α-1,2-glycosidic* linkage, as mentioned by Edwards and Hayashi (1965) [195]. The proportion of mono- and di-rhamnolipids in the produced rhamnolipids is mainly dependent on bacterial strains [196], different environmental conditions such as pH, temperature, salinity and medium compositions, including C and $N^2$ sources [197]. The hydrophobic tails of rhamnolipids are *β-hydroxy* fatty acid chains with a chain length of $C_8H$ to $C_{16}H$, and the fatty acid chains are linked by an ester bond [191, 198, 199]. According to Abdel-Mawgoud et al. (2010) [200], there are 60 numbers of rhamnolipid congeners and homologues reported so far. Different rhamnolipid congeners produced by different bacterial strains are summarised in Table 9.2.

### 9.6.1.2 Trehalolipids

*Mycobacterium, Nocardia, Corynebacterium* and *Rhodococcus* are well-known bacterial genera for the production of trehalolipids [201]. Like rhamnolipids, the structures of trehalolipids also vary by the origin of microorganisms, size of the *α-branched-β-hydroxy* fatty

**Table 9.2** Different rhamnolipid congeners produced by different biosurfactant-producing bacterial strains

| S. no. | Bacterial strains | Rhamnolipid congeners | ST(mN/m),IFT(mN/m), CMC (mg/l) | Ref. |
|---|---|---|---|---|
| 1 | *Pseudomonas aeruginosa UG2* | Rha-$C_{10}$-$C_{10}$, Rha-Rha-$C_{10}$-C8, Rha-Rha-$C_{10}$- $C_{10}$, and Rha-Rha-$C_{10}$-$C_{12}$ | ST: 36, CMC: 37 | [278] |
| 2 | *Pseudomonas aeruginosa LBI* | Rha-$C_{10}$-$C_{10}$, Rha-Rha-$C_{10}$-$C_{10}$, Rha-Rha-$C_{10}$-$C_{12}$, and Rha-Rha-$C_{10}$-$C_{12:1}$ | ST: 24, IFT: 1.31,CMC: 120 | [279] |
| 3 | *Pseudomonas aeruginosa DAUPE 614* | Rha-$C_{10}$-$C_{10}$, Rha-$C_{10}$-$C_8$, Rha-$C_8$-$C_{10}$, Rha-$C_{10}$-$C_{12:1}$, Rha-$C_{12}$-$C_{10}$, Rha-$C_{10}$-$C_{12}$, Rha-Rha-$C_{10}$-$C_{10}$, Rha-Rha-$C_{10}$-$C_8$, Rha-Rha-$C_{10}$-$C_{12:1}$, Rha-Rha-$C_{12}$-$C_{10}$, and Rha-Rha-$C_{10}$-$C_{12}$ | ST: 27.3, CMC: 13.9 | [280] |
| 4 | *Pseudomonas aeruginosa D* | Rha-Rha-$C_{10}$-$C_{10}$, Rha-$C_{12}$-$C_{10}$, and Rha-$C_{10}$-$C_8$ | ST: 24.02 | [242] |
| 5 | *Pseudomonas aeruginosa FIN2* | Rha-$C_{10}$-$C_{10}$, Rha-$C_8$-$C_{10}$, Rha-$C_{10}$-$C_{12:1}$, and Rha–Rha-$C_{10}$-$C_{10}$ | ST: 28.6, CMC: 195 | [281] |
| 6 | *Pseudomonas aeruginosa PG1* | Rha-($C_{10}$-$C_{10:1}$)-$CH_3$, Rha-$C_{12:2}$, Rha-$C_8$-$C_8$, Rha-Rha-$C_8$-$C_8$, and Rha-Rha-$C_{10}$-$C_8$ | ST: 29.6, CMC: 56 | [282] |
| 7 | *Pseudomonas aeruginosa ST5* | Rha–Rha–$C_{10}$–$C_{10}$, Rha–$C_{10}$–$C_{10}$, Rha–Rha–$C_8$–$C_{10}$,Rha–$C_8$–$C_{10}$, Rha–$C_{12}$–$C_{10}$, and Rha–Rha–$C_{12}C_{10}$ | ST: 32.3 | [283] |
| 8 | *Pseudomonas aeruginosa* | Rha-$C_{10}$-$C_8$, Rha-$C_{10}$-$C_{10}$, Rha-$C_{10}$-$C_{12}$, Rha-$C_{10}$-$C_{12:1}$, Rha-Rha-$C_{10}$-$C_{10}$, and Rha-Rha-$C_{10}$-$C_{12}$ | IFT: 0.57, CMC: 55.87 | [284] |
| 9 | *Serratia rubidaea SNAU02* | Rha-$C_{10}$-$C_{10}$ and Rha-Rha-$C_{10}$-$C_8$ | ST: 34.4 | [93] |
| 10 | *Burkholderia pseudomallei* | Rha-Rha-$C_{14}$-$C_{14}$ | – | [285] |
| 11 | *Pseudomonas aeruginosa ATCC 10145* | Rha-$C_{10}$-$C_{10}$ and Rha-Rha-$C_{10}$-$C_{10}$ | ST: 28, CMC: 50 | [286] |
| 12 | *Lysinibacillus sphaericus strain IITR51* | Rha-$C_{10}$-$C_{10}$ and Rha-Rha-$C_{10}$-$C_{10}$ | ST: 52 | [287] |
| 13 | *Burkholderia thailandensis* | Rha-Rha-$C_{14}$-$C_{14}$ | ST: 42, CMC: 225 | [288] |
| 14 | *Achromobacter sp. TMB1* | Rha-$C_{8:2}$, Rha-$C_{10}$, Rha-$C_{12:2}$, Rha-$C_8$-$C_8$, Rha-$C_{10}$-$C_{10}$, Rha-$C_{12}$-$C_8$, Rha-$C_{10}$-$C_{12:1}$, Rha-Rha-$C_{12}$, Rha-Rha-$C_{10}$-$C_{10}$, and Rha-Rha-$C_{10}$-$C_{12:1}$ | ST: 30.7, CMC: 81 | [289] |

acid chain (mycolic acids) and the number of carbon atoms present in the disaccharide. White et al. (2013) reported *Rhodococcus* PML026 from seawater in the United Kingdom with trehalolipid-producing ability, which reduced the maximum ST of water to 29 mN/m at the CMC of 250 mg/l. *Rhodococcus wratislaviensis* BN38 was also reported for trehalolipid production with ST-reducing capacity from 72 mN/m to 24.4 mN/m and IFT reduction between water and n-hexadecane to 1.3 mN/m at the CMC value of 5 mg/l [202, 203].

### 9.6.1.3 Sophorolipids

Sophorolipids are another type of glycolipids which are reported from yeast, for example *Torulopsis apicola* [204], *Candida bogoriensis* [205], *Candida bombicola* [206] and *Wickerhamiella domericqiae* [207]. The hydrophilic head of sophorolipids is made up of disaccharide sophorose ($C1_{12}H_{22}O_{11}$), whereas the hydrophobic tail is the fatty acid chain with a carbon length of $C_{16}$ or $C_{18}$. According to Van Bogaert, sophorolipids lower the ST of water from 72.8 mN/m to the range of 40–30 mN/m at the CMC range of 40 mg/l to 100 mg/l [208].

## 9.6.2 Lipopeptides

Lipopeptides, made up of a lipid and a polypeptide chain, are differentiated on the basis of the length of the polypeptide chain, along with the types of amino acids [209]. *Bacillus* sp.-produced surfactin is the well-studied lipopeptide BS, first reported by Arima et al. in 1968 from *B. subtilis* growth medium [210]. Generally, extracellularly produced surfactins are made up of seven amino acids containing peptide loop incorporation with the *β-hydroxy* fatty acid chain [211]. According to Ohno et al., surfactin is the most powerful BS with the water ST-reducing capacity from 72 mN/m to 27 mN/m [212]. Table 9.3 lists some of the reported bacterial strains along with the produced lipopeptide BSs and their properties.

**Table 9.3** List of lipopeptide producing bacteria with their properties

| S. no. | Bacterial strains | Lipopeptide | ST (mN/m), IFT (mN/m), CMC (mg/l) | Ref. |
|---|---|---|---|---|
| 1 | *Bacillus subtilis MG495086* | Surfactin | ST: 29.85, CMC: 40 | [290] |
| 2 | *Bacillus atrophaeus strain AKLSR1* | Cyclic lipoheptapeptides | ST: 28,CMC: 0.9 | [291] |
| 3 | *Bacillus stratosphericus strain FLU5* | Surfactin and pumilacidin | ST: 28, CMC: 30 | [292] |
| 4 | *Aneurinibacillus aneurinilyticus SBP–11* | Aneurinifactin | ST: 26, CMC: 26 | [293] |
| 5 | *Paenibacillus dendritiformis CN5* | Lipopeptide | ST: 30, CMC: 185 | [294] |
| 6 | *Acinetobacter junii B6* | Anionic lipopeptide (surfactin, iturin A, and fengycin) | ST: 36, CMC: 300 | [295] |
| 7 | *Arthrobacter species strain MIS38* | Arthrofactin | ST: 24, CMC: $10^{-4}$ M | [296] |
| 8 | *Bacillus circulans* | Surfactin | ST: 28 | [297] |
| 9 | *Bacillus licheniformis JF-2* | Lichenysin B | ST: 27, IFT: 0.43 | [217] |

## 9.6.3 Phospholipids, Fatty Acids and Neutral Lipids

Phospholipid and fatty acid surfactants are produced extracellularly by bacteria and yeast in the presence of n-alkanes, and the phospholipid structures differ with the length of the hydrocarbon (HC) chain [213, 214]. *P. putida* BD2 was reported by Janek et al. with rhamnolipids and phosphatidylethanolamine (a phospholipid)-producing ability [215]. *Rhodococcm erythropolis* was also reported for phosphatidylethanolamine production during growth in n-alkane, which reduced the IFT between water and hexadecane to $<1$ mN/m at the CMC value of 30 mg/l [213]. *Lactobacillus pentosus*-produced BS was reported with a fatty acid chain, which was composed of linoelaidic acid ($C_18H_{32}O_2$), elaidic acid ($C_18H_{34}O_2$), palmitic acid ($C_16H_{32}O_2$) and stearic acid ($C_16H_{32}O_2$) as per Vecino et al. (2015) [216].

## 9.7 Biosurfactant-Mediated MEOR Study at the Laboratory Level

BS-mediated EOR has been studied extensively at the laboratory level as compared to field studies. Due to IFT reduction capabilities, BSs overcome the capillary effect and mobilise the entrapped oil with the water flooding to the surface. Lichenysin B was reported as the most powerful BS produced by *B. licheniformis* JF-2, lowering the IFT between oil and water to $10^{-2}$ mN/m at a concentration range

**Table 9.4** Biosurfactant-mediated in situ MEOR investigations at the lab scale

| S. no. | Bacterial strains | Mode of action | Oil recovery (%) | Ref. |
|---|---|---|---|---|
| 1 | *Chelatococcus daeguensis HB-4* | Viscosity reduction | Core flooding: 20.5 | [266] |
| 2 | *Bacillus amyloliquefaciens 702* | IFT reduction | Core flooding: 7.04 | [298] |
| 3 | *Pseudomonas aeruginosa DQ3* | ST reduction, emulsification | Core flooding: 5.22 | [299] |
| 4 | *Bacillus stearothermophilus SUCPM14* | IFT reduction | Core flooding: 11.54 | [300] |
| 5 | *Enterobacter cloacae* | IFT reduction | Core flooding: 6.9 | [300] |
| 6 | *Alcaligenes faecalis* | IFT reduction, emulsification, wettability alteration | Shut in after core flooding: 8.2, quick flooding in cores: 5.2 | [301] |
| 7 | *Enterobacter Cloacae* | Surfactin: IFT reduction, biofilm formation, wettability alteration | Shut in after core flooding: 4; quick flooding in cores: 6.3 | [302] |
| 8 | *Pseudomonas aeruginosa SG* | Rhamnolipid: ST and IFT reduction | Additional recovery in core flooding: 8.33 | [303] |
| 9 | *Pseudomonas aeruginosa WJ-1* | Rhamnolipid: ST and IFT reduction, wettability alteration, emulsification | Additional recovery in sand pack column: 7.46 to 7.32 | [304] |

of 10–60 mg/l [217]. Reports say this method recovered around 10%–85% of extra oil from core floods or sand pack column tests [176, 218]. There are numerous investigations on BS-based EOR technology at the lab scale, and some of the reported BSs for both in situ and ex situ MEOR are summarised in Tables 9.4 and 9.5, respectively.

**Table 9.5** Biosurfactants mediated ex situ MEOR investigations at the lab scale

| S. no. | Bacterial strains | Mode of action | Oil recovery in (%) | Ref. |
|---|---|---|---|---|
| 1 | *C. tropicalis MTCC230* | Biosurfactant: ST and IFT reduction | Additional recovery in sand pack column: 7.46 | [272] |
| 2 | *Paenibacillus alvei ARN63 and Bacillus mycoides SH2* | Lipopeptide: ST reduction | Core flooding: 31 | [305] |
| 3 | *Bacillus licheniformis DSM* | Lipopeptide: ST reduction and emulsification index | Sand pack column: 116.6 | [306] |
| 4 | *Candida albicans IMRU 3669e* | Sophorolipids: ST reduction and emulsification index | Sand pack column: 8.6 | [126] |
| 5 | *Bacillus subtilis* | Surfactin: IFT reduction, wettability alteration | Core flooding: 1.7 to 5 | [259] |
| 6 | *Enterobacter Cloacae* | Surfactin: IFT reduction, biofilm formation, wettability alteration | Shut in after core flooding: 4; quick flooding in cores: 14 | [300] |
| 7 | *Bacillus subtilis K1* | Surfactin: ST and IFT reduction, emulsification index | Sand pack column: 43 | [307] |
| 8 | *Pseudomonas aeruginosa MM1011* | Rhamnolipid: ST and IFT reduction, emulsification | Sand pack column: 27 | [308] |
| 9 | *Pseudomonas aeruginosa NCIM 5514* | Rhamnolipid: ST reduction, emulsification | Core flooding: 27 | [309] |
| 10 | *Pseudomonas aeruginosa WJ-1* | Rhamnolipid: ST and IFT reduction, wettability alteration, emulsification | Sand pack column: 4.49 to 4.64 | [304] |

## 9.8 Biosurfactant-Mediated MEOR in Field Studies

Successful field studies have been carried out by Czechoslovakia, Hungary, the Netherlands, Poland, Romania, the United States, Russia, China, Australia, Argentina, Bulgaria, Hungary, India, Malaysia, Peru, Poland and Romania with significant oil recovery from depleted reservoirs [144, 166]. Hitzman et al. reported that approximately 78% of MEOR field trials have been successful with the increasing efficiency of 25% [219]. Moreover, Youssef et al. reported 96% of the success rate of MEOR implementations from 403 laboratories and field applications [220]. An in situ field test was conducted with lipopeptide-producing *Bacillus* strain RS-1 and *B. subtilis* subsp. *spizizenii NRRL B-23049* on a limestone petroleum reservoir, where oil production was increased by 10% [221]. *B. licheniformis* ACO1–, *B. subtilis*– [222] and *B. subtilis* B20 [223]–produced biosurfactants were also reported with successful MEOR field trials. According to Lazar et al., China is leading in MEOR field applications and around 70% of MEOR field trails have shown a positive response with significant oil recovery [143, 178]. More than 4600 MEOR field trials have been conducted in China, where microbial flooding recovery field tests recovered an additional 4.95% of oil from OOIP [224]. She et al. also reported the influence of reservoir fluid characteristics, salinity and temperature on MEOR field applications [224]. A bacterial consortium was developed from the Baolige Oilfield, China, with *B. subtilis*, *Arthrobacter*, *G. subterraneus*, *P. aeruginosa*, *B. licheniformis* and *Rhodococcus* sp., which recovered an additional 210,000 tons of crude oil from 169 wells after 43 months of operation [225]. In 2019, Wang et al. implemented the laboratory test results into the reservoir well of the Liaohe high pour-point oil reservoir, China [226]. Exogenous BS-producing and nitrate-reducing bacterial consortiums, including *Lysinibacillus*, *Acinetobacter junnii* and *Sulphurihydrogenibium*, were stimulated inside the reservoir and it was reported that bioemulsification played a more important role in enhancing oil mobility than biodegradation [226]. In the Xinjiang No. 6 oilfield of China also, the indigenous microbes present in the reservoir were stimulated and an additional 1872 tons of oil was recovered from the heavy oil reserves [227]. An exogeneous BS-producing consortium including *P. aeruginosa* DN1 and *B. subtilis* QHQ110 was injected two times into the Hujianshan oilfield during 2010–2013 at specific time intervals. After the first treatment, the production of cumulative oil increased from 3.15 bbl/days to 14.77 bbl/days with a maximum yield of 38.55 bbl/day. Moreover, the production increased to 13.28 bbl/days from 3.38 after the second treatment of the mentioned consortium [228]. In this process, the water cut was also reduced from 93.7% to 72.9% and 94.4% to 77.6% after the first and second treatments of the consortium [228].

## 9.9 Why Biosurfactant over Chemical Surfactant in MEOR

In recent decades, BSs have gained more recognition than chemical surfactants due to their structural diversity and functional properties. The environment-friendly nature of microbial-origin BSs makes them even more acceptable than chemical surfactants. Other advantages include bioavailability, biodegradability and their functional stability at diverse temperature, pH and salinity [144, 166, 176, 229].

The most essential properties of BSs for MEOR applications are ST reduction, IFT reduction, emulsification and wettability alteration abilities. BSs exhibit better efficiency for ST and IFT reduction than chemically synthesised surfactants. Moreover, the CMC of the BSs is also very much lower than that of chemical surfactants. The ST and IFT reduction capacity and CMC values of different microbial-origin BSs are already summarised in Tables 9.2 and 9.3. Tergitol 15-s-7, a non-ionic chemical surfactant reduced he maximum ST of an aqueous solution and IFT between crude oil and aqueous solution to 29.2 mN/m and 2.03 mN/m, respectively [67], whereas rhamnolipid produced by *P. aeruginosa* UCP0992 reduced the ST to 27.4 mN/m and IFT between oleic and aqueous phase to 2 mN/m at a CMC of 700 mg/l [230]. Moreover, the CMC of SDS was reported to be 2100 mg/l [231].

## 9.10 Production of Biosurfactants

The production costs of BSs are very high than those of commercially synthesised surfactants because of their low yield from the bacteria species [232, 233]. Some strategies were developed to reduce production costs, as follows:

(1) Usage of waste or cheap substrates for BS production in the culture media [234–236]
(2) Optimisation of culture conditions and media components for maximum BS production at the lab scale [145, 237]
(3) Development of efficient bioprocess technologies [145, 237, 238]
(4) Development of mutant or recombinant bacterial strains to maximise the yields [145, 239, 240]

### 9.10.1 Usage of Waste or Cheap Substrates

Use of low-cost and renewable substrates in the BS production medium is one of the ways to reduce the production cost since almost 30%–50% of their production cost has been influenced by the production medium [241]. There are various reports for BS production using renewable and waste substrates such as waste frying coconut oil used by *P. aeruginosa* D [242], kitchen waste oil used by *P. aeruginosa*, molasses waste used by *B. subtilis* B20 [223], sugarcane molasses and sugarcane juice alcohol stillage used by *B. subtilis* [243], olive mill waste used by *P. aeruginosa* and *B. subtilis* [244], agroindustrial waste (crude glycerol and corn steep liquor) used by *Rhizopus arrhizus* UCP 1607 [245] and agroindustrial waste (soybean waste frying oil and corn steep liquor) used by *Streptomyces* sp. DPUA1566 [246].

### 9.10.2 Optimisation

Optimisation of culture conditions provides organisms the most suitable conditions for maximum growth and BS production at the laboratory scale. There are various reports on optimisation of BS production using classical methods such as the one factor at a time

(OFAT) method and statistical methods such as response surface methodology (RSM) and the Taguchi method, as summarised in Table 9.6 [246]. In the classical method, one variable is changed at a time, keeping other variables constant, which is known as the OFAT method. But the RSM and fractional factorial designs are the two most widely used optimisation techniques in recent times, where multiple factors can be changed at a time.

### 9.10.3 Bioprocess Engineering

As process intensification approaches, bioreactors are used for the scale-up of BS yield, where aeration and agitation rates are optimised. Production of excessive foam is a serious challenge during BS production in aerobic conditions due to the aeration and agitation rates as well as the foaming properties of BSs [247, 248]. Many chemical antifoaming agents like silicon oil, polyethylene glycol and polypropylene can be used, but these agents may have negative effects on product quality and productivity [249]. In these circumstances, the production of excessive foam can be used as an advantage for in situ product recovery during batch fermentation [250, 251]. Generally in in situ product recovery, stirred tank bioreactors are most commonly used, where a foam-separated column is attached additionally to the lead of the reactor vessel and collects the surfactant-containing foam during operation. In 2006, Chien-Yen Chen et al. integrated a batch mode bioreactor with a foam fractionation column for surfactin production from *B. subtilis* BBK006, and the production was 41.67% increased in the reactor than in shake-flask conditions [252]. Moreover, Yeh et al. designed another integrated bioreactor for BS scale-up from *B. subtilis* ATCC 21332. Aeration and agitation rates were optimised at 1.5 vvm and 300 rpm, respectively, and the maximum production rate of 190 mg/l/h was reported with an overall production rate of 106 mg/l/h, surfactin concentration of 6.45 g/l and surfactin yield of 161 mg/g glucose in the optimised conditions [253]. In the case of *Candida lipolytica* UCP 0988, the BS production was found to be 10 g/l in a 2 l bioreactor at an agitation rate of 200 rpm. But this BS production was increased to 40 g/l in a 50 l bioreactor with the same agitation rate

**Table 9.6** Optimum conditions for maximum biosurfactant production by selected bacterial strains

| S. no. | Bacterial strains | Optimum conditions | Optimisation method | Ref. |
|---|---|---|---|---|
| 1 | *Planococcus sp. MMD26* | Temp.: 30°C and pH 7 | Response surface methodology | [310] |
| | | Incubation time: 48 h, Carbon source: Glucose, Nitrogen source: Ammonium nitrate | Classical method | [310] |
| 2 | *Bacillus brevis* | Temp.: 33°C, pH: 8, Incubation time: 10 days, Carbon source: Glucose (8.5 g/l) | Response surface methodology | [311] |
| 3 | *Acinetobacter junii* B6 | Aeration rate: 300 rpm, Temp.: 25°C, ILCO percentage (v/v): 5%, Inoculum size (v/v): 2%, and $NaNO_3$: 2 g/l | Two-level fractional factorial design | [295] |
| 4 | *Pseudomonas aeruginosa* | Carbon: Kitchen waste oil, pH: 8, and Nitrogen: Yeast extract (2 g/l) | Classical method | [284] |
| 5 | *Pseudomonas aeruginosa* | Carbon: kitchen waste oil, pH: 8, and $N_2$: Yeast extract (2 g/l) | Classical method | [284] |
| 6 | *Burkholderia glumae* | Carbon: Mannitol and $N_2$: Urea | Classical method | [312] |
| | | Carbon (w/v): Mannitol (0.5%), Urea (M): 0.01, $KH_2PO_4$ buffer (mM): 100, and $CaCl_2 2H_2O$ (g/l): 0.01 | Response surface methodology | [312] |
| 7 | *Bacillus subtilis* MJ01 | $N_2$(g/l): Yeast extract (1.49), $KH_2PO_4$ (g/L): 7.62, $KH_2PO_4$ (g/l): 33.68, and Carbon (g/l): 11.9 | Response surface methodology | [313] |
| 8 | *Pseudomonas stutzeri* SSASM1 | Temp.: 35°C, pH: 9, salinity : 15 psu, Carbon: Sucrose (1%), and $N_2$: Yeast extract | Classical method | [ [314] |
| 9 | *Serratia marcescens* | pH: 8, Temp.: 30°C, salinity: 1%, Agitation rate: 200 rpm, Incubation time: 120 h, Carbon: Glycerol and $N_2$: Peptone and $(NH_4)_2SO_4$ | Classical method | [315] |
| | | Carbon: 5%, Nitrogen: Peptone(4 g/l), and $(NH_4)_2SO_4$(5 g/l) | Central composite design | [315] |
| 10 | *Achromobacter* sp. TMB1 | pH: 5.5–7.2, Temp.: 30°C, Carbon: Glucose, $N_2$: $NaNO_3$ and Incubation time: 48 h | Classical methods | [289] |
| | | $N_2$: $NaNO_3$ (0.5%), Carbon: Glucose (1%), and Inducer: Yeast extract (0.05%) | Taguchi method | [289] |

(200 rpm) [254]. This technology is extensively studied for surfactin production from *B. subtilis*; however, there are very limited reports on rhamnolipid scale-up through these methods. In 2016, Beuker et al. used these methods for rhamnolipid production from the genetically engineered strain *P. putida* KT2440, where also the rhamnolipid concentration was found higher in the foammate than the reactor vessel with enrichment up to 200 [255].

### 9.10.4 Genetic Engineering

In the modern biotechnological approach, genetic engineering has emerged as a single-window solution for many of the problems. Rhamnolipids and surfactins are two widely studied BSs, which are generally known to be produced by *Pseudomonas* sp. [256–258] and *Bacillus* sp. [259, 260], respectively. In rhamnolipid biosynthesis, *RhlA*, *RhlB* and *RhlC* genes are mainly responsible as per Wittgens et al. (2018), whereas A, B, C and D are responsible for surfactin biosynthesis [261, 262]. *RhlA* is responsible for the synthesis of 3-(3-hydroxyalkanoyloxy)alkanoic acids (HAAs), whereas *RhlB* and *RhlC* are involved in mono- and di-rhamnolipid formation, respectively [240]. Wittgens et al. expressed the rhamnolipid synthesis genes of *P. aeruginosa* in *P. putida* KT2440 and revealed that *RhlA* and *RhlB* are involved independently in rhamnolipid synthesis, not as a *RhlAB* heterodimer complex [240].

One major success of the genetic engineering approach for scale-up of rhamnolipid production was reported by Tavares et al. in 2013 [263]. Rhamnolipid synthesis genes *RhlA* and *RhlB* from *Pseudomonas* sp. were expressed in *Burkholderia kururiensis* and increased the rhamnolipid production by 600% in the engineered strain than in the wild-type strain [263]. A lipopeptide-producing recombinant strain, *P. stutzeri* DQ1 was constructed through protoplast fusion from *B. mojavensis* JF-2 for in situ MEOR applications [264]. Moreover, *RhlAB* genes were inserted into the chromosomes of the nonrhamnolipid-producing *P. aeruginosa* PAO1 and *E. coli* BL21 strains and two recombinant strains were constructed, named *P. aeruginosa* PEER02 and *E. coli* TnERAB, respectively. Both of the recombinant strains produced rhamnolipids with varying rhamnolipid congeners. The rhamnolipid production was 56.64%

and 50.2% increased by the recombinant strain *P. aeruginosa* PEER02 in glucose- and soybean oil–containing MSM medium than by the wild-type strain *P. aeruginosa* PAO1. However, the recombinant *E. coli* TnERAB strain produced 175.3 mg/l and 75.6 mg/l of rhamnolipids in glucose-containing LB and MSM medium [265].

## 9.11 Conclusion

MEOR is an extensively studied tertiary oil recovery method, where microbes, particularly bacteria and their by-products, such as biosurfactants, biopolymers and biogases, are used for enhancing recovery of oil from depleted reservoirs. Mechanically, biosurfactant-containing fluids displace the trapped oil from the reservoir as displacing fluids, whereas the high-permeable zones formed during conventional recovery methods are blocked by biopolymers to recover the trapped oil present in the low-permeable zones. From an economical point of view also, MEOR poses an advantage over EOR due to the use of microbes and their produced metabolites. Although biosurfactant-mediated core-flooding studies at the laboratory scale showed 10%–85% enhancement in recovery, the method still has not received due attention in the petroleum industries. One major cause of unpopularity implies the low product yield, which can be mitigated by approaches like media optimisation, process intensification and genetic manipulation of microorganisms. There are two different approaches of biosurfactant-based EOR technology: in situ MEOR has been preferred more to ex situ MEOR in field applications. The second approach needs an additional production cost for production in bioreactors and subsequent downstream processing. While both methods look attractive for their potential use in tertiary oil recovery, a reservoir has unique properties to influence overall recovery either way. The properties influencing the success story of MEOR have also been described in this chapter. Therefore, it is essential to study biosurfactant-based EOR applications in simulated conditions prior to actual field application, which can help in predicting the oil recovery efficiency of the BSs to be used in the candidate reservoir.

## Acknowledgements

The authors gratefully acknowledge the financial assistance extended in the form of a project junior research fellowship, entitled 'Strengthening of Biotechnology Teaching, Training and Research in Universities and Colleges in the North-East', by the Department of Biotechnology, Ministry of Science and Technology, Government of India, to Saurav Haloi.

# Chapter 10

# A Preliminary Theoretical Study on Microbial Enhanced Oil Recovery

**Saurav Bhattacharjee and Borkha Mech Das**
*Department of Petroleum Technology, Dibrugarh University, Assam, India*
bhattacharjeesourav342@gmail.com

In recent times, vast amounts of the energy arising from crude oil is still left in the pore spaces of rocks after regular enhanced oil recovery (EOR) methods. Only one-third to one-half of the oil is recovered from the original oil in place (OOIP). This reason has led the researchers to develop a method which is cost-effective and environmentally safe which can help to recover the left oil efficiently. Microbial enhanced oil recovery (MEOR) is a biological-based technology to improve the recovery of oil entrapped in porous media with increasing economic profits. Several decades of research and successful applications claim MEOR as a mature technology. MEOR is a method in which microorganisms are activated using specially prepared nutrients, which helps them to produce metabolic products such as biopolymers, biosurfactants (BSs), etc. BSs will help in emulsification and de-emulsification through reduction of interfacial tension (IFT), and biopolymers will aid in injectivity profile and viscosity modification and selective plugging. The selection of nutrients is a very crucial factor for

*Advances in Petroleum Technology*
Edited by Subrata Borgohain Gogoi

ISBN 978-981-4877-19-0 (Hardcover), 978-1-003-04993-7 (eBook)
www.jennystanford.com

the success of MEOR. Microorganisms are introduced with a food source so that the desired bioproducts get produced, and the survival of bacteria is completely dependent on what nutrients are fed to them. The best way to select nutrients is done after a detailed study of the subsurface properties like pH, salinity and trace minerals. The selection of a microbe species is also another crucial factor in MEOR. They should adapt to the reservoir conditions and produce the desired bioproducts. Different species of bacteria have different adaptabilities, so the best way to get the most suitable bacterial strain is after careful laboratory studies and test on core samples. MEOR technology can help in 20%–30% more additional oil recovery during the abandonment of an oilfield.

## 10.1 Introduction

In recent times, crude oil is the main source of the world's energy. After the application of conventional oil extraction methods, this valuable and nonrenewable energy is still left in the pore spaces of rocks. Only one-third to one-half of the oil is recovered from the original oil in place (OOIP). This has led the researchers to develop a method which is cost-effective and can help to recover the left oil efficiently. Hence, meeting the rising demand for crude oil has led to the evaluation of old and new enhanced oil recovery (EOR) processes [316].

One such method known as microbial enhanced oil recovery (MEOR) is a process in which microorganisms are used to recover the trapped oil in the rock pore spaces. MEOR is a method in which a combination of nutrients and microbes helps to produce several metabolic products either in situ or ex situ which can lead to the recovery of left oil efficiently [317]. Several theories have developed about MEOR, but very few of them have been accepted worldwide [318–320]. Later, in 1947, Zobell was the first to perform several experiments and came to the conclusion that the release of crude oil is possible with the help of microbes [166]. The first MEOR field test was conducted in Lisbon Field, Vision Country, AR, USA in 1954 [144]. Two crucial factors for the success of MEOR is the selection of a microbe species and the nutrients designed for them. The selection of microbes can be done using formation fluids and tests performed in cores [321]. The selection of nutrients is also another crucial factor in the success of MEOR. The nutrients are designed on the basis of microorganisms' characteristics so that they can produce sufficient metabolic products [322]. MEOR is an inexpensive EOR method as compared to other oil recovery methods, especially the thermal recovery method, where a large amount of energy is consumed. MEOR is a multidisciplinary field incorporating geology, chemistry, microbiology, fluid mechanics, petroleum engineering, environmental engineering and chemical engineering [323].

## 10.2 Process

There are mainly two types of MEOR processes, namely in situ and ex situ MEOR. In situ MEOR is a process where nutrients are injected into the reservoir directly to stimulate the indigenous reservoir microorganisms. In this process a combination of nutrients and microbes is injected together or one after another into the reservoir. Injection of microbes with genetic modifications can be effective in reservoir conditions, and resultant products would lead to incitation of oil. In fact in reservoirs, microbes are already residing and thriving on nutritional components. The metabolic products are formed inside the reservoir in these processes; hence it is known as in situ MEOR [324].

Ex situ MEOR is a process where only metabolic products are injected into the reservoir. The microbes are filtered out, and only specific metabolic products are injected into the reservoir. In this method, microbial products such as polysaccharides and surfactants would be produced in fermenters and would be added to the reservoir after separation and purification. In this process, metabolic products are formed in the surface; hence it is known as ex situ MEOR [322]. Among the above methods, the best method is in situ MEOR because in ex situ MEOR, metabolic products may be decomposed by reservoir microbes. Injection of exogenous microbes is also not desirable, because characteristics of reservoir microbes are unknown, so compatible studies must be conducted before using exogenous microbes.

## 10.3 Mechanisms

There are various mechanisms by which bacteria might enhance the recovery of oil from petroleum reservoirs [320]:

(1) Microbes can reduce the viscosity of the oil and increase the flow rate of the oil by breaking large hydrocarbon (HC) molecules into smaller ones.
(2) Microbes often tend to attach to solid surfaces of the rocks. This process might crowd oil off the surfaces.
(3) Microbes often produce surface-active agents, detergents or wetting agents when grown in aqueous culture media, which helps to reduce the interfacial tension (IFT) between oil and water, releasing oil from sand particles.
(4) Microbes can produce several gases like carbon dioxide, hydrogen and methane, which can increase reservoir pressure and liberate oil from micro traps.
(5) Microbes can produce several water-soluble polymers like polysaccharides, which can help to increase the viscosity of the water.

## 10.4 Factors under Consideration

Several factors need to be considered before applying MEOR in the field. The selection of nutrients and microbe species is the most crucial factor for the success of MEOR.

### 10.4.1 Selecting the Reservoir

Formation fluids should be collected from any ongoing water flooding fields, and later these fluids samples will be tested in a laboratory. Formation water (FW) analysis will be done to measure the pH of the fluid sample and salinity. To select the candidate reservoir, petrophysical analyses are to be done using core samples. Porosity, permeability and grain size are measured in petrophysical analysis. Organic and inorganic compounds present in formation fluids are also measured.

### 10.4.2 Selecting Microbes

The selection of a microbe species is a crucial factor for the success of MEOR. Microbes should adapt to the reservoir conditions and produce the desired bioproducts. Different species of bacteria have different adaptabilities based on reservoir conditions [325]. FW should be tested for the identification of indigenous reservoir microorganisms. The DNA sequencing of microbes must be done to identify the pure culture of bacterial strain present in formation fluids. The most suitable mixed-culture species used in MEOR are *Clostridium, Mycobacterium, Pseudomonas, Micrococcus, Escherichia, Bacillus, Arthobacter* and *Peptococcus* [326]. Essentially, the most suitable strain of bacteria can only be determined after careful laboratory analyses and tests on cores.

### 10.4.3 Selecting Nutrients

The selection of nutrients is another crucial factor for the success of MEOR. Microorganisms are introduced with a food source so that the desired bioproducts get produced and the survival of bacteria is completely dependent on what nutrients are fed to them [323]. The nutrient package mainly consists of seven components: water, an energy source, a carbon source, an electron acceptor, essential minerals, a nitrogen source, a growth factor and inorganic salts [144]. Water will be a source of a medium where all these chemicals and vitamins will be added to prepare the right nutrients for the microbes. Nutrients must be prepared on the basis of microorganism characteristics. The combination of nutrients which gives the best bioproducts must be chosen for oil recovery. The best way to select nutrients is done after a detailed study of the subsurface properties like pH, salinity and trace minerals [322].

### 10.4.4 Preparation of a Microbial Slug

A microbial slug will be prepared with both selected nutrients and microbes, and afterwards the microbial slug will be flooded in core samples to recover the left oil. Reservoir microbes with a target salinity of FW should be placed in a glass septa bottle. Since these live cultures contain a mixture of both anaerobic and aerobic microbes, respiratory gases must be supplied to the closed container [327]. All fluids before being introduced to a closed system should be filtered through a 0.45 μm. These cultures are incubated in a 35°C oven and closely monitored for nutrient consumption. The microbial slug will be flooded in core samples to recover the left oil [324].

## 10.5 Advantages

The most outstanding advantages of MEOR over other EOR technologies are listed below [323]:

(1) The injected nutrients and bacteria are inexpensive and easy to obtain and handle in the field.
(2) MEOR processes are suitable alternatives before the abandonment of marginal wells and are economically attractive for marginally producing oilfields.
(3) MEOR agents can be produced with very little input of energy. MEOR technologies which are cost-effective to install need less modification of the existing field.
(4) Where some EOR technologies cannot be applied efficiently, especially in carbonate oil reservoirs, the MEOR process can be applied.
(5) MEOR products will be not accumulated in the environment, as all the products are biodegradable and therefore are environmentally compatible.

## 10.6 Problems

MEOR methods face some common problems which are outlined as follows [322]:

(1) Injectivity loss due to microbial plugging of the wellbore is the main problem faced by MEOR technology. Some actions must be taken to avoid wellbore plugging, such as avoiding biopolymer production, filter before injection and minimise

microbial adsorption to the rock surface by using dormant cell forms, spores or ultramicrobacteria.

(2) Transportation or dispersion of all necessary components to the target zone.

(3) Optimisation of the desired in situ bioproducts due to the effect of variables such as pH, temperature, salinity and permeability for any in situ MEOR operation.

(4) Isolation of bacterial strains, adaptable to the extreme reservoir conditions pH, pressure, temperature and salinity.

(5) The low in situ concentration of bacterial metabolites; application of genetic engineering techniques might be the solution to this problem.

## 10.7 Conclusion

MEOR is a well-proven technology to improve oil recovery, especially in mature oil wells. MEOR is an eco-friendly and cost-effective process which shows several advantages over other EOR processes. MEOR has great potential to become a viable alternative to the conventional CEOR method. In spite of the various advantages of MEOR over other EOR methods, it has gained little attention in the oil industry, because the value of MEOR is mostly determined by the result of field trials. Most MEOR literature is based on laboratory data and not enough field trials can be seen. Although MEOR is a highly attractive method in the field of oil recovery, there are still uncertainties in meeting the engineering design criteria required by the application of microbial processes in the field. Optimisation of nutrients and testing the microbes and their by-products compatible with reservoir conditions are required.

## Acknowledgements

The authors gratefully acknowledge the experimental facilities extended by the Department of Petroleum Technology, Dibrugarh.

# Chapter 11

# Revisiting Evaluation of Crude Oil in a Cost-Effective Way

**Pankaj Boruah,[a] Miranda Kakoty,[b] Walid M. Alakhdar[b] and Subrata Borgohain Gogoi[b]**

[a] *Department of Chemical Engineering, Indian Institute of Technology, Guwahati, Assam, India*

[b] *Department of Petroleum Technology, Dibrugarh University, Assam, India*

pankajboruah1@iitg.ac.in

This study is based on the characterisation of crude oil samples obtained from the Upper Assam Basin on the basis of the conversion of American Standard for Testing of Materials (ASTM) D-86 data which is obtained from distillation apparatus to true boiling point (TBP) distillation data by Edmister's method. However, for obtaining TBP data expensive laboratory equipment is required, and it is also costly and time consuming, which is inappropriate for quick estimation of crude oil distillation characteristics. It was found that the behaviour of crude oils ranges from class A to E. The characterisation factor ($K$) and correlation index (CI) showed that the crude oil behaviour is mixed based and inclined both towards naphthenic and aromatic sides. This study is aimed to tentative determination of crude oil products (petroleum fractions) from the TBP data. The TBP analyses also indicate that the likely products

*Advances in Petroleum Technology*
Edited by Subrata Borgohain Gogoi

ISBN 978-981-4877-19-0 (Hardcover), 978-1-003-04993-7 (eBook)
www.jennystanford.com

from these crude oils are mainly naphtha fractions and lesser amounts of kerosene and gasoline fractions and least amounts of lube oil base stock (LOBS) fraction.

## 11.1 Introduction

Crude oil or petroleum is a mixture which occurs naturally which consists mainly of HCs with a small amount of other elements such as sulphur (S), nitrogen ($N_2$) and oxygen ($O_2$) appearing in the form of organic compounds or in some cases complexes with metals. The chemical utilisation of crude oil and its refined product has been on the increase globally over the years. In the current economic condition, refineries are at pace to improve the efficiency of individual units and producing products of different ranges. Different technological advancements have been done to increase the efficiency. Refineries also go for refining, 'opportunity crude', as an option for improving economics. One of the feasible ways to improve refinery profitability is diversifying the crude oil slate [328]. However, in frequent changing of crude oil to the refineries, process engineers face major challenges in determining the optimum operating conditions for the new feed. Therefore, crude oil needs to be analysed in the laboratory in order to cope with these problems. Refinery product qualities may vary widely due to the lack of crude oil distillation data [329, 330].

According to the American Standard for Testing of Materials (ASTM) D-2892 standard true boiling point (TBP) is one of the most important laboratory analyses, which separates individual components from the hydrocarbon mixtures and thus gives a very good approximation of the possible product which can be obtained from the crude oil [331]. However, expensive laboratory equipment is required to obtaining TBP data. A single TBP distillation can take up to 48 h, and it is also costly and time-consuming. Therefore, it is impractical to monitor the crude oil quality for refining. But we can convert the crude oil distillation data which is obtained from the Engler distillation (ASTM D-86) by Edmister's method. This method is a fast and low-cost method for measuring distillation characteristics of oils. The present study includes how the data obtained from ASTM D-86 is converted to TBP distillation data, and the volume of different petroleum fractions corresponding to different boiling ranges obtained from a particular crude can be determined from the TBP curve, and this will lead to the determination of the percentage of specific cut fractions (part of the TBP range).

## 11.2 Experimental

### 11.2.1 Materials and Instruments

Calcium chloride fused having a water absorption capacity $\geq$20.0% was bought from Merck Specialties Private Limited with purity 98% (molecular weight [MW] 110.98 g/mol), Mumbai. An oil-soluble de-emulsifier was procured from Imperial Oilfield Chemicals Pvt. Ltd. Baroda, India. All the chemicals used were of analytical grade. Crude oil samples were obtained from different oilfields of the Assam Basin from the wellhead and subjected to ASTM distillation. ASTM distillation apparatus manufactured by Toshniwal Instrument Manufacturing Pvt. Ltd., Ajmer, India, was used for determination of ASTM distillation data.

### 11.2.2 Methodology

#### 11.2.2.1 ASTM distillation

ASTM distillation was done by taking 100 ml of a crude oil sample in a distillation flask. It was distilled under the prescribed condition of heat input and rate of distillation. The volume of distillate obtained was recorded at a fixed interval of time, up to a maximum temperature at which there was no rise in temperature. If the crude contains water it makes the distillation difficult, so the water present in the crude oil was separated by using a de-emulsifier to remove the emulsified water and using $CaCl_2$ to remove small amounts of free water, which acts as an absorbent for water. The temperature at which the first drop comes is the initial boiling point (IBP) and at which the last drop falls is the final boiling point (FBP). The ASTM data thus obtained was then plotted as the volume of

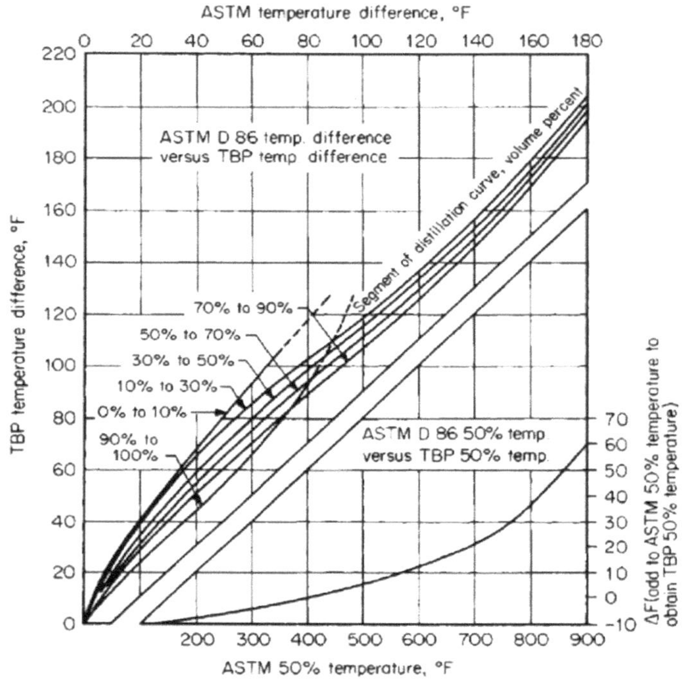


**Figure 11.1** Relationship between ASTM D-86 and TBP distillation curves developed by Edmister et al. (1948) [334].

distillate collected versus temperature rise, and then this data was converted to TBP data and equilibrium flash vaporisation (EFV). Different empirical approaches are available for converting ASTM D-86 distillations to TBP distillations. The most applicable methods for conversion in engineering purposes are Edmister's method, Riazi–Daubert method and Daubert's method [332, 333].

The earliest method for available conversion of the distillation curves is Edmister's method [334]. This method was developed in the period of 1948–1961. It is a graphical approach in which the first step is making a graphical definition of the TBP temperature differences (△TBP) from ASTM D-86 temperature difference (△ASTM) from the graph illustrated in Fig. 11.1, and then the next step is definition of the TBP 50% as a function of ASTM D-86 50% in °F using the same graph. This value of TBP 50% obtained is added to 50% of ASTM D-86 to obtain the TBP temperature at which 50% of distillate was obtained. Then the TBP data above 50% was obtained by adding $\triangle$TBP to the corresponding TBP temperature and subtracting $\triangle$TBP to obtain a TBP temperature below 50%. By means of all the TBP data obtained, it is easy to draw the TBP curve. The relationship between ASTM D-86 and TBP distillation curves developed by Edmister et al. are in Fig. 11.1 [333].

Another method for conversion of ASTM distillation to TBP distillation is the Riazi–Daubert method, which is given by the following correlation in the form of Eq. 11.1 [333]:

$$\text{TBP} = a(\text{ASTM} - D86)^{\text{b}}, \tag{11.1}$$

where TBP and ASTM temperatures are for the same vol.% distilled and are in Kelvin. Constants $a$ and $b$ at various points along the distillation curve with the range of application are given in Table 11.1. Daubert and his group developed a different set of equations to convert ASTM to TBP through modifying Riazi–Daubert (2005) correlations, which is also known as Daubert's method [333]. The following Eq. 11.2 is used to convert an ASTM D-86 distillation at 50% point temperature to a TBP distillation 50% point temperature:

$$\text{TBP (50\%)} = 255.4 + 0.8851[\text{ASTM} - D86\ (50\%) - 255.4]^{1.0258}, \tag{11.2}$$

where ASTM (50%) and TBP (50%) are temperatures at 50% volume distilled in °K. Equation 11.2 can also be used in reverse

**Table 11.1** Correlation constants in Eq. 11.1

| S. no. | Vol% | *a* | *b* | ASTM D-86,°C * |
|---|---|---|---|---|
| 1 | 0 | 0.9177 | 1.0019 | 20–320 |
| 2 | 10 | 0.5564 | 1.0900 | 35–305 |
| 3 | 30 | 0.7617 | 1.0425 | 50–315 |
| 4 | 50 | 0.9013 | 1.0176 | 55–320 |
| 5 | 70 | 0.8821 | 1.0226 | 65–330 |
| 6 | 90 | 0.9552 | 1.0110 | 75–345 |
| 7 | 95 | 0.8177 | 1.0355 | 75–400 |

*Source*: [338].

*Temperatures are approximated to the nearest 5.

**Table 11.2** Correlation constants in Eq. 11.2

| S. no. | Cut point range (%) | *A* | *B* | Maximum allowable $X_i$ °C* |
|---|---|---|---|---|
| 1 | 0 | 0.9177 | 1.0019 | 20–320 |
| 2 | 10 | 0.5564 | 1.0900 | 35–305 |
| 3 | 30 | 0.7617 | 1.0425 | 50–315 |
| 4 | 50 | 0.9013 | 1.0176 | 55–320 |
| 5 | 70 | 0.8821 | 1.0226 | 65–330 |
| 6 | 90 | 0.9552 | 1.0110 | 75–345 |
| 7 | 95 | 0.8177 | 1.0355 | 75–400 |

*Source*: [335, 336].
*Temperatures are approximated to the nearest 5.

form to estimate ASTM from TBP. The following Eq. 11.3 is used to determine the difference between two cut points:

$$Y_i = AX_i^B, \tag{11.3}$$

where $Y_i$ is the difference in TBP temperature between two cut points, °K; $X_i$ is the observed difference in ASTM D 86 temperature between two cut points, °K; and $A$ and $B$ are constants varying for each cut point and are given in Table 11.2 [335, 336].

To determine the true boiling point temperature at any percent distilled, calculation should begin with 50% TBP temperature and addition or subtraction of the proper temperature difference Yi as in Eqs. 11.4, 11.5, 11.6, 11.7, 11.8 and 11.9.

$$\text{TBP }(0\%) = \text{TBP }(50\%) - Y4 - Y5 - Y6 \tag{11.4}$$

$$\text{TBP }(10\%) = \text{TBP }(50\%) - Y4 - Y5 \tag{11.5}$$

$$\text{TBP }(30\%) = \text{TBP }(50\%) - Y4 \tag{11.6}$$

$$\text{TBP }(70\%) = \text{TBP }(50\%) + Y3 \tag{11.7}$$

$$\text{TBP }(90\%) = \text{TBP }(50\%) + Y3 + Y2 \tag{11.8}$$

$$\text{TBP }(100\%) = \text{TBP }(50\%) + Y3 + Y2 + Y1 \tag{11.9}$$

The data obtained ASTM D-86 distillation obtained for the crude oil sample taken was converted to TBP data by Edmister's method. From the TBP curve we can estimate the percentage of specific cut fractions of crude oil [337]. Table 11.3 illustrates the ASTM D-86 data and the corresponding TBP data for four different crude oil samples of the Upper Assam Basin.

**Table 11.3** ASTM D-86 data with corresponding TBP data for large different crude oil samples of the Upper Assam Basin

| Jorajan | | | Lakwa | | | Galeki | | | Bhogpara | | |
|---|---|---|---|---|---|---|---|---|---|---|---|
| Vol. (cc)* | ASTM (°C) | TBP (°C) | Vol. (cc)* | ASTM (°C) | TBP (°C) | Vol. (cc)* | ASTM (°C) | TBP (°C) | Vol. (cc)* | ASTM (°C) | TBP (°C) |
| 0 | 66 | −20.5556 | 0 | 90 | −32.667 | 0 | 120 | −1.8889 | 0 | 124 | −31.6667 |
| 10 | 101 | 32.7778 | 9 | 136 | 35.1111 | 4 | 170 | 71.44 | 3 | 130 | −17.7778 |
| 16 | 110 | 51.667 | 14 | 165 | 79 | 4.2 | 190 | 107 | 6 | 144 | 8.8889 |
| 23 | 117 | 67.2222 | 16 | 180 | 106.222 | 7.4 | 216 | 149.22 | 8 | 146 | 15 |
| 29 | 126 | 86.1111 | 17 | 196 | 135.667 | 9.6 | 234 | 181.44 | 10 | 149 | 21.667 |
| 35 | 133 | 100 | 23 | 235 | 188.444 | 10.4 | 250 | 210.33 | 12 | 154 | 33.8889 |
| 39 | 143 | 117.222 | 27 | 250 | 215.667 | 18 | 274 | 248.667 | 14 | 160 | 47.222 |
| 44 | 149 | 129.444 | 32 | 260 | 235.111 | 29 | 295 | 283.11 | 15 | 166 | 60.55 |
| 48 | 156 | 143.333 | 36 | 282 | 267.333 | 32 | 310 | 313.667 | 15.6 | 174 | 77.22 |
| 50 | 167 | 161.667 | 40 | 292 | 287.333 | 36 | 312 | 317.556 | 16 | 179 | 88.889 |
| 53 | 173 | 171.667 | 45 | 310 | 313.444 | 38 | 313.5 | 320.33 | 16.2 | 180 | 90 |
| 54 | 180 | 183.333 | 48 | 316 | 327.889 | 42 | 315 | 323.11 | 17.4 | 185 | 101.11 |
| 54.5 | 182 | 187.778 | 48.5 | 317 | 330.111 | 44 | 317 | 327 | 19 | 192 | 115.556 |
| 56 | 191 | 202.222 | 49.5 | 319 | 335.667 | 46 | 319 | 330.889 | 20 | 200 | 133.33 |
| 57.5 | 198 | 213.889 | 52 | 326 | 346.778 | 48 | 321 | 334.778 | 22 | 217 | 163.33 |
| 58.5 | 202 | 221.8889 | 54 | 330 | 353.444 | 49 | 325 | 341.44 | 26 | 229 | 185.556 |
| 59.5 | 208 | 232.222 | 59 | 337 | 365.111 | 50 | 327 | 345.33 | 28 | 240 | 206.667 |
| 60 | 211 | 236.667 | 62 | 340 | 369.556 | 53 | 328 | 347 | 30 | 247 | 221.667 |
| 61 | 214 | 241.1111 | 63 | 346 | 379.556 | 56 | 328.9 | 348.556 | 34 | 251 | 228.33 |
| 61.5 | 218 | 248.889 | 63.5 | 351 | 388.444 | 57.5 | 330 | 350.27 | 35 | 255 | 235 |
| 62.5 | 224 | 259.444 | – | – | – | 59 | 331 | 351.9 | 38 | 272 | 261.11 |
| – | – | – | – | – | – | – | – | – | 40 | 281 | 276.667 |
| – | – | – | – | – | – | – | – | – | 45 | 289 | 290.556 |
| – | – | – | – | – | – | – | – | – | 46 | 296 | 302.778 |
| – | – | – | – | – | – | – | – | – | 50 | 315 | 331.667 |
| – | – | – | – | – | – | – | – | – | 54 | 325 | 347.222 |
| – | – | – | – | – | – | – | – | – | 58 | 336 | 363.8889 |
| – | – | – | – | – | – | – | – | – | 60 | 341 | 373.8889 |

*Volume of crude oil collected collected during ASTM distillation.

## 11.3 Results

The percentage of different petroleum fractions obtained from four different crude oils from Jorajan, Lakwa, Geleki and Bhogpara oilfields are given in Figs. 11.1 to 11.4 and Table 11.4.

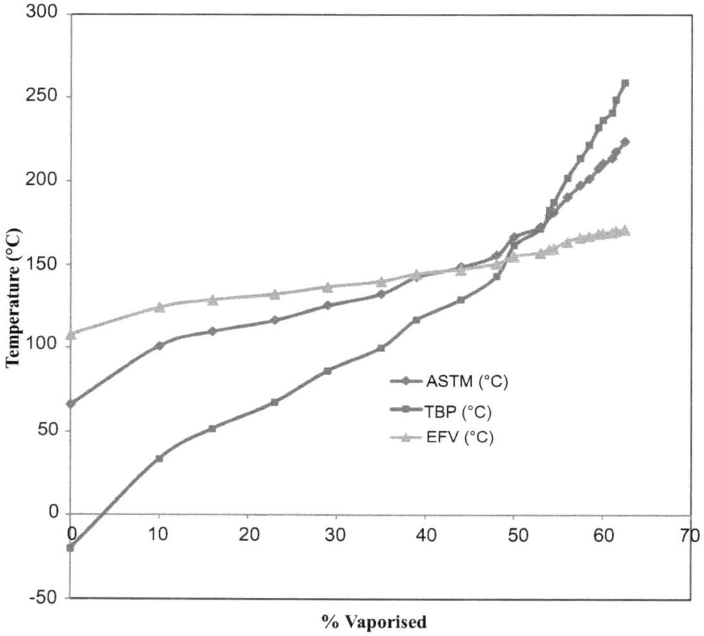


**Figure 11.2** ASTM, TBP and EFV curves of Jorajan crude oil.

**Table 11.4** Percentage of different product fractions from crude oils

| S. no. | Crude oil | Petrol fr*. (%) | Naphtha fr*. (%) | Kerosene & Diesel fr*. (%) | LOBS fr*. (%) |
|---|---|---|---|---|---|
| 1 | Jorajan | 46 | 41 | 13 | 0 |
| 2 | Lakwa | 22 | 30 | 27 | 21 |
| 3 | Galeki | 8 | 11 | 46 | 35 |
| 4 | Bhogpara | 27 | 16 | 32 | 25 |

*Fraction.

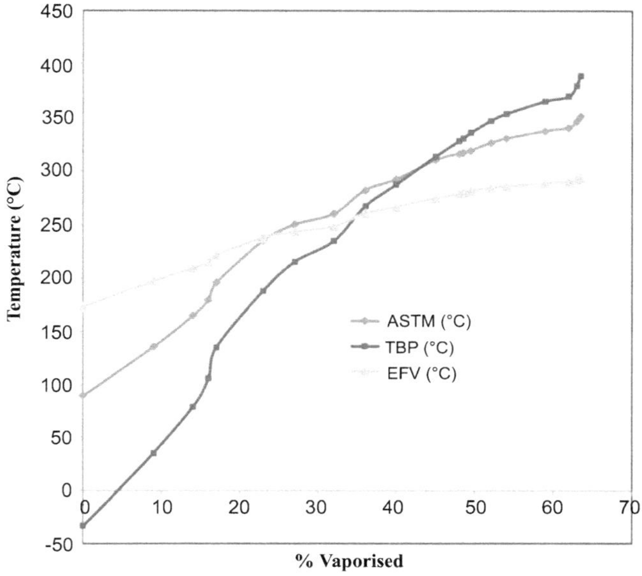


**Figure 11.3** ASTM, TBP and EFV curves of Lakwa crude oil.

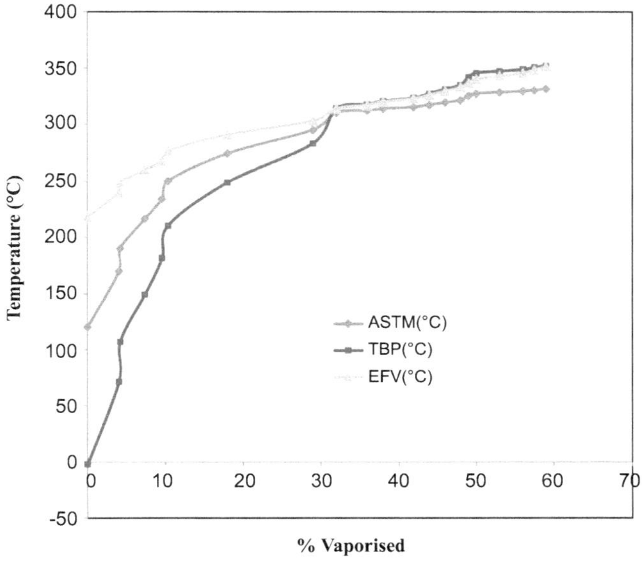


**Figure 11.4** ASTM, TBP and EFV curves of Galeki crude oil.

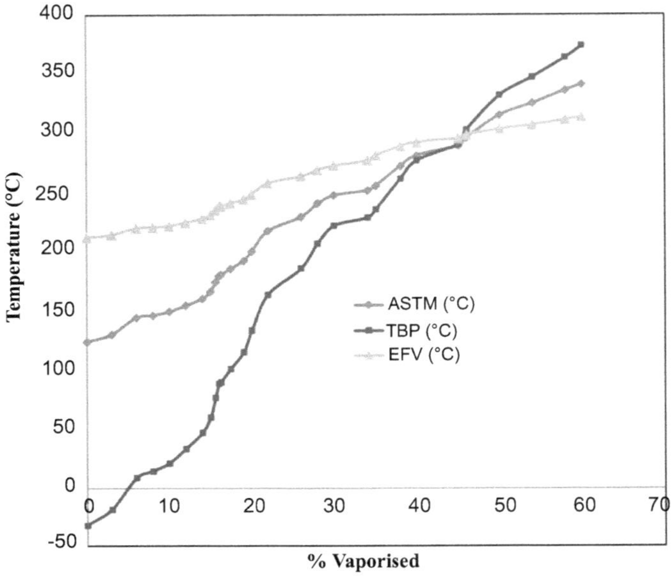


**Figure 11.5** ASTM, TBP and EFV curves of Bhogpara crude oil.

## 11.4 Conclusion

TBP distillation analysis has been proven to be very useful for petroleum characterisation and design and operation of refinery units. TBP distillation analysis has contributed to petroleum science and technology by giving information on the class of petroleum and petroleum property correlations. The TBP curve is very important for the oil industry and is used to understanding the behaviour of crude oil before distillation, for when the crude oil is subjected to a distillation tower on an industrial scale the percentage of distillate obtained at a specific temperature is already known. The results help us to better define the strategies and operating conditions for oil processing and achieving better economy due to its better characterisation. The other major use of these data is in deciding refinery processes needed to refine a given crude oil. TBP distillation can also be used as a method to isolate a specified fraction from crude oil for testing.

## Acknowledgements

We would like to acknowledge the Indo–US Project No. 194-1/2009(IC) dated 20.2.15, titled 'Foam-Assisted $CO^2$ Flooding for the Depleted Reservoirs of Upper Assam Basin and Candidate Reservoir in Louisiana', of the 21st Century Knowledge Initiative programme under University Grant Commission.

## Chapter 12

# A Study on the Effects of Iron Nanoparticles in Water-Based Drilling Mud

**Debashree Dutta and Borkha Mech Das**
*Department of Petroleum Technology, Dibrugarh University, Assam, India*
rs_debashreedutta@dibru.ac.in

The study of the science of materials at the nanoscale helps in enhancing the performance of various chemical processes. The surface-area-to-volume ratio of nanoparticles (NPs) is very high, which makes it superior as compared to other additives used in drilling fluids. The performance of drilling fluids in a high-pressure high-temperature (HPHT) environment along with conventional additives such as polymers and potassium chloride poses a great challenge. The performance of the conventional additives used in drilling fluids is very challenging, especially in HPHT formations. NPs as an additive have proved to be very efficient in drilling fluids. But the high cost of NPs is creating a hindrance towards their use, especially in the oil industries. In this chapter, iron NPs have been synthesised by the chemical reduction method using ferric chloride ($FeCl_3$) as a precursor and sodium borohydride ($NaBH_4$) as a reducing agent. Characterisation was done after

*Advances in Petroleum Technology*
Edited by Subrata Borgohain Gogoi

ISBN 978-981-4877-19-0 (Hardcover), 978-1-003-04993-7 (eBook)
www.jennystanford.com

synthesis using a UV-visible spectrophotometer and a particle size analyser (PSA). Filtration properties were studied for 0.1 M and 0.2 M $FeCl_3$. Another comparative study on rheological properties was carried out for 0.1 M and 0.2 M $FeCl_3$. The effect of iron NPs on water-based mud (WBM) with increasing concentrations was then observed. A low-pressure low-temperature (LPLT) filter press, M3600 Fann viscometer and methylene blue test (MBT) kit were used for determining the various properties of nanoenhanced drilling mud. In addition, 1% bentonite and varied concentrations of iron NPs were used for the experiment. Reduction in filtrate loss and improvement in rheological properties were observed on addition of iron NPs. It was observed that with pure WBM (without NPs), filtrate loss was recorded to be 95 ml with a mud cake thickness (MCT) of 1.4 mm. On addition of NPs, the filtrate loss reduced by 43%. A subsequent reduction in filtrate loss was observed with further addition of NPs. After a certain concentration the filtrate loss remained constant even after addition of NPs. The rheological properties such as plastic viscosity, apparent viscosity, yield point (YP) and gel strength (GS) were studied, and it was observed that on addition of iron NPs, the PV and AV increased by 33% as compared to the base mud. Moreover, it was found that on increasing the molarity of $FeCl_3$ the optimum concentration increased from 20% to about 25%.

## 12.1 Introduction

Drilling fluid is basically a circulating fluid used in oil and gas well drilling operations. The various kinds of drilling fluids available are water-based mud (WBM), oil-based mud (OBM), synthetic-based mud (SBM), etc., of which WBM is most widely used. In the Upper Assam Basin, WBM is mostly widely used. But drilling in complex formations, say for example, high-pressure high-temperature (HPHT) formations and shale formations, is what poses a challenge to these conventional drilling fluids. Using WBM along with its regular additives to drill in such complex formations is practically not viable. WBM completely swells the shale formations, making it prone to swelling and spalling. Such prevailing conditions trigger the need for the use of OBM, which is not only highly compliant with shale but also is best for HPHT formations. Although the oil and gas industry today prefers the use of OBM and SBM for most of the complex formations, yet the economy associated with them is something that cannot be ignored. A large number of regular additives are used to meet the standards of various functions of drilling fluid and thereby increase their efficiency. Although the addition of these additives serves the present purpose, but in the long run it builds up complications as the weight of the drilling mud increases. This in turn affects other properties such as viscosity and gel strength. Bore hole cleaning with such viscous fluids becomes very challenging, which again calls for the need of using some other additives. Polymers especially start degrading at high temperature and pressure. Hence, maintaining an optimum concentration of these additives is very crucial and challenging as well.

All these complications call for the use of a technology which can single-handedly deal with such issues. Nanotechnology today is an emerging trend with which wide-scale laboratory experiments are being carried out by researchers. Nanofluids are formulated by adding nanosized particles in a very lesser amount to the fluid [339]. Laboratory experiments show that inoculating nanoparticles (NPs) in drilling fluid can replace most of the regular additives used. It results in the formation of a smart drilling fluid which encompasses most of the challenges faced by the conventional drilling fluid. Due to the cost of NPs, the oil and gas industry is not using them practically in oilfields. But here, an important matter that needs to be highlighted is that although the initial cost will be high, the overall cost will be highly reduced. This is because of the fact that a very small amount of NPs will replace most of the additives and will result in a fluid which is compatible in HPHT conditions. A small amount of NPs will serve most of the functions of the regular additives. Researchers have stated that about 1l lb/gal of NPs can replace 10 lb/gal of other materials [340]. A high surface-area-to-volume ratio of NPs increases the interaction with the matrix or surrounding fluid, which serves as bonding sites with functional groups [341]. NPs have a great potential for their application in drilling fluid. The present formations call for the use of smart drilling fluids, and nanoenhanced drilling mud are one bearing high potential.

## 12.2 Experimental

### 12.2.1 Materials

$NaBH_4$ and $FeCl_3$ were used for synthesising iron NPs. Bentonite powder was used for formulating the drilling fluid. Distilled water was used for both mud formulation and synthesis of NPs.

Hamilton beach mixture was used for the formulation of bentonite-based drilling fluid. A filter press was used for determining the filtrate loss of the mud. Mud balance was used for determining the density and specific gravity. Rheological properties were determined by a M3600 Fann viscometer. A magnetic stirrer was used for synthesising iron NPs. A UV-visible spectrophotometer and a particle size analyser (PSA) were used for characterisation of iron NPs. A methylene blue test kit was used to determine the cationic exchange capacity (CEC).

### 12.2.2 Methods

#### 12.2.2.1 Formulation of bentonite-based drilling fluid

Bentonite-based drilling mud was formulated by using 1% bentonite powder and distilled water. The formulation was done by using a Hamilton beach mixture, which mixes the mud homogeneously by vigorously mixing it for around 15 min.

#### 12.2.2.2 Synthesis of iron NPs

The chemical reduction method was employed for synthesising iron NPs. In this method, $NaBH_4$ is the reducing agent and $FeCl_3$ is the precursor. The required concentration of $FeCl_3$ was stirred vigorously in distilled water in a magnetic stirrer for 20 min at room temperature. Under the condition of vigorous stirring, $NaBH_4$ was added drop-wise to the mixture of $FeCl_3$. The final mixture was stirred vigorously for around 25 min so that a homogeneous solution of iron nanofluid was formed without any deposition at the bottom and around the magnetic beads.

#### 12.2.2.3 Characterisation of iron NPs

After synthesising iron NPs, their characterisation was done using a UV-visible spectrophotometer and a PSA. The UV-visible spectrophotometer gave the maximum absorbance of iron NPs, which was compared with the standard. The sample was then tested for particle size in a PSA, which gave the average size of the particles.

#### 12.2.2.4 Formulation of nanobased mud

Nanobased mud (NBM) was formulated by adding the synthesised NPs to bentonite-based mud. For two different molarities (0.1 M, 0.2 M) of iron NPs synthesised, varying concentrations of iron NPs were added to the base mud. Concentrations of iron NPs were varied from 5% to 35% for each molarity. After addition of the iron NPs, the resultant mud mixture was again mixed in for 20 min.

## 12.3 Results and Discussion

### 12.3.1 Characterisation of Iron NPs

The maximum absorbance was checked for two different concentrations of iron NPs, that is, 0.1 M and 0.2 M, as shown in Figs. 12.1 and 12.2. The maximum absorbance of iron NPs was observed to be in the range of 216–219 nm. The standard maximum absorbance of iron NPs occurs in the range of 200–300 nm. From the results it is observed that on increasing the concentration of NPs, the peak of maximum absorbance also increases from 216 to 219 nm. This can be attributed to the reason that on increasing concentration, the absorbance peak also increases. However, since the absorbance reading is within the permissible range of the standard, it can be concluded that the formation of iron NPs has occurred.

#### 12.3.1.1 Particle size determination

The average diameter of iron NPs was determined by a PSA. The average size was reported to be 29 nm. From Fig. 12.3 it is clearly seen that the peak is within the range of 10–100 nm.

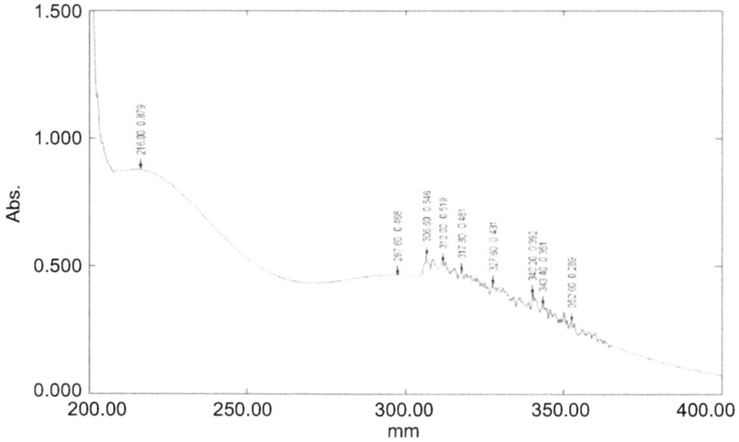


**Figure 12.1** Absorbance of iron NPs with 0.1 M $FeCl_3$.

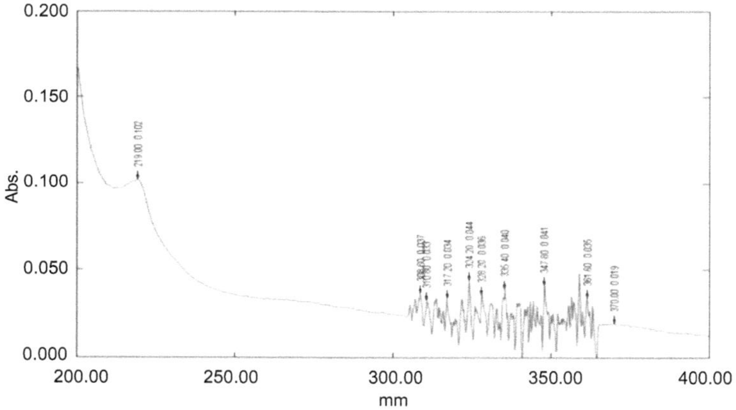


**Figure 12.2** Absorbance of iron NPs with 0.2 M $FeCl_3$.

## 12.3.2 Rheological Properties

### 12.3.2.1 Rheological properties of base mud without NPs

Figure 12.4 shows the effect of plastic viscosity on increasing bentonite concentration. The bentonite concentration was varied from 1% to 5% to observe the changes. Addition of bentonite

| | | | Size (d.nm): | % Number: | St Dev (d.n... |
|---|---|---|---|---|---|
| Z-Average (d.nm): | 50.06 | Peak 1: | 29.29 | 100.0 | 7.178 |
| PdI: | 0.446 | Peak 2: | 0.000 | 0.0 | 0.000 |
| Intercept: | 0.944 | Peak 3: | 0.000 | 0.0 | 0.000 |
| Result quality : | Refer to quality report | | | | |

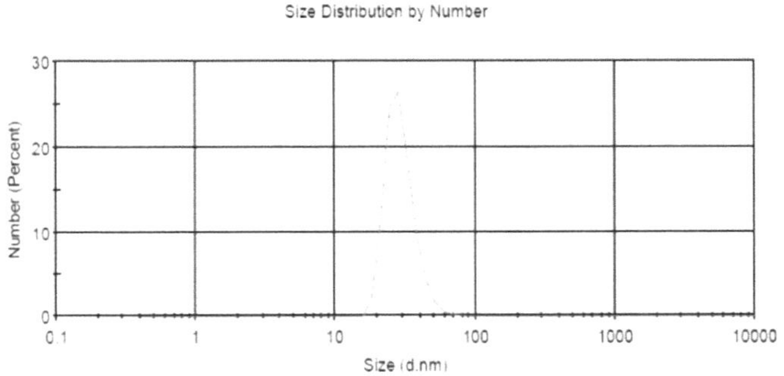


**Figure 12.3** Particle size of iron NPs.

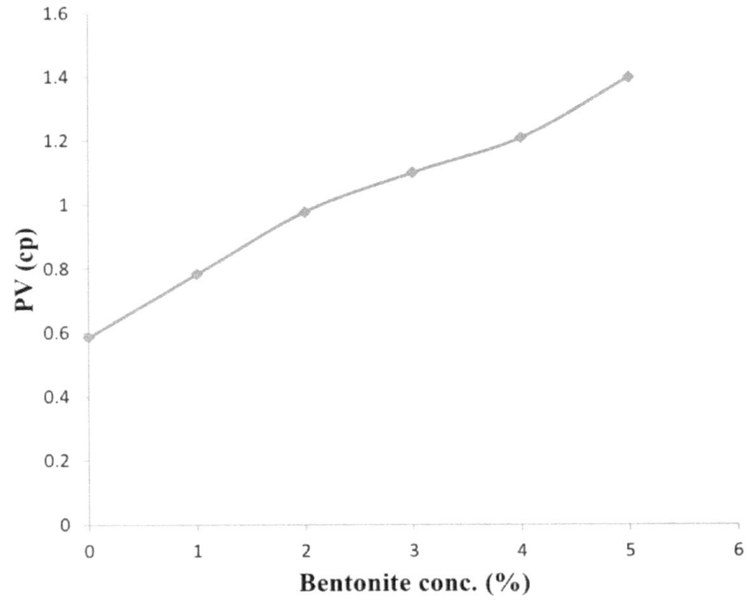


**Figure 12.4** Effect on plastic viscosity by increasing bentonite concentration.

makes the drilling fluid viscous, and hence it can be concluded that bentonite acts a viscosifier. But it also increases the density of the drilling fluid, which may give rise to serious complications later. Since it adds to the weight and solid concentration of the drilling fluid, an optimum concentration of bentonite is permitted.

### 12.3.2.2 Effect of iron NPs on rheological properties

From Figs. 12.5 and 12.6, it is observed that rheological properties such as plastic viscosity, yield point (YP) and gel strength (GS) increase up to a certain concentration of NPs and then not much effect is observed as the graph shows a straight line. It is clearly seen that iron NPs improve the properties of the drilling fluid. Improvement up to an optimum concentration is observed for both the molarities of NPs. Maintaining proper rheology of drilling fluid is

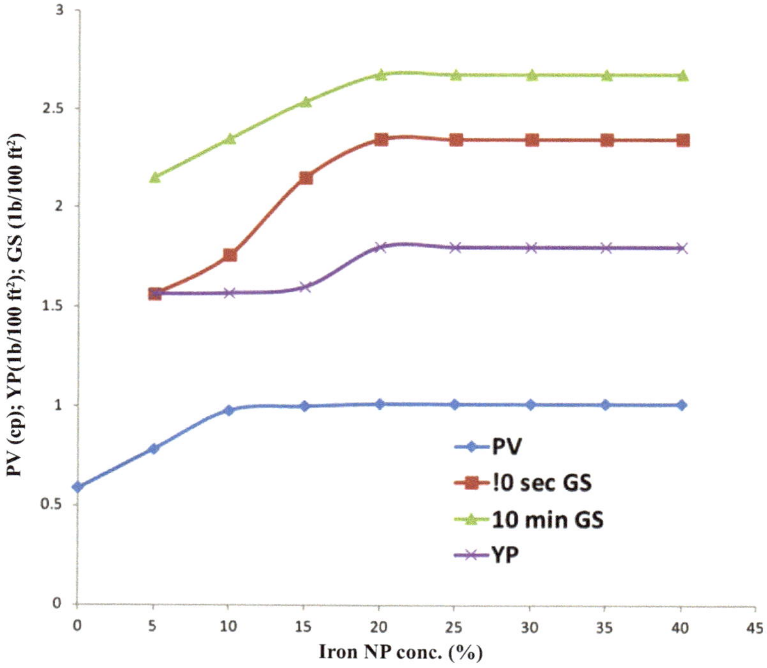


**Figure 12.5** Rheological properties of nanomud with 0.1 M iron NPs.

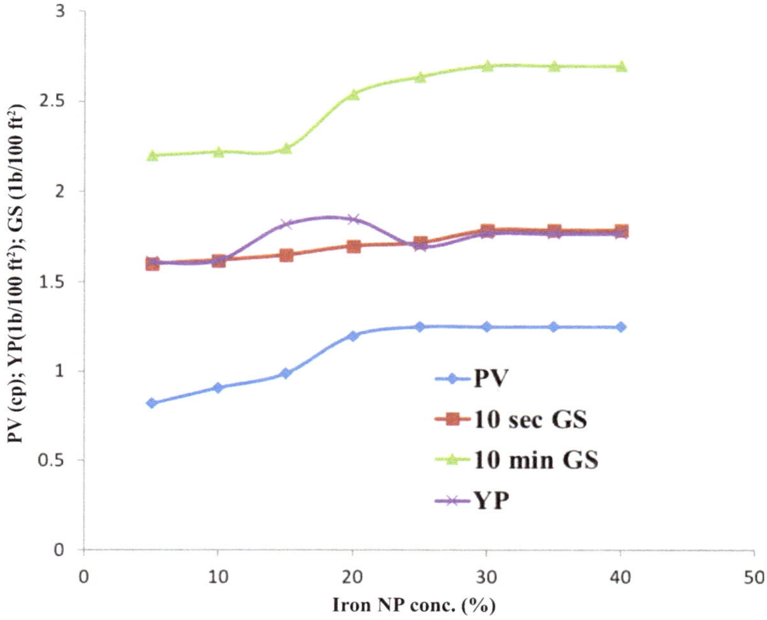


**Figure 12.6** Rheological properties of nanomud with 0.2 M iron NPs.

very important for a successful drilling operation. It is to be noticed that improvement in properties takes place on addition of only 5% of iron NPs. Having a high surface-area-to-volume ratio, the iron NPs interact with the mud much effectively as compared to other macro- and microadditives added to drilling fluid. Another observation is the ability of the iron NPs to improve all the rheological properties on addition of a just little amount. Generally different additives are added to control different rheological properties, such as plastic viscosity, YP and GS. This implies that complications which arise due to high density of the drilling fluid can easily be avoided.

#### 12.3.2.3 Comparative analyses of rheological properties of 0.1 M and 0.2 M iron NPs in drilling fluid

For 0.1 M iron NPs the optimum concentrations were observed to be 20% as no changes were reported after that. On the other hand the optimum concentration varies between 25% and 30% for 0.2 M of iron NPs. On observing the plastic viscosity graphs in both Fig. 12.5 and Fig. 12.6, it is observed that with drilling fluid with 0.2 M iron NPs, the graph shifts upwards, that is, the plastic viscosity increases with increasing concentration of NPs. Also, it is seen that 10 min GS results in a more stable mud than 10 s GS. On studying the YP graphs, it is seen that the graph is consistent in the case of 0.1 M iron NPs, while a ripple is observed for 0.2 M iron NPs. On comparing both Fig. 12.5 and Fig. 12.6, it can be concluded that although NBM with 0.2 M iron NPs shows better results and is an efficient mud, it is less consistent. Results of 0.1 M iron NPs give a consistent and stable drilling mud without any ripples. However, it cannot be denied that by increasing the concentration of iron NPs, the properties get better and the mud more efficient.

## 12.3.3 Filtration Properties

### 12.3.3.1 Filtration properties of base mud without NPs

Static and dynamic fluid losses usually increase with an increase in temperature and are also affected by gelation and thermal degradation of polymers [342]. Preventing loss of circulation by using micro- and macro-loss-circulating materials is limited. NPs have the potential to provide effective sealing of porous and permeable zones [343].

Properties of base mud, that is, bentonite-based mud without NPs, was first observed. From Fig. 12.7, it is clear that on increasing the concentration of bentonite, the filtrate loss reduces by almost 38% initially and subsequently by around 20%. The MCT increases on increasing concentration. Here, it is very important to maintain an optimum concentration of bentonite, because although the filtrate loss is reduced, the MCT increases simultaneously. An increase in the MCT will lead to various drilling complications such as pipe sticking and inefficient bore hole cleaning.

### 12.3.3.2 Effect of iron NPs on filtration properties

Filtration properties were experimentally studied for 0.1 M and 0.2 M iron NPs by varying the concentration of iron NPs from 5% to 35%. It was observed that the filtrate loss reduced subsequently

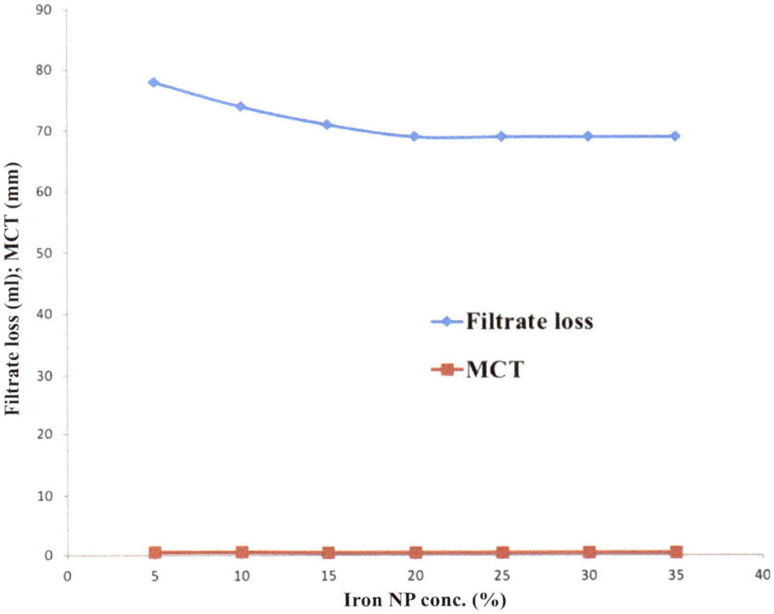


**Figure 12.7** Effect on filtrate loss and MCT by increasing bentonite concentration.

by around 5%–7% on increasing the concentration of iron NPs. Observing the graph for the MCT in Figs. 12.8 and 12.9, it is clear that the MCT decreases from 0.5 mm to 0.4 mm. From Figs. 12.8 and 12.9, it is observed that both filtrate loss and MCT decrease till 20% and after that remain constant. A similar decreasing trend was observed for 0.2 M iron NPs. A thin and tough mud cake is always desirable during drilling operations. It is clearly seen from the figures that on adding iron NPs, the MCT decreases and remains constant, unlike conventional mud where the MCT goes on increasing.

#### 12.3.3.3 Comparative analyses of filtration properties of 0.1 M and 0.2 M iron NPs in drilling fluid

From Fig. 12.10, it is observed that on increasing the concentration of iron NPs from 0.1 M to 0.2 M, the optimum concentration also increases from 20% to 25%. While the filtrate loss decreases till

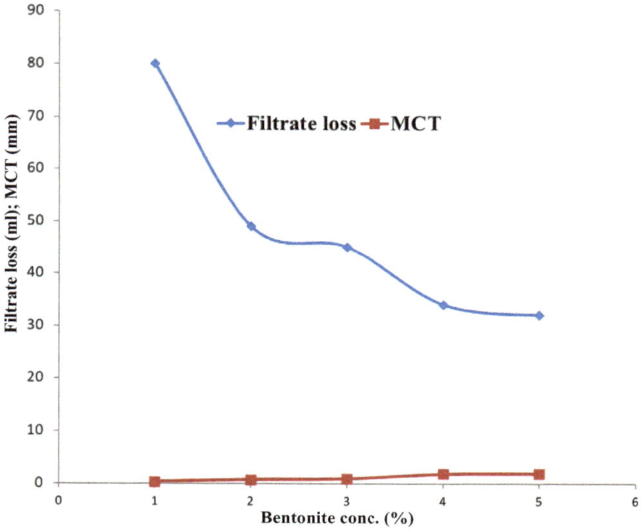


**Figure 12.8** Effect of iron NPs (0.1 M) on filtrate loss and MCT.

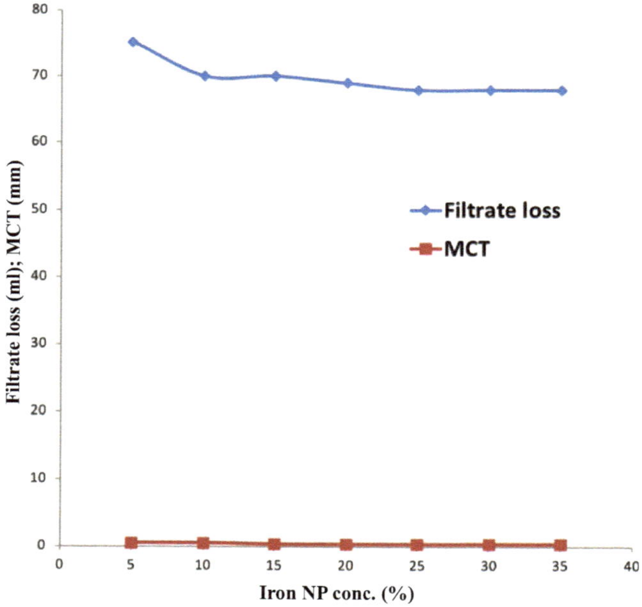


**Figure 12.9** Effect of iron NPs (0.2 M) on filtrate loss and MCT.

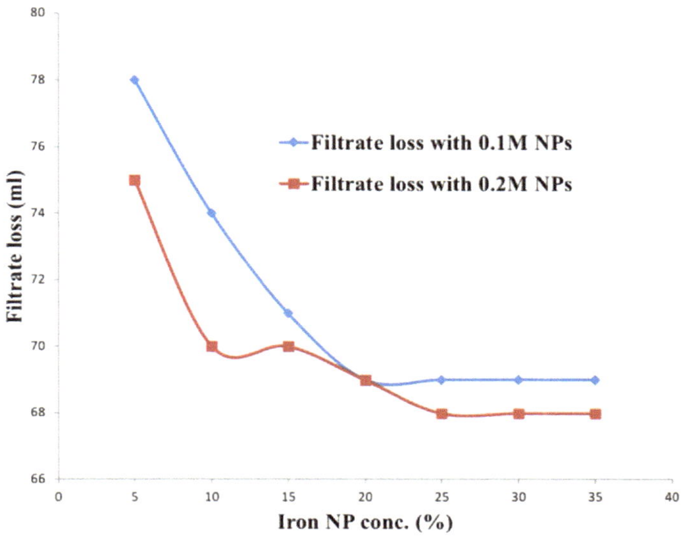


**Figure 12.10** Comparison of effect of 0.1 M and 0.2 M iron NPs on filtrate loss.

addition of 20% of iron NPs for 0.1 M NPs, it decreases till addition of 2% of iron NPs for 0.2 M NPs. This is due to the fact that higher concentrations of NPs show better results with filtrate loss and at the same time maintain a thin and tough mud cake. Figures 12.11 and 12.12 show the overall effect of iron NPs on WBM without NPs and with NPs (0.1 M and 0.2 M).

## 12.3.4 Cationic Exchange Capacity of NBM

To study the reactiveness of clays, the cationic exchange capacity (CEC) was determined by MBT. An MBT kit was used for used for the same. Experiments were conducted by varying the concentration of iron NPs from 5% to 25%. Reactiveness increased with increasing concentration of iron NPs. From Fig. 12.13, it is also clear that on increasing the molarity from 0.1 M to 0.2 M, the graph shifted upwards, indicating the fact that injecting more concentration of iron NPs will result in a more reactive mud. Because of this, the interaction of NPs with the drilling fluid will be more efficient simply

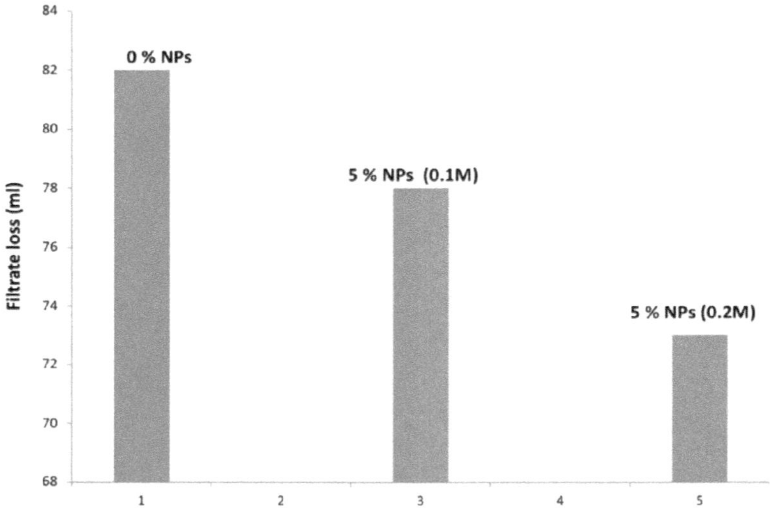


**Figure 12.11** Effect of NPs on filtrate loss with and without NPs with change in concentration.

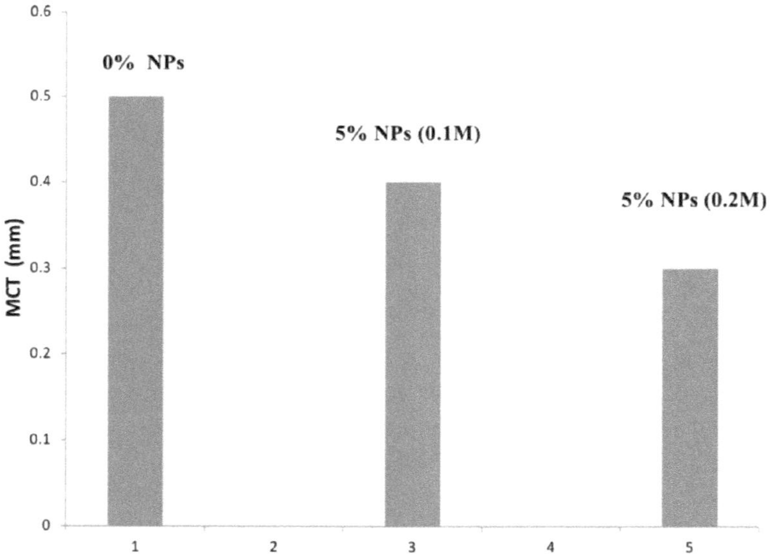


**Figure 12.12** Effect of NPs on MCT with and without NPs with change in concentration.

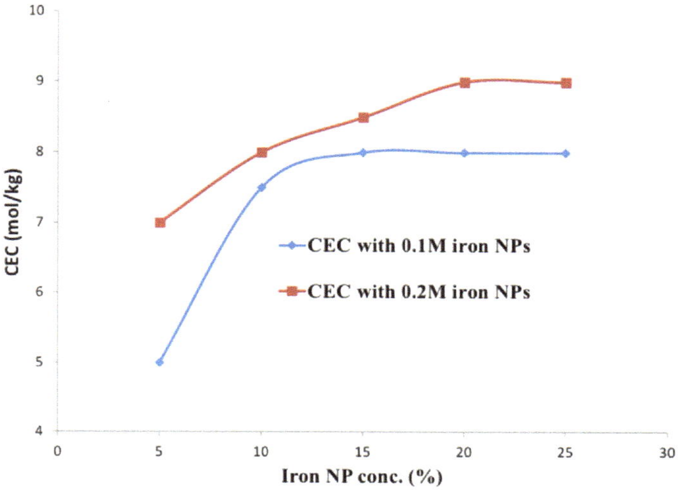


**Figure 12.13** Effect of iron NPs on CEC.

by injecting a small amount of iron NPs. This can be attributed to the fact that NPs have a high surface-area-to-volume ratio.

## 12.4 Conclusion

The experimental results clearly show the efficiency of NPs in drilling fluid. While different additives are added to deal with various properties of drilling fluid, here iron NPs alone deal with both filtration properties as well as rheological properties. On observing the results of the filtration properties, it is clear that iron NPs are highly efficient in reducing filtrate loss by maintaining a thin and tough mud cake. Decrease in filtrate loss will prevent formation damage, and its quality of maintaining appropriate MCT will prevent most of the downhole complications and pipe-sticking problems. Improvement in rheological properties has also been observed. On increasing the concentrations of iron NPs, plastic viscosity, YP and GS increase up to a substantial level. Analysis of the results of 0.1 M and 0.2 M of iron NPs shows that drilling fluid with 0.2 M iron NPs is more efficient as properties are improved more as compared to

0.1 M of iron NPs. Maintaining a stable drilling mud is also very crucial on the part of the driller. From this pretext, drilling mud with 0.1 M iron NPs shows more consistency in results. NPs have a high surface-area-to-volume ratio. Because of this property, even on adding only 5% of iron NPs, a considerable improvement in filtration and rheological properties was observed. So the cost associated with a large number of regular additives that are added in drilling fluid can be reduced to a greater level. Reactiveness of NBM is much more than the base mud. Increase in reactiveness of clays will provide better bonding of NPs with the drilling fluid and thereby increase the number of reactive sites. This is another reason why small amounts of NPs can replace large amounts of conventional additives. The method which was used for synthesising iron NPs was very economic as not much chemicals were used and only a precursor and a reducing agent were used. The method employed was also user-friendly without any use of highly sophisticated instruments that demands for more economy. Purchasing NPs directly from the market will cost very high. This is one of the prime reasons why oil industries are not employing NPs in oilfields despite the fact that NPs possess great potential in the field of drilling technology. Synthesising NPs instead of purchasing directly will definitely deal with this economy prospect to a considerable level. Use of NPs in drilling fluid will thus result in a smart drilling fluid, which will be efficient enough to deal with most of the drilling complications.

## Acknowledgements

At the very outset, the authors would like to acknowledge all the researchers mentioned in the references for enlightening us with their experiments and results. We would also like to extend our gratitude to the Drilling Fluid and Rheology Laboratory, Department of Petroleum Technology, Dibrugarh University.

## Chapter 13

# Use of Environment-Friendly Additives for Improving the Properties of Water-Based Mud for Drilling Reactive Formations

**Bichakshan Borah and Borkha Mech Das**
*Department of Petroleum Technology, Dibrugarh University, Assam, India*
rs_bichakshanborah@dibru.ac.in

Due to an increase in environmental awareness, there is a need for drilling companies to come up with improvised drilling mud with more environment-friendly additives. Drilling mud generally consists of an aqueous clay suspension, containing weighting agents to increase the hydrostatic head, and frequently also contains concentrated colloidal suspending and conditioning agents. Drilling mud additives play an important role in improving the mud rheological properties and filtration capability by allowing bits to penetrate heterogeneous geological formations. In this chapter, the effect of polyethylene glycol (PEG) in water-based drilling fluid (WBDF) is determined. PEGs are hydrophilic compounds and are environmentally safe and commercially available over a wide range of molecular weights from 300 g/mol to 10,000,000 g/mol. The effect has been determined by analysing the rheological and

*Advances in Petroleum Technology*
Edited by Subrata Borgohain Gogoi

ISBN 978-981-4877-19-0 (Hardcover), 978-1-003-04993-7 (eBook)
www.jennystanford.com

filtration properties of drilling fluid with varied concentrations of PEG, and it has been compared with a commercial product carboxy methyl cellulose (CMC). The experimental results have shown that the use of PEG in drilling fluid shows a good reduction in filtrate loss. Apart from the filtrate loss reduction, its addition also helps in improving the viscosity and density a bit. But the degree of improvement is less as compared to the improvement with CMC, which is a known commercial additive. So, a mixed additive was prepared using PEG and a very less concentration of the commercial product CMC to analyse the combined effect on the properties of drilling fluid. It was observed that the PEG+CMC mixture as an additive helps in improving the rheological properties and reducing the filtrate loss in a considerable amount. The mixture proved to be a more effective, economically efficient and environment-friendly additive for improving the properties of drilling fluids and thereby helped in obtaining an efficient drilling operation. The formulated drilling mud with the prepared mixed additive was finally analysed to determine its ability to drill reactive shale formations.

## 13.1 Introduction

Drilling fluids are heterogeneous fluids used to drill heterogeneous geological formations of the earth, which consists of different clay particles, weighing materials, chemicals and polymers [344]. The use of a drilling fluid to drill different formations has been employed since the early 1900s [345]. Water was the first drilling fluid used in history, which was poured in wells to soften and carry the cuttings of the rocks to the surface. Water is the main constituent in most of the drilling fluids [346, 347]. The main functions of drilling fluids are to cool and lubricate the drill bit and drill string, to provide hydrostatic pressure to prevent formation fluids from entering into the wellbore carrying out drill cuttings and to suspend the drill cuttings while the drilling operation is at rest [348, 349]. With the maturation of technology to overcome various difficulties in drilling, different chemicals, polymers, were introduced into the drilling fluid. These additives provide sanguine properties to the drilling fluid to operate its functions efficiently. The chemical properties of the drilling fluid and its mud properties should provide a stable wellbore. A highly important property of drilling fluids is the ability to form an impervious filter cake upon the permeable walls of the bore hole to prevent filtrate loss. Excessive fluid loss from drilling fluids can cause severe problems. An ideal drilling fluid is the one which shows less filtrate loss and forms a thin and tough mud cake. A drilling fluid with high fluid loss is particularly a damaging fluid and proves to be inefficient in maintaining properties like cleaning the bottom of the hole and transporting the cutting particles to the surface [350–352]. Mud is, in varying degrees, a toxic substance as numerous chemicals and polymers are added to it to improve its properties. Many of the chemicals and polymers are harmful. It is also difficult and expensive to dispose of this type of drilling fluid in an environment-friendly manner. It is because when this toxic fluid discharges to an open surface, it may percolate through the pore spaces of the crust and may contaminate the ground water. WBDFs have little toxicity, although they are made from water, as some additives used to prepare the drilling fluid are corrosive and toxic. With water-based mud, chemical differences can cause interactions between mud and shale, which lead to the softening of the native rock, and the highly fractured, dry, brittle reactive shale formations can become extremely unstable, which can lead to differential pipe sticking and fluid loss. Due to an increase in environmental awareness, nowadays in this 21st century, to overcome these difficulties there is a need for drilling companies to come up with an improvised drilling mud with more environment-friendly additives [353, 354].

Many types of polymers and chemicals can be used to prepare drilling fluids which are environment-friendly and more effective. One of the polymers which is already commercialised is carboxy methyl cellulose (CMC). CMC polymer is a nontoxic, anionic, water-soluble, colourless and odorless, high-viscosity sodium CMC which is designed to control filtration loss which provides viscosity in WBDFs, both in fresh water and in saturated salt water, and is absorbed by clay [355, 356]. This long-chained molecule can be polymerised to produce different molecular weights, and they are available in different viscosity grades. They are not subjected to bacterial effects and are stable up to temperatures of 121°C [357]. PEG is a polyether compound which is environment-friendly with many applications from industrial manufacturing to medicine. The structure of PEG is commonly expressed as $HOCH_2CH_2)n{-}OH$. PEGs are prepared by polymerisation of ethylene oxide and are commercially available over a wide range of molecular weights from 300 g/mol to 10,000,000 g/mol. Depending on their molecular weights PEG are liquids, semiliquids or solids. Due to difference in structure, different PEGs have different physical properties and are used in different applications. High-molecular-weight PEGs are synthesised by polymerisation suspension. PEGs have also a significant impact on the gas hydrate phase stability [358]. PEGs also reduce mud maintenance cost, reduce fluid loss, are less toxic and are easily biodegradable. In this chapter, we used PEG4000 [359]. The main objective of the study was to prepare and analyse CMC and PEG4000 as environmentally safe additives to improve the properties of water-based drilling mud and for drilling reactive shale formations.

## 13.2 Experimental

### 13.2.1 Materials

Materials used in this experiment were distilled water, bentonite, CMC, PEG4000 and shale of the Champhai district of Mizoram, India.

### 13.2.2 Instruments

Instruments which were used in this experiment were a weight balance, beaker, stirrer, stand mixer (Hamilton Beach Mixer), Ofite mud balance, Marsh funnel viscometer, stop watch, M3600 viscometer, Grace 3600 software, Ofite low-pressure low-temperature (LPLT) filter press, filter paper, graduated cylinder, scale, mortar and test tubes.

### 13.2.3 Methodology

#### 13.2.3.1 Preparation of base mud

The raw materials which were required in this experiment were measured with the help of a weight balance. We took 3% bentonite and 1000 ml of distilled water to prepare the base mud. The distilled water was poured into a beaker, and the bentonite was mixed in the distilled water by a stirrer. Then the mud was firmly mixed by the stand mixer, that is, Hamilton beach mixer, for about 10 to 15 min.

#### 13.2.3.2 Preparation of CMC mud and PEG4000 mud

In this step different amounts of CMC and PEG4000 were mixed separately with the base mud. The amounts of CMC used in this experiment were 1 to 5 g and a total of five samples were prepared. Similarly, for the preparations of PEG 4000 mud, 1 to 5 g of PEG 4000 were used and five samples were prepared.

#### 13.2.3.3 Preparation of CMC-PEG4000 mud

After the preparation of CMC and PEG4000 muds separately, here we combined CMC and PEG4000 to prepare drilling mud. Taking the minimum amount (1 g) of CMC as a constant, 1 to 5 g of PEG4000 was mixed with the base mud and accordingly another five mud samples were prepared.

### 13.2.4 Property Analyses of Prepared Mud Samples

#### 13.2.4.1 Density

The densities of the prepared mud samples were measured with the help of a mud balance. The mud balance method is the simplest, quickest and most reliable method for determining the density of a mud sample instantly. It consists of a graduated beam having a bubble level to make sure that the balance is completely horizontal when the bubble is in mid-position. The graduated beam of the balance has a slider weight which can move along the beam, and it has a cup on one end which can hold a fixed amount of mud in it.

#### 13.2.4.2 Funnel viscosity

Funnel viscosity or Marsh funnel viscosity is used to determine the viscosity of mud instantly in the field. It consists of a cone-shaped mud having a mesh at the top to remove any solid particles entering the funnel or mug and a cup. We poured 950 ml of drilling mud through the mesh in the funnel, and the time required to complete discharge the mud was measured by a stopwatch. In this process, by considering the density and the time, three types, that is, effective viscosity as in Eq. 13.1, apparent viscosity (considering time) and apparent viscosity (considering both time and viscosity), can be calculated using Eqs. 13.2–13.3, respectively, as follows:

$$\text{effective viscosity} = \rho(t - 25)\ \text{cp} \tag{13.1}$$

$$\text{Apparent viscosity} = -0.0118t^2 + 1.6175t - 32.168 \tag{13.2}$$

$$\text{Apparent viscosity} = \rho(t - 28)\ \text{cp}, \tag{13.3}$$

where $\rho$ is the density of the mud, $t$ is time and cp is the unit.

#### 13.2.4.3 Rheological properties

The deformation of fluids and flow of matter are known as rheological properties which include plastic viscosity and yield point (YP), as in Eqs. 13.4 and 13.5. These properties were measured in the M3600 viscometer and the Grace 3600 software. The main formulae which were required to obtain these properties were in [356, 360].

$$\text{Plastic viscosity} = \theta_{600} - \theta_{300} \tag{13.4}$$

$$\text{Yield point} = \theta_{300} - \text{Plastic viscosity} \tag{13.5}$$

#### 13.2.4.4 Filtrate loss and mud cake thickness

The filtrate loss of the mud samples was determined using the Ofite LPLT filter press. A certain amount of drilling mud was kept in the cup of the filter press, and pressure was given to it by a hydraulic dead weight. Each set of experiments was done for 30 min. The amount of filtrate loss was measured with the help of a graduated cylinder, and the thickness of the mud cake produced was measured with the help of a scale.

#### 13.2.4.5 Shale stability

The shales which were to observe the effect of the mud on its stability were taken from the Champhai district, Mizoram, India. With the help of a mortar the shales were first crushed into powered form, and then equal amounts, that is, 8 g of powdered shales, were put in three test tubes, which were, later, filled by three different types of mud, that is, CMC mud, PEG4000 mud and CMC-PEG4000 mud. This setup was kept for 48 h and the swelling property of the shale was observed after every 2 h.

## 13.3 Results and Discussion

### 13.3.1 Density

The measurement of densities of the three types of mud is shown in Fig. 13.1. With the addition of varying concentrations of CMC and PEG4000 it shows almost negligible effects in the change in mud

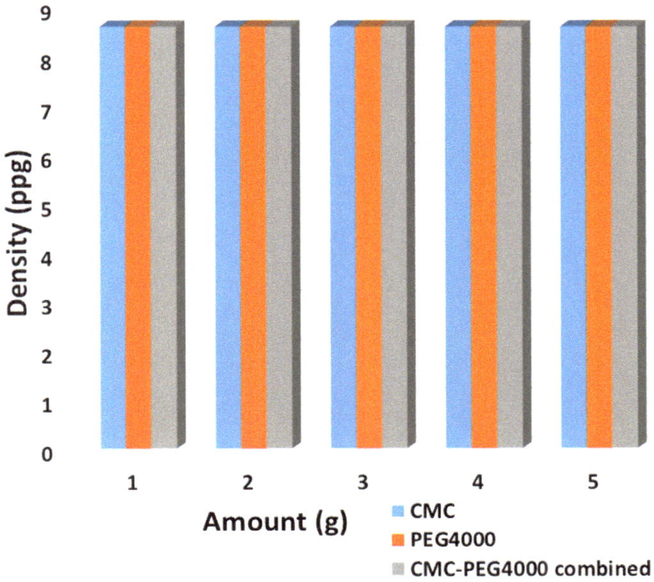


**Figure 13.1** Comparison of density of muds.

density. All the 15 samples showed almost 8.60 lb/gal of mud weight. Therefore addition of these polymers has no effect on density.

### 13.3.2 Funnel Viscosity

The values of effective viscosity and apparent viscosity of CMC and PEG4000 mud, which were determined by a Marsh funnel viscometer, and their variations with an increase in concentration are shown in Fig. 13.2. From Fig. 13.2, it is clearly seen that with increasing amounts of CMC, PEG4000 and CMC-PEG4000, the viscosities increased. However, with PEG4000, the viscosities increased in higher amounts as compared to CMC, and the combined CMC-PEG4000 mud showed a greater value than CMC mud and lower value than PEG4000 mud, which proved that the addition of PEG4000 with CMC to a drilling fluid is a better viscosifier.

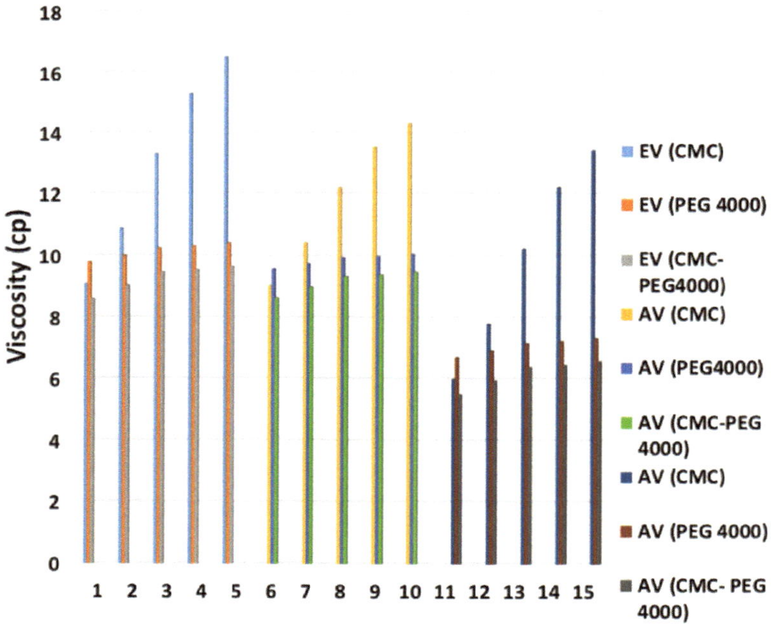


**Figure 13.2** Comparison of funnel viscosity of muds.

### 13.3.3 Rheological Properties

The rheological properties of the drilling fluid are a very important factor regarding the property enhancement of the drilling mud. The variation and comparison of plastic viscosity, YP and gel strength (GS) are in Figs. 13.3–13.8. It has been found from the figures that plastic viscosity and YP increase with an increase in the concentration of both CMC and PEG4000. Initially when 1 g each of CMC and PEG4000 was added to the mud, PEG4000 showed better results in plastic viscosity than CMC, while with increasing concentration CMC showed better improvement in plastic viscosity than PEG4000 [356]. In the case of YP, CMC shows much better results [356, 361] than PEG4000 mud as from 1 to 5 g of CMC, the YP increases from about 4 lb/100 ft$^2$ to more than 10 lb/100 ft$^2$, while the range of YP for the same amount of PEG4000 is less than 6 lb/100 ft$^2$. Like YP, the same trend of improvement in GS is observed

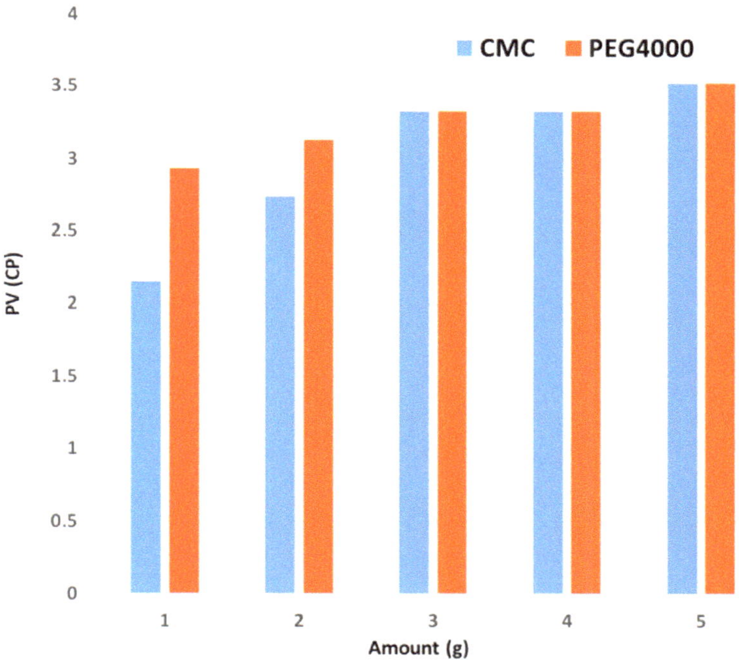


**Figure 13.3** Comparison of plastic viscosity of CMC and PEG muds.

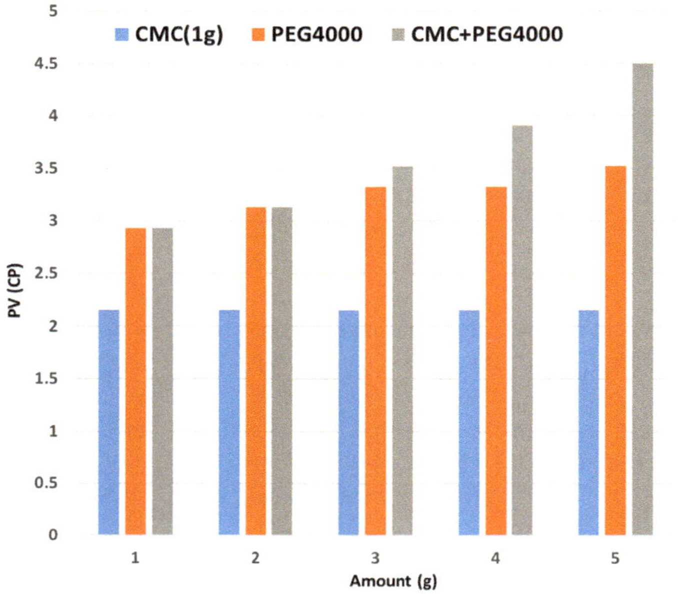


**Figure 13.4** Comparison of plastic viscosity of CMC, PEG and CMC+PEG muds.

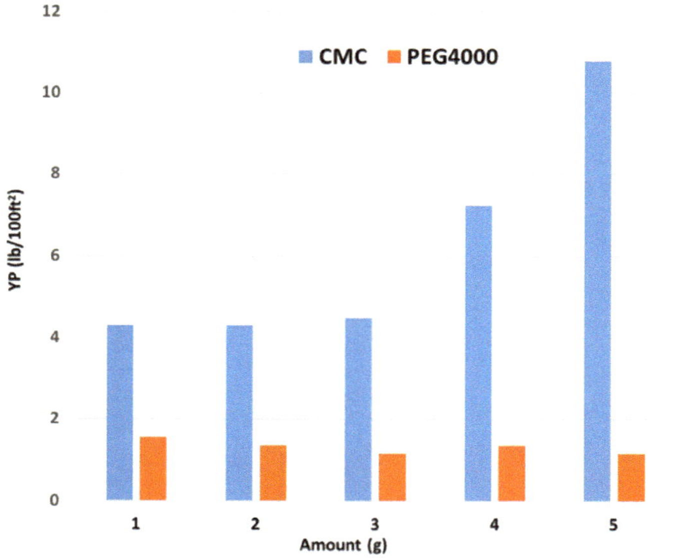


**Figure 13.5** Comparison of yield point of CMC and PEG muds.

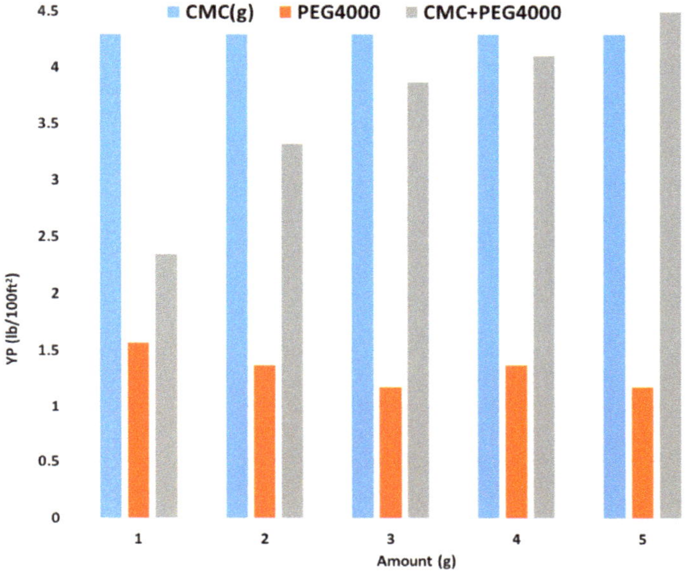


**Figure 13.6** Comparison of yield point of CMC, PEG and CMC+PEG muds.

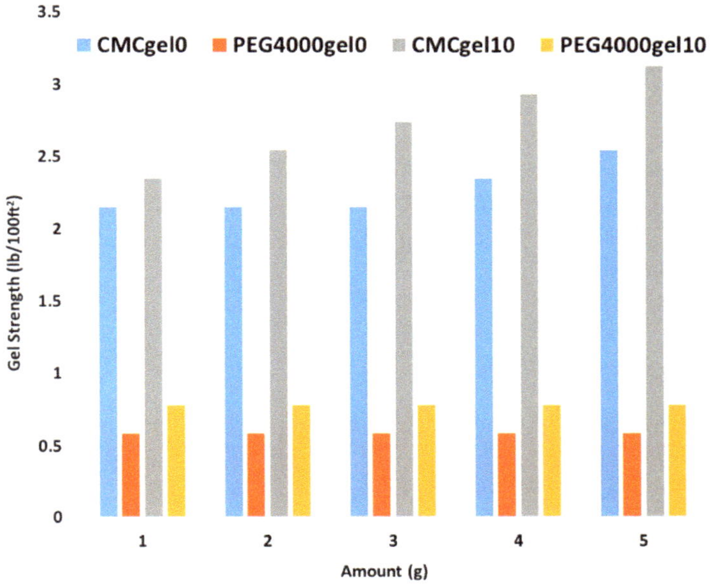


**Figure 13.7** Comparison of gel strength of CMC and PEG muds.

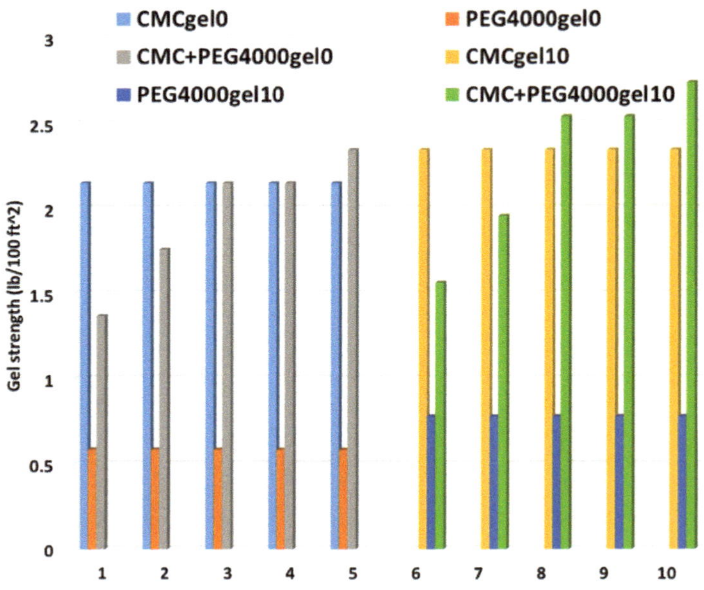


**Figure 13.8** Comparison of gel strength of CMC, PEG and CMC+PEG muds.

for CMC and PEG4000 muds [356]. Now if we look upon the combined mud of CMC-PEG4000, it can be seen that it shows better improvement in all the three rheological properties, that is, plastic viscosity, YP and GS of the mud, with an increase in concentration [361]. Therefore addition of PEG4000 with CMC proves to be a better polymer for improving the rheological properties.

### 13.3.4 Filtration Properties

A highly important property of drilling fluids is the ability to form an impervious filter cake upon the permeable walls of the bore hole to prevent filtrate loss, which can be obtained by the filter press method. In Figs. 13.9 and 13.10, the filtrate loss of CMC, PEG4000 and combined CMC-PEG4000 muds is shown, followed by the thickness of the mud cake in Figs. 13.11 and 13.12.

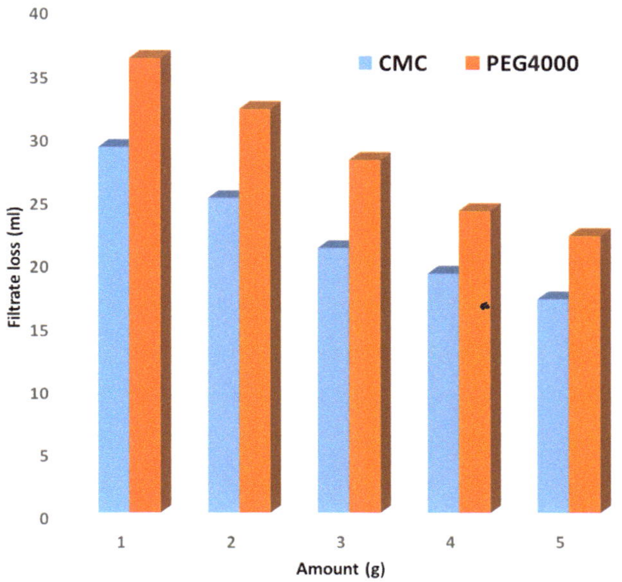


**Figure 13.9** Comparison of filtrate loss of CMC and PEG muds.

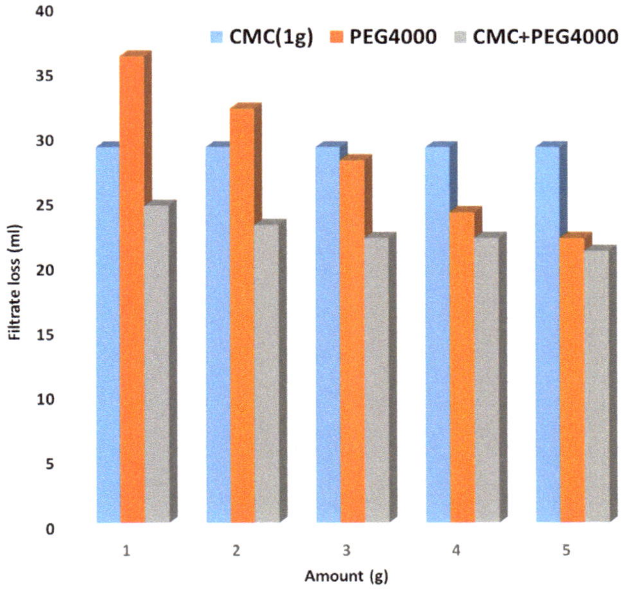


**Figure 13.10** Comparison of filtrate loss of CMC, PEG and CMC+PEG muds.

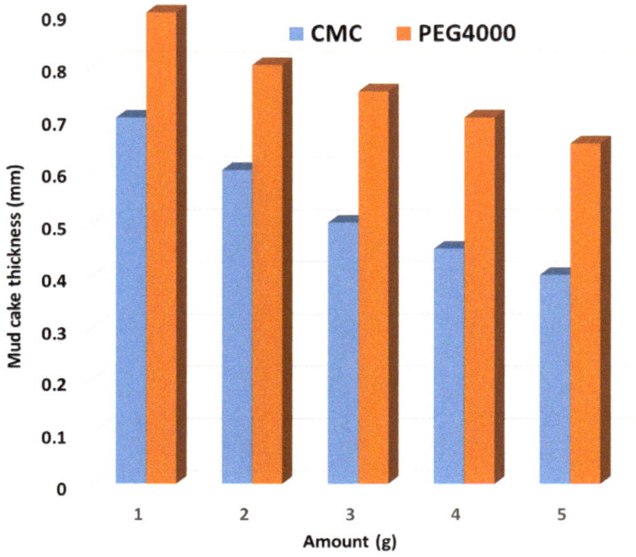


**Figure 13.11** Comparison of mud cake thickness of CMC and PEG muds.

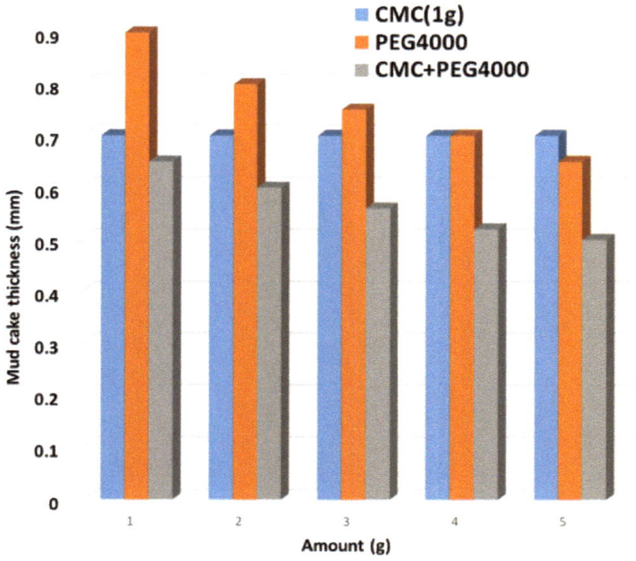


**Figure 13.12** Comparison of mud cake thickness of CMC, PEG and CMC+PEG muds.

CMC is an excellent filtrate loss reducer than PEG4000, as shown in the figures, and the amount of filtrate loss reduces with increasing amounts of the polymer. CMC also produces a very thin and tough mud cake than PEG4000, and its thickness further reduces with increasing amounts of the polymer [356, 361]. But when CMC and PEG4000 are combined together to observe their effect on filtration properties, it is seen that the combined mud is more effective in reducing filtrate loss and mud cake thickness than the individual effect of CMC and PEG4000, which can be clearly seen in the figures. That is why PEG4000 proves to be a better filtrate loss reducer when combined with CMC.

### 13.3.5 Shale Stability

The shale stability test was done to check the resistance of the shale particle to swelling when they are in contact with different types of mud. The swelling did not occur up to 8 h from the starting of the test in any of the three specimens. From 8 h the swelling of the shale started to occur, as shown in Fig. 13.13. From Fig. 13.13 it is clearly observed that the resistance of shale to swelling was less in the combined CMC-PEG4000 mud than individual muds. The clay particle swells more in CMC mud than PEG4000 mud. This experiment proves that PEG4000 is more environment-friendly than CMC as it shows less swelling of the clay particle. It is because when a drilling fluid passes through a well it encounters different shale formations.

## 13.4 Conclusion

CMC and PEG4000 show excellent results in enhancement of drilling fluid properties. They have very negligible effect on density, that is, the density almost remains constant. PEG4000 shows better results in improving the viscosity of the drilling fluid than CMC, as it shows better results in effective viscosity and apparent viscosities. However, in the case of rheological properties, addition of PEG4000 with CMC to a drilling fluid proves to be a better option as the combined mud shows excellent results in rheological properties.

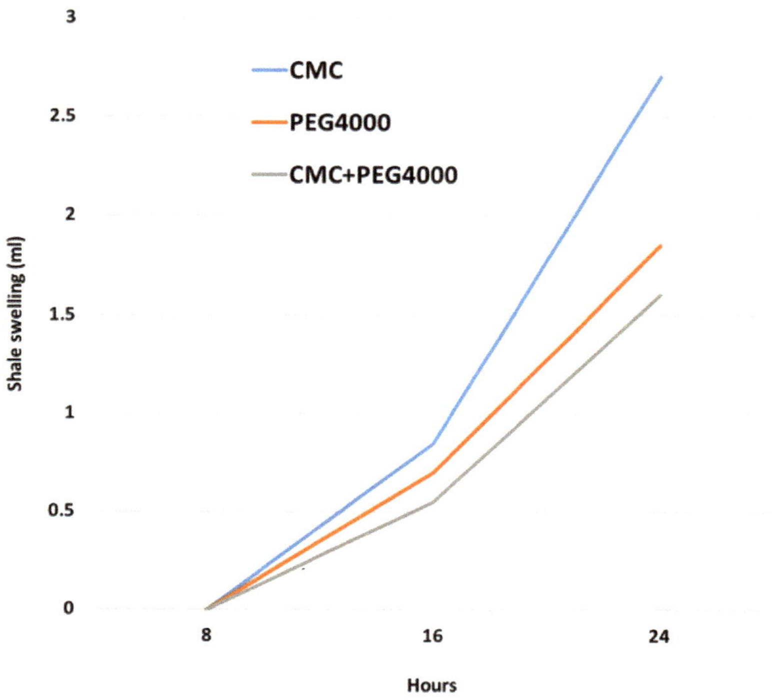


**Figure 13.13** Shale stability of CMC, PEG and CMC+PEG muds.

While comparing filtration properties, CMC works better than PEG4000. But when PEG4000 is combined with CMC it shows better results than CMC alone. This combined mud has better filtration properties than CMC mud, which proves that PEG4000 can be used as an active fluid loss–reducing agent and it can contribute to wellbore stability by forming thin and tough mud cakes which will also prevent pipe sticking. Comparing CMC and PEG4000 in the shale stability test, PEG4000 shows excellent results than CMC as the shale takes a comparatively long time to swell and it shows better results in case of the combined CMC-PEG4000 mud as it shows less amount of shale swelling compared to the others. So concerning all the properties of the drilling fluid discussed in this research work, the addition of PEG4000 with CMC proves to be more reliable, economically efficient and environment-friendly and helps in obtaining an efficient drilling operation.

## Acknowledgements

We would like to acknowledge Dibrugarh University for the laboratory facilities.

## Chapter 14

# Effect of Different Plant Extracts on Enhancing Rheological Properties of Water-Based Drilling Fluids

**Deepjyoti Mech,[a] Borkha Mech Das,[b] Rahul Kumar M.,[a] Meet Priyavadan Patel,[a] Sandra Jayan[a] and Rishabh Pardeshi[a]**

[a]*Department of Petroleum Engineering, Presidency University, Bengaluru, Karnataka, India*
[b]*Department of Petroleum Technology, Dibrugarh University, Assam, India*
deepjyotimech@presidencyuniversity.in

A mixture of viscous fluid or drilling fluid is often known as drilling mud, which helps to drill a borehole into the earth to produce hydrocarbons (HCs) or crude oil or natural gas. A lot of research work has been done to improve the characteristics and properties of drilling fluids to perform the necessary functions in wellbore formations at various depths. Through years, efforts have been made to make drilling fluids economic and eco-friendly using a number of natural and man-made additives for efficient drilling operations. In various research works, several natural additives were used to improve the properties of water-based drilling mud. In this work, different plant extracts such as mango leaves and neem leaves have been used for rheological studies. The effects of these plant extracts on drilling fluid properties of water-based drilling fluid are

*Advances in Petroleum Technology*
Edited by Subrata Borgohain Gogoi

ISBN 978-981-4877-19-0 (Hardcover), 978-1-003-04993-7 (eBook)
www.jennystanford.com

determined and analysed. Mud samples of various concentrations were formulated in the presence and absence of extracts of mango leaves and neem leaves. The mud samples were investigated for rheological properties. For controlling the corrosion of subsurface equipment, a corrosion-inhibiting additive was added, and it can only be added when it improves the properties of the drilling fluid. All the required experiments were carried out according to API standards. After proper revised analyses of the results, it was observed that mango leaf extracts highly enhance the mud properties, followed by neem leaf extract, and thus can be used as an additive to enhance the mud properties and make the drilling mud perform required functions optimally. In addition, the use of leaf extract as a drilling fluid additive may aid the oil and gas industry as it is economically efficient and environment-friendly.

## 14.1 Introduction

Drilling fluid is a lubricating fluid used predominantly in the oil and gas well–drilling activities. In geotechnical engineering, drilling fluid is used to aid the drilling of boreholes into the earth. Often used while drilling oil and natural gas wells and on exploration drilling rigs, drilling fluids are also used for much simpler boreholes, such as water wells. Various kinds of drilling muds, such as water-based muds (WBMs), oil-based muds (OBMs) and their formatives are being used along with appropriate polymer and clay additives for drilling various oil and gas formations. Liquid drilling fluid is often called drilling mud. The three main categories of drilling fluids are WBMs (which can be dispersed and nondispersed); non-aqueous muds, usually called OBMs; and gaseous drilling fluid, in which a wide range of gases can be used [362, 363]. On a drilling rig, mud is pumped from the mud pits through the drill string where it sprays out of nozzles on the drill bit, cleaning and cooling the drill bit in the process. The mud then carries the crushed or cut rock (cuttings) up the annular space (annulus) between the drill string and the sides of the hole being drilled, up through the surface casing, where it emerges back at the surface. Cuttings are then filtered out with either a shale shaker or the newer shale conveyor technology, and the mud returns to the mud pits. The mud pits let the drilled 'fines' settle; the pits are also where the fluid is treated by adding chemicals and other substances. The returning mud can contain natural gases or other flammable materials which will collect in and around the shale shaker/conveyor area or in other work areas. Because of the risk of a fire or an explosion if they ignite, special monitoring sensors and explosion-proof certified equipment is commonly installed, and workers are advised to take safety precautions. The mud is then pumped back down the hole and further re-circulated. After testing, the mud is treated periodically in the mud pits to ensure properties which optimise and improve drilling efficiency and borehole stability [362, 363]. It has observed that cashew leaf extract and mango leaf extract (MLE) are suitable additives for the production of WBM. MLE showed higher improvement of the rheological properties of the drilling mud (yield stress increased with temperature; gel strength [GS] increased). Use of plant leaf extracts as drilling mud additives will encourage local content development [364]. Henna extract was found to be compatible with common WBDF additives [365]. Henna extract enhanced the lubricity of WBDFs and showed an incredible significance in directional drilling. Advantages such as environmental friendliness, low cost, accessibility and the countercorrosion property made the applicability of henna extract in WBDFs suitable [365, 366]. Kosun (2014) has studied about the advantages and disadvantages of WBM. Some of the advantages are that WBMs use water as the continuous phase. Additives are used to change density and viscosity and to control filtrate losses. Also, chemical compatibility with the rock is important. Water-based drilling fluids (WBDFs) are cheap. They are good for curing mud losses. By pumping coarse bridging materials, mud losses often are cured. It is believed that the main mechanism is that in water-wet formations, filtrate losses occur, leaving dense particles in the mud in the fracture [367]. Some of the disadvantages observed are that WBM can swell shale formation, collapse boreholes and impact drilling outcomes in the drilling operations; gases produced among shale cracks whose non-organic part is possibly an aqueous wetting phase can be easily displaced by water. WBM can easily block the layers of very low permeability and influence the capability [367]. Scientists have carried out an experiment where grass was introduced as the sustainable drilling fluid additive. Experiments were conducted on different samples of grass to study the characteristics and behaviour of the newly developed drilling fluid at room temperature. And the experiment showed that the grass samples with varying particle size and concentration improved the viscosity, GS and filtration of the drilling fluid. These observations recommended the use of grass as a rheological modifier, filtration control agent and pH control agent to substitute the toxic agent from drilling fluids [368, 369].

## 14.2 Experimental

As the base fluid, water is the major constituent of the WBM. All other constituents are regarded as additives. The additives are for proper viscosity, filtration and fluid loss control. The drilling mud additives also serve as weighing agents and corrosion inhibitors. Most chemicals used as corrosion-inhibitive substances are harmful to the environment. Studies revealed that plant extracts are potential eco-friendly corrosion inhibitors. Plant-derived substances are of great interest owing to their versatile applications. As bioresources, plant extracts have been used for the production of drugs and corrosion inhibitors [364].

### 14.2.1 Materials

An accurate weighing machine (Mettler Toledo PB303-S) was used to weigh all the materials with accuracy up to 0.02 g. All the raw materials used in this study are mentioned in Table 14.1. Distilled water was used to prepare various water-based drilling muds with various concentrations of bentonite, barite and plant extracts.

### 14.2.2 Experimental Setup

Various following equipment or experimental setups were used in this study. A marsh funnel is a simple device for measuring viscosity by observing the time taken by a known volume of liquid to flow from a cone through the short tube. A mud balance is a device used to

**Table 14.1** Raw materials used in this study and their suppliers

| S. no. | Material | Source |
|---|---|---|
| 1 | Bentonite powder | Himedia, Mumbai |
| 2 | Barite powder | Himedia, Mumbai |
| 3 | Filter papers (90 mm) | Whatman (GE Healthcare) |
| 4 | Polyacrylamide (PAM) | Himedia, Mumbai |
| 5. | Methanol | Emplura, Mumbai |
| 6. | Hydrogen peroxide | Emplura, Mumbai |

measure the density and specific gravity of a drilling fluid. It consists of a graduated scale beam with a bubble level and a weight slider along its length and a cup with a lid on one end. The cup is used to hold a fixed amount of fluid so it can be weighed. A slider weight can be moved along the beam, and a bubble indicates when the beam is level. Density and specific gravity are read at the point where the slider weight sits on the beam at level. A Fann VG viscometer is a device used to measure the viscosity and GS of a drilling fluid. It has a rotating cylinder and bob instrument. The cylinder can be rotated at a speed of 600 rpm, 300 rpm and 3 rpm.

A water analyser is an instrument used to determine the pH of drilling mud. It has a knob on which the diode is connecting to measure the pH. A low-pressure low-temperature (LPLT) filter press is a pressurised cell fitted with a filter medium used for evaluating the filtrate loss and mud cake thickness of a drilling fluid. A Soxhlet apparatus is extraction equipment. It has three main sections: a percolator which circulates the solvent, a thimble which retains the solid to be extracted and a siphon mechanism which periodically empties the thimble. A fractional distillation unit was used to separate miscible liquids which have a boiling point difference of less than 25°C. It includes a burner, a round-bottom flask and a fractional column. A thermometer will be placed above the flask to indicate the temperature of the fluid. A methylene blue test was carried out to determine the cation exchange capacity and bentonite equivalent (amount of bentonite in a specific volume of drilling fluid).

| | | |
|---|---|---|
| Methylene blue (ml) | = | No. of drops of methylene blue * 0.5 |
| Methylene blue capacity | = | [Methylene blue (ml)]/[Drilling fluid (ml)] |
| Bentonite equivalent (lb/bbl) | = | [5 * Methylene blue (ml)]/[Drilling fluid (ml)] |
| Bentonite equivalent ($kg/m^3$) | = | 2.85 * bentonite equivalent (lb/bbl) |

### 14.2.3 Experimental Procedure

Water-based drilling mud samples (1000 ml) were prepared with different concentrations of bentonite (1%, 1.5%, 2%, 2.5%, 3%, 3.5%, 4%, 4.5%, 5%). Mixers were used to prepare samples. Firstly, a marsh funnel was cleaned using water and dried using a clean cloth, and then 950 ml of the prepared base mud was poured into the funnel, simultaneously covering the outlet of the funnel with a finger. The time taken by the base mud to flow out of the funnel was measured using a stopwatch, and required calculations were carried out to determine the effective viscosity and apparent viscosity.

$$\text{Effective viscosity} = \rho(\text{t} - 25)$$
$$\text{Apparent viscosity} = \rho(\text{t} - 28)$$

where $\rho$ = density of the mud in g/ml and $t$ = average time taken by the mud to flow out of marsh funnel in seconds [370].

Mud balance apparatus was used to determine the density and specific gravity of the base mud and also water-based drilling mud with leaf extract. The sample was poured in the mud cup, and the cup was closed airtight. Moving arm on the scale of the mud balance was adjusted till the bubble settled down in the centre and the readings were noted down. A viscometer was used to determine plastic viscosity, yield point (YP) and GS (GS at 0 s [gel 0] and GS at 10 min [gel 10]) of the base mud and water-based drilling mud with a leaf extract. The container was filled with the sample and set at a certain height as marked in the viscometer. Firstly, it was stirred at 600 rpm for 120 s in order to obtain $\theta_{600}$ and then again stirred at 300 rpm for 120 s to obtain $\theta_{300}$. For gel 0 and gel 10, the bob was rotated at 600 rpm for 20 s and 003 rpm for 120 s, and then results were observed after waiting for 10 s and 10 min, respectively.

Water Analyser 371 (Siemens) was used to determine the pH of various base muds and the pH of various water-based drilling muds with a leaf extract. Electric connectivity was given to the equipment and required connections were made with a pH diode. The diode was then dipped in a flask filled with sample mud. Then the Enter button on the dial was pressed after showing the sample number. Readings were noted down from the digital screen on the instrument. The LPLT filter press had been used for the study of filtrate loss and mud cake thickness. The pressure which was applied on the drilling mud is produced by the force of the water stream which was pressurised by metal weight when it was suspended in air. a Filter paper was placed on the outlet of the mud container before it was filled up with drilling mud, which resulted in the formation of mud cake. Pressure inside the container during the experiment may vary from 0 psi to 200 psi (average pressure of 80 psi to 120 psi). Soxhlet apparatus was used for the extraction of various leaf extracts. Methanol was used as an extraction solvent here and the extraction chamber was filled with leaves, where extraction takes place. Continuous water supply was given to the condenser, and the solvent was heated at 65°C as methanol's boiling point is 64.15°C. The setup is continuously kept for more than 20 h as the extraction process gets completed. A fractional distillation unit was used to separate methanol from plant extracts as there was a fine gap between their boiling points. A water tank was filled with ice water and the mixture of methanol and extract was poured in a round-bottom flask. A thermometer had been placed inside the flask and the top outlet was sealed using a cork. The heater was adjusted to a low temperature (5°C–10°C). Methanol evaporated and then condensed while passing through the water tank which had a temperature of 0°C to 10°C. Methanol was collected from the outlet and the plant extract was left in the flask.

## 14.3 Results and Discussion

### 14.3.1 Formulation of Base Mud

Bentonite was mixed in water (1000 ml) for determination of the optimum concentration for the base mud. Nine different concentrations (1%, 1.5%, 2%, 2.5%, 3%, 3.5%, 4%, 4.5%, 5%) of bentonite were tested for different properties of drilling fluid, and from the results, three different concentrations (3%, 3.5%, 4%) of bentonite were chosen as the optimum concentration. Five different concentrations (1%, 2%, 3%, 4%, 5%) of barite were mixed with the above-mentioned predetermined concentrations of bentonite for determining the optimum concentration of barite for the base mud. From the results, three concentrations of bentonite+barite were determined to be optimum for the base mud, (a) 3%–4%, (b) 3.5%–1% and (c) 4%–2% of bentonite and barite. Hamilton beach mixer was used for the proper mixing of bentonite and barite with water. The samples were tested for determining the rheological properties of drilling fluid. Mud balance was used to determine the density and specific gravity. Effective and apparent viscosities were determined using marsh funnel equipment. A water analyser was used to determine the pH of the drilling fluid. A Fann VG viscometer was used to determine the plastic viscosity, GS and YP. Mud cake thickness and filtrate loss of the drilling fluid was determined using a filter press. Table 14.2 shows the variation of drilling fluid properties

**Table 14.2** Rheological properties of various concentrations of bentonite

| Conc. of bentonite (%) | $\rho$ (ppg) | Sp. gr. | Effective $\mu$ (cP) | Apparent $\mu$ (cP) | pH | Filtrate loss (ml) | Mud cake thickness (mm) |
|---|---|---|---|---|---|---|---|
| 1 | 1.014 | 1.02 | 5.41 | 2.37 | 8.51 | 88 | 1.5 |
| 1.5 | 1.02 | 1.02 | 6.29 | 3.23 | 8.55 | 72 | 2 |
| 2 | 1.02 | 1.02 | 6.76 | 3.70 | 8.58 | 65 | 2 |
| 2.5 | 1.032 | 1.03 | 7.017 | 3.75 | 8.76 | 61 | 2 |
| 3 | 1.032 | 1.03 | 7.92 | 3.92 | 8.8 | 56 | 2 |
| 3.5 | 1.032 | 1.03 | 8.01 | 4.51 | 8.92 | 52 | 2 |
| 4 | 1.034 | 1.03 | 8.4 | 5.83 | 8.98 | 46 | 3 |
| 4.5 | 1.038 | 1.04 | 11.42 | 8.32 | 9.01 | 44 | 3 |
| 5 | 1.044 | 1.05 | 11.88 | 8.74 | 9.08 | 41 | 3.5 |

**Table 14.3** Plastic $\mu$ of various concentrations of bentonite at different temperatures

| Conc. of bentonite (%) | Plastic $\mu$ (70°F) | Plastic $\mu$ (75°F) | Plastic $\mu$ (80°F) | Plastic $\mu$ (85°F) | Plastic $\mu$ (90°F) |
|---|---|---|---|---|---|
| 1 | 7.24 | 7.15 | 6.92 | 5.7 | 4.5 |
| 1.5 | 8.2 | 7.45 | 7.02 | 6.21 | 5.2 |
| 2 | 8.61 | 7.92 | 7.31 | 6.42 | 5.38 |
| 2.5 | 8.9 | 8.01 | 7.48 | 6.82 | 5.5 |
| 3 | 8.96 | 7.98 | 7.53 | 6.89 | 5.67 |
| 3.5 | 9.01 | 8.12 | 7.68 | 6.92 | 5.7 |
| 4 | 9.17 | 8.18 | 7.73 | 7.03 | 5.77 |
| 4.5 | 9.28 | 8.23 | 7.86 | 7.21 | 5.9 |
| 5 | 9.73 | 8.35 | 7.98 | 7.53 | 6.18 |

with different concentrations of bentonite. With an increase in the concentration of bentonite, density, specific gravity, effective and apparent viscosity and pH of the mud increased while filtrate loss decreased.

Table 14.3 represents the variation of the plastic viscosity of mud with temperature at different concentrations of bentonite. With an increase in temperature the plastic viscosity decreased but with an increase in the concentration of bentonite, plastic viscosity increased. The optimum concentration of bentonite was determined to be 3% (30 g), 3.5% (35 g) and 4% (40 g), because higher concentrations of bentonite may cause damage to the formation and a lower concentration of bentonite will not give accurate results. Different concentrations of barite were mixed with predetermined concentrations of bentonite and the variations in the drilling fluid properties were recorded. Table 14.4 shows the variation of drilling fluid properties in bentonite+barite mud. The variation of plastic viscosity with temperature at different concentrations of barite in predetermined concentrations of bentonite is shown in Table 14.5. The plastic viscosity of the drilling mud decreased with an increase in temperature. Barite is used in the petroleum industry as weighting material in the formulation of base mud. It also increases the hydrostatic pressure of the drilling mud, allowing it to compensate for higher-pressure zones. Therefore, viscosity

**Table 14.4** Rheological properties of various concentrations of barite in predetermined concentrations of bentonite

| Conc. of bentonite (%) | Conc. of barite (%) | $\rho$ (lb/gal) | Sp. gr. | Effective $\mu$ (cP) | Apparent $\mu$ (cP) | pH | Filtrate loss (ml) | Mud cake thickness (mm) |
|---|---|---|---|---|---|---|---|---|
| | 1 | 8.6 | 1.03 | 7.360 | 4.27 | 8.72 | 45 | 2 |
| | 2 | 8.7 | 1.04 | 8.040 | 4.742 | 9.26 | 43 | 2 |
| 3 | 3 | 8.72 | 1.05 | 8.136 | 4.89 | 9.18 | 40 | 2 |
| | 4 | 8.9 | 1.07 | 8.504 | 5.5 | 9.25 | 38 | 2.5 |
| | 5 | 8.9 | 1.07 | 8.66 | 5.65 | 9.27 | 37 | 3 |
| | 1 | 8.6 | 1.03 | 8.67 | 5.58 | 9.18 | 48 | 2 |
| | 2 | 8.7 | 1.05 | 9.76 | 6.63 | 9.25 | 39 | 2.5 |
| 3.5 | 3 | 8.8 | 1.06 | 8.54 | 5.37 | 9.48 | 38 | 2.5 |
| | 4 | 8.85 | 1.065 | 8.19 | 5.01 | 9.40 | 34 | 2.5 |
| | 5 | 8.9 | 1.07 | 15.74 | 12.53 | 9.49 | 32 | 3 |
| | 1 | 8.7 | 1.04 | 9.98 | 6.84 | 9.14 | 39 | 2 |
| | 2 | 8.75 | 1.05 | 11.07 | 7.921 | 9.37 | 37 | 2 |
| 4 | 3 | 8.9 | 1.07 | 12.18 | 9.32 | 9.32 | 35 | 2.5 |
| | 4 | 8.9 | 1.07 | 11.23 | 8.03 | 9.28 | 33 | 3 |
| | 5 | 8.95 | 1.08 | 11.86 | 8.66 | 9.25 | 30 | 3 |

**Table 14.5** Variation of plastic $\mu$ with different temperatures of bentonite+barite mud

| Plastic $\mu$ (70°F) | Plastic $\mu$ (75°F) | Plastic $\mu$ (80°F) | Plastic $\mu$ (85°F) | Plastic $\mu$ (90°F) | | |
|---|---|---|---|---|---|---|
| | 1 | 1 3.91 | 3.91 | 3.522 | 3.13 | 2.55 |
| | 2 | 7.04 | 6.07 | 5.28 | 5.09 | 0.98 |
| 3 | 3 | 11.15 | 8.99 | 7.05 | 5.87 | 4.89 |
| | 4 | 9 | 7.43 | 7.04 | 6.46 | 5.67 |
| | 5 | 9.59 | 8.02 | 7.83 | 7.04 | 4.10 |
| | 1 | 11.15 | 8.79 | 8.22 | 7.24 | 5.86 |
| | 2 | 9.39 | 8.61 | 7.44 | 6.07 | 5.02 |
| 3.5 | 3 | 7.85 | 7.24 | 6.65 | 6.26 | 5.72 |
| | 4 | 11.16 | 7.26 | 6.46 | 6.06 | 5.68 |
| | 5 | 9.8 | 7.62 | 6.43 | 5.86 | 4.11 |
| | 1 | 9.84 | 8.03 | 7.63 | 6.46 | 5.89 |
| | 2 | 11.55 | 9 | 8.03 | 6.06 | 5.87 |
| 4 | 3 | 8.99 | 7.23 | 6.45 | 4.5 | 4.08 |
| | 4 | 11.53 | 9.77 | 9.38 | 6.84 | 5.27 |
| | 5 | 12.03 | 10.58 | 9.4 | 7.3 | 5.87 |

plays an important role in the formulation of drilling mud, and the optimum concentration of barite in drilling mud was chosen according to the plastic viscosity and density. The concentrations of barite which were determined to be optimum are 4% (40 g) in 3% (30 g) of bentonite, 1% (10 g) in 3.5% (35 g) of bentonite and 2% (20 g) in 4% (40 g) of bentonite.

## 14.3.2 Plant Extract

Table 14.6 shows the variation of density and specific gravity after using the two plant extracts (mango and neem leaves) in the base mud. Comparing the results, MLE showed values which are much higher than neem leaf extract (NLE). For all the plant extracts, the density and specific gravity decreased with an increase in concentration of the plant extract. With an increase in concentration of the plant extract, the filtrate loss also increased and the mud cake thickness decreased. This was the same for all the plant extracts. Even if the values do not have much variation in between them, MLE has much lesser filtrate loss than neem plant extract. A comparison

**Table 14.6** Comparison of $\rho$ and specific gravity of MLE and NLE

| Conc. of bentonite (%) | Conc. of barite (%) | Plant extract (ml) | Density of MLE (lb/gal) | Sp. gr. of MLE | $\rho$ of NLE (lb/gal) | Sp. gr. of NLE |
|---|---|---|---|---|---|---|
| | | 2 | 8.65 | 1.04 | 8.25 | 0.99 |
| | | 4 | 8.40 | 1.02 | 8.10 | 0.96 |
| 3 | 4 | 6 | 8.15 | 0.98 | 7.8 | 0.95 |
| | | 8 | 8 | 0.975 | 7.7 | 0.94 |
| | | 10 | 7.85 | 0.85 | 7.6 | 0.93 |
| | | 2 | 8.6 | 1.03 | 7.9 | 0.98 |
| | | 4 | 8.35 | 0.9 | 7.85 | 0.97 |
| 3.5 | 1 | 6 | 8.1 | 0.975 | 7.8 | 0.965 |
| | | 8 | 8 | 0.96 | 7.7 | 0.96 |
| | | 10 | 7.9 | 0.95 | 7.65 | 0.95 |
| | | 2 | 8.7 | 1.05 | 8.2 | 1.03 |
| | | 4 | 8.4 | 1.04 | 8.15 | 1.02 |
| 4 | 2 | 6 | 8.15 | 1.02 | 8 | 0.98 |
| | | 8 | 7.95 | 0.95 | 7.95 | 0.95 |
| | | 10 | 7.9 | 0.9 | 7.9 | 0.94 |

of the filtrate loss and mud cake thickness by using the two plant extracts is shown in Table 14.7. Table 14.8 through Table 14.10 show the variation of plastic viscosity with temperature at different concentrations of plant extracts. The plastic viscosity increased from the base mud after using the plant extracts. With an increase in concentration of the plant extract, plastic viscosity increased and with an increase in temperature, it decreased. MLE was identified to be much more effective as it showed a more optimum value than NLE. The variation of GS (gel 0 and gel 10) and YP with temperature of the three plant extracts has been recorded in Table 14.11 and Table 14.12. With an increase in the concentration of the plant extract the GS and YP increased. The variation of GS and YP with temperature was found to be inverse as both GS and YP decreased with an increase in temperature. The results which we got from using MLE were found to be higher than NLE. MLE shows higher values of GS and YP.

**Table 14.7** Comparison of filtrate loss and mud cake thickness of MLE and NLE

| Conc. of bentonite (%) | Conc. of barite (%) | Plant extract (ml) | MLE filtrate loss (ml) | MLE mud cake thickness (mm) | NLE filtrate loss (ml) | MLE mud cake thickness (mm) |
|---|---|---|---|---|---|---|
| | | 2 | 34 | 2 | 35 | 2 |
| | | 4 | 35 | 2 | 36 | 2 |
| 3 | 4 | 6 | 37 | 2 | 37 | 2 |
| | | 8 | 38 | 1.5 | 39 | 1.5 |
| | | 10 | 40 | 1.5 | 40 | 1.5 |
| | | 2 | 30 | 3 | 36 | 2 |
| | | 4 | 32 | 2.5 | 38 | 2 |
| 3.5 | 1 | 6 | 35 | 2 | 39 | 1.5 |
| | | 8 | 36 | 2 | 41 | 1.5 |
| | | 10 | 37 | 2 | 42 | 1.5 |
| | | 2 | 33 | 2 | 35 | 2 |
| | | 4 | 34 | 2 | 37 | 2 |
| 3.5 | 2 | 6 | 36 | 2 | 38 | 2 |
| | | 8 | 38 | 1.5 | 40 | 1.5 |
| | | 10 | 40 | 1.5 | 41 | 1.5 |

**Table 14.8** Comparison of plastic viscosity $\mu$ of MLE and NLE at (70°F)

| Conc. of bentonite (%) | Conc. of barite (%) | Temp. (°F) | Plant extract (ml) | Plastic viscosity $\mu$ of MLE (cP) | Plastic viscosity $\mu$ of NLE (cP) |
|---|---|---|---|---|---|
| | | | 2 | 11.53 | 11.73 |
| | | | 4 | 11.83 | 14.87 |
| 3 | 4 | 70 | 6 | 12.52 | 15.68 |
| | | | 8 | 12.95 | 16.82 |
| | | | 10 | 13.42 | 17.01 |
| | | | 2 | 13.69 | 113.13 |
| | | | 4 | 15.84 | 13.58 |
| 3.5 | 1 | 70 | 6 | 15.95 | 13.79 |
| | | | 8 | 17.22 | 14.05 |
| | | | 10 | 18.19 | 14.97 |
| | | | 2 | 11.74 | 10.59 |
| | | | 4 | 12.13 | 11.38 |
| 3.5 | 1 | 70 | 6 | 13.69 | 11.97 |
| | | | 8 | 14.67 | 12.35 |
| | | | 10 | 16.21 | 13.81 |

**Table 14.9** Comparison of plastic viscosity $\mu$ of MLE and NLE at 75°F

| Conc. of bentonite (%) | Conc. of barite (%) | Temp. (°F) | Plant extract (ml) | Plastic viscosity $\mu$ of MLE (cP) | Plastic viscosity $\mu$ of NLE (cP) |
|---|---|---|---|---|---|
| | | | 2 | 9.9 | 9.97 |
| | | | 4 | 10.12 | 10.75 |
| 3 | 4 | 75 | 6 | 10.89 | 11.74 |
| | | | 8 | 11.24 | 13.68 |
| | | | 10 | 11.49 | 14.67 |
| | | | 2 | 10.75 | 10.75 |
| | | | 4 | 12.08 | 11.05 |
| 3.5 | 1 | 75 | 6 | 12.32 | 11.38 |
| | | | 8 | 14.67 | 11.97 |
| | | | 10 | 15.05 | 12.62 |
| | | | 2 | 10.16 | 9.18 |
| | | | 4 | 10.56 | 9.25 |
| 3.5 | 1 | 75 | 6 | 11.54 | 10.54 |
| | | | 8 | 12.71 | 11.01 |
| | | | 10 | 13.1 | 11.97 |

**Table 14.10** Comparison of plastic viscosity $\mu$ of MLE and NLE at 80°F

| Conc. of bentonite (%) | Conc. of barite (%) | Temp. (°F) | Plant extract (ml) | Plastic viscosity $\mu$ of MLE (cP) | Plastic viscosity $\mu$ of NLE (cP) |
|---|---|---|---|---|---|
| | | | 2 | 8.79 | 8.79 |
| | | | 4 | 9 | 9.58 |
| 3 | 4 | 80 | 6 | 9.87 | 9.77 |
| | | | 8 | 10.35 | 11.14 |
| | | | 10 | 10.94 | 12.32 |
| | | | 2 | 8.21 | 8.41 |
| | | | 4 | 9.97 | 8.64 |
| 3.5 | 1 | 80 | 6 | 10.56 | 8.95 |
| | | | 8 | 12.12 | 9.17 |
| | | | 10 | 13.49 | 10.57 |
| | | | 2 | 8.6 | 7.95 |
| | | | 4 | 19.19 | 8.06 |
| 3.5 | 1 | 80 | 6 | 10.95 | 8.26 |
| | | | 8 | 11.14 | 8.59 |
| | | | 10 | 12.13 | 9.08 |

**Table 14.11** Comparison of plastic viscosity $\mu$ of MLE and NLE at 85°F

| Conc. of bentonite (%) | Conc. of barite (%) | Temp. (°F) | Plant extract (ml) | Plastic viscosity $\mu$ of NLE (cP) | Plastic viscosity $\mu$ of NLE (cP) |
|---|---|---|---|---|---|
| | | | 2 | 7.63 | 7.63 |
| | | | 4 | 7.83 | 8.6 |
| 3 | 4 | 85 | 6 | 8.02 | 9.19 |
| | | | 8 | 8.54 | 9.97 |
| | | | 10 | 8.97 | 10.75 |
| | | | 2 | 7.03 | 7.24 |
| | | | 4 | 8.22 | 7.36 |
| 3.5 | 1 | 85 | 6 | 9.00 | 7.94 |
| | | | 8 | 9.59 | 8.15 |
| | | | 10 | 10.19 | 9.01 |
| | | | 2 | 7.23 | 6.87 |
| | | | 4 | 7.63 | 7.04 |
| 3.5 | 1 | 85 | 6 | 7.81 | 7.48 |
| | | | 8 | 9.78 | 7.93 |
| | | | 10 | 10.56 | 8.21 |

**Table 14.12** Comparison of plastic viscosity $\mu$ of MLE and NLE at 90°F

| Conc. of bentonite (%) | Conc. of barite (%) | Temp. (°F) | Plant extract (ml) | Plastic viscosity $\mu$ of NLE (cP) | Plastic viscosity $\mu$ of NLE (cP) |
|---|---|---|---|---|---|
| | | | 2 | 4.38 | 4.49 |
| | | | 4 | 4.42 | 4.63 |
| 3 | 4 | 90 | 6 | 4.65 | 5.08 |
| | | | 8 | 4.79 | 4.88 |
| | | | 10 | 4.92 | 5.08 |
| | | | 2 | 5.81 | 4.55 |
| | | | 4 | 5.86 | 4.64 |
| 3.5 | 1 | 90 | 6 | 5.94 | 4.83 |
| | | | 8 | 5.89 | 4.91 |
| | | | 10 | 5.58 | 5.06 |
| | | | 2 | 5.08 | 4.09 |
| | | | 4 | 5.09 | 4.18 |
| 3.5 | 1 | 90 | 6 | 5.1 | 4.29 |
| | | | 8 | 5.86 | 4.37 |
| | | | 10 | 6.06 | 4.65 |

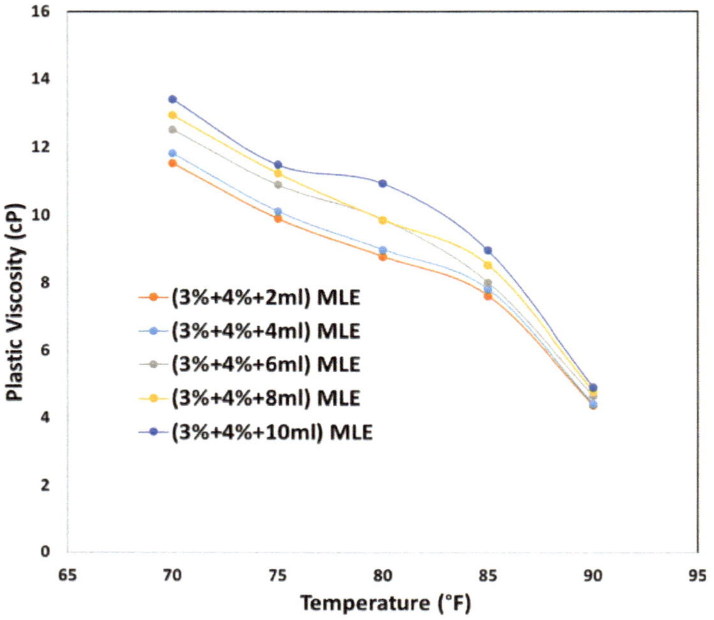


**Figure 14.1** Variation of plastic viscosity with temperature for different concentrations of MLE.

## 14.3.3 Plastic Viscosity

Figures 14.1 and 14.2 show the variation of plastic viscosity with temperature at different concentrations of MLE and NLE, respectively. As the concentration of the plant extract increases, the plastic viscosity of the drilling mud also increases. For all the plant extracts which were being used, a similar trend was observed. With an increase in temperature the plastic viscosity decreased. MLE showed high values than NLE, but if the plastic viscosity is more, then more pump pressure has to be applied for the circulation of mud, and thus, the plastic viscosity need not increase further.

## 14.3.4 Yield Point

Figures 14.3 and 14.4 show the variation of YP with temperature at different concentrations of MLE and NLE, respectively. The variation

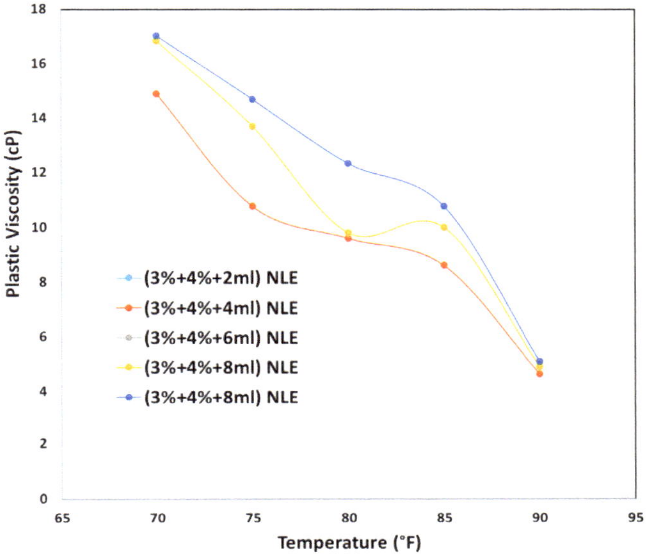


**Figure 14.2** Variation of plastic viscosity with temperature for different concentrations of NLE.

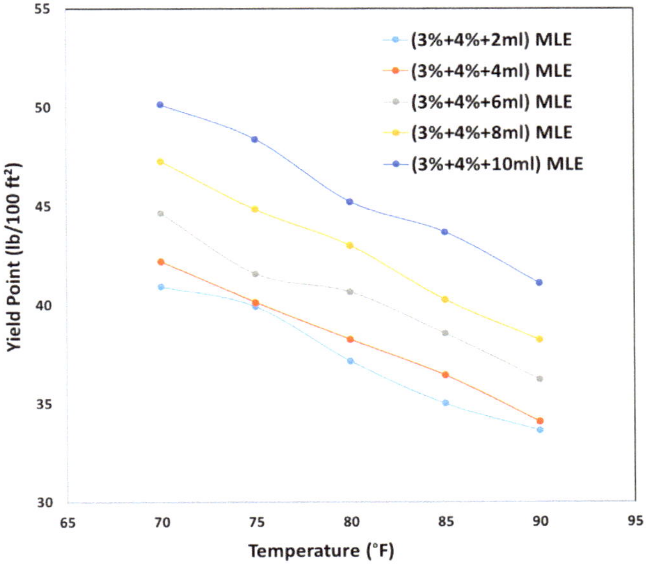


**Figure 14.3** Variation of yield point with temperature for different concentrations of MLE.

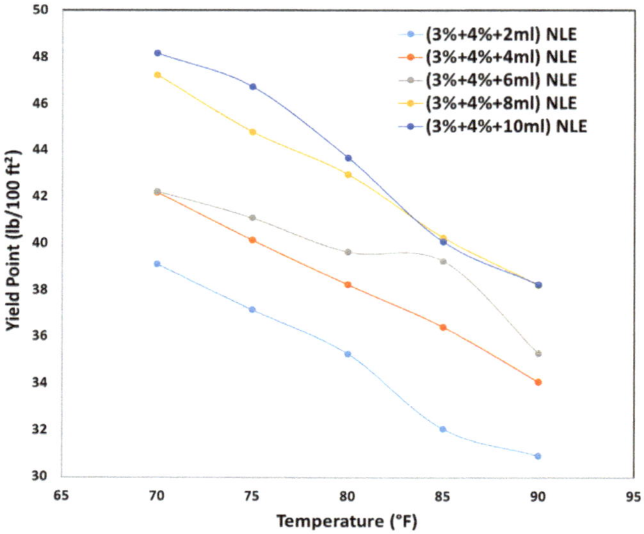


**Figure 14.4** Variation of yield point with temperature for different concentrations of NLE.

of YP was also similar in the case of each plant extract. As the concentration of the plant extract increases, YP also increases, and it decreases with an increase in temperature.

## 14.4 Conclusion

The use of plant extracts as an additive can enhance the properties of drilling fluids. The effect of a plant extract increases as the concentration of the particular extract increases. MLE shows more effects than other leaves which were used (neem leave). After analysing the results, the density of the drilling fluid decreased, filtrate loss increased, while mud cake thickness decreased, plastic viscosity of the drilling fluid increased and the GS and YP of the drilling fluid increased with an increase in the plant extract's concentration. With the addition of locally sourced MLE and NLE, there were improvements in the rheological properties of the drilling mud samples. Use of plant extracts as drilling mud additives will encourage local content development. The effect of temperature on the rheological properties of drilling mud samples is also noticed to be similar. For all samples with MLE and NLE, plastic viscosity of the drilling fluid increases as the concentration increases and decreases with an increase in temperature. The YP also followed the same trend. From the two plant extracts which were used, MLE shows higher improvement in enhancing the properties of drilling mud samples. The use of locally available natural sources will help to curb the cost to the petroleum industry.

## Acknowledgements

We would like to acknowledge the Department of Petroleum Technology, Dibrugarh University, for laboratory facilities.

## Chapter 15

# Effects of Temperature on the Transportation of Assam Crude Oil through Pipelines

**Bondita Robidas and Subrata Borgohain Gogoi**
*Department of Petroleum Technology, Dibrugarh University, Assam, India*
bonditarodidas@dibru.ac.in

Crude oil is transported through a pipeline over a long distance, and temperature plays a vital role during this transportation process. The pour point is an important factor in the pipeline transportation system as the pour point is the temperature below which there is no flow. Crude oil containing a high amount of wax exhibits a high pour point and non-Newtonian viscosity behaviour which creates difficulty in transportation. Viscosity is the resistance to flow. Again, at lower temperatures the paraffin may cause gel formation which consists of wax crystals which affects the flow properties. Generally, pour point depressants or wax crystal modifiers are used to lower the pour point, viscosity and yield stress of crude oil so that we can easily transport crude oil at lower temperatures through the pipeline. In my work, five crude oil samples were collected from five different wells of the Upper Assam Basin and laboratory experimental work was done. The M3600

*Advances in Petroleum Technology*
Edited by Subrata Borgohain Gogoi

ISBN 978-981-4877-19-0 (Hardcover), 978-1-003-04993-7 (eBook)
www.jennystanford.com

Grace Viscometer apparatus was used to carry out the rheological experiments. Effects of temperature on flow behaviour were studied by considering the sample at 30°C, 40°C, 50°C, 60°C, 70°C and 80°C. This chapter pertains to the determination of viscosity, shear stress and shear rate at different temperatures of 30°C, 40°C, 50°C, 60°C, 70°C and 80°C. So, investigating the rheological study of crude oil at 30°C, 40°C, 50°C, 60°C, 70°C and 80°C was carried out. From the study it was observed that temperature plays an important role in the crude oil transportation system through a pipeline. It was also observed from the results that crude oil exhibits non-Newtonian flow behaviour at a low shear rate and Newtonian flow behaviour at a high shear rate. As a conclusion, the rheological study showed that temperature, shear stress and shear rate have great impacts on the viscosity of crude oil, and it is important to understand these factors to make the transportation process continuous through the pipeline.

## 15.1 Introduction

It has been observed that wax deposition in the pipeline interrupts the continuous flow of the pipeline transportation system. Wax crystallisation usually takes place when the pressure and temperature values start dropping at the onset of crystallisation [371, 372]. It has been observed that wax deposition is one of the major problems in the pipeline transportation system. The presence of a small percentage of waxes would increase the pour point temperature of waxy crude oil, and this may cause an early blockage of pipelines in the presence of a temperature gradient between the fluid and the surrounding weather [373]. On exposure to cold conditions, wax molecules crystallise and become a solid [374]. The outer environment of the pipeline also has significant effects on wax deposition, where an unfavourable surrounding temperature results in amorphous crystallisation. Kasumu et al. (2013) reported an opposite trend between the wax precipitation temperature and the cooling rate [375]. Lashkarbolooki et al. (2010) observed significant effects of the temperature difference between crude oil and ambient

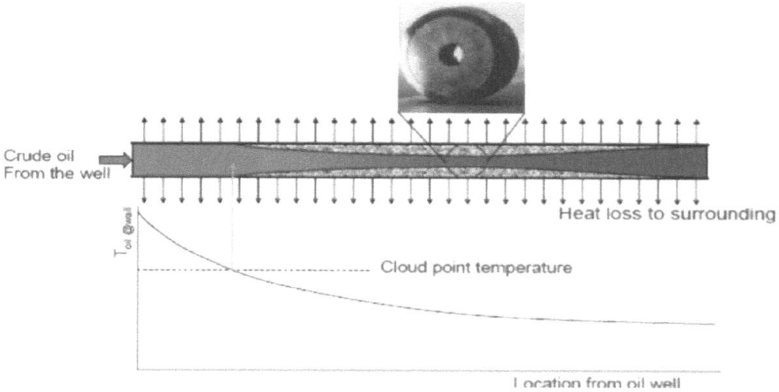


**Figure 15.1** Wax deposition when inner temperature is lower than cloud point temperature in a pipeline. Reprinted with permission from Muhammad Theyab, 2018, London South Bank University [391].

temperature on the wax deposition thickness in the pipelines [376]. Wax deposition decrease with increasing temperature was reported in [376–378]. Creek et al. (1999) observed the effects of temperature and flow rate on the wax deposition rate [379]. The deposition of waxes on the wall of pipelines usually occurs below the wax appearance temperature, followed by gelation over lower temperatures.

Figure 15.1 shows the stages of wax formation and deposition in a pipeline. While coming out from the well, the waxes within the liquid crude oil precipitate and deposit on the wall of pipelines when the temperature of the crude oil drops below the cloud point temperature due to the heat loss to surroundings [380]. Isaac Newton discovered that the viscosity of most fluids is only affected by temperature. But in some cases, viscosity is also affected by factors other than temperature, and these fluids are termed as non-Newtonian fluids. The viscosity of a non-Newtonian fluid will change due to agitation or pressure, that is, technically known as shear stress. The thermal behaviour, which mainly relies on the temperature, has been a key parameter in understanding the wax-phase behaviour of a waxy crude oil [381, 382]. The relationship between temperature and viscosity, for instance, has been an important factor for defining the Newtonian and non-Newtonian regions of waxy crude oils. Carnahan et al. (2007) [383] stated that the rheology of waxy oil and its physical properties could be defined by considering the effects of the pour point and apparent viscosity over the ranges of temperatures and the static yield stress. When the crude oil temperature is higher than the wax appearance temperature, waxy crude oil shows Newtonian fluid behaviour, below which it shows non-Newtonian fluid characteristics [384]. Thixotropy is a time-dependent change of viscosity of a waxy crude oil in the non-Newtonian region [385, 386]. Waxy crude oil changes from low viscous Newtonian to high viscous shear thinning or thixotropic when paraffin particles precipitate out during the cooling process [387]. Crude oil becomes a liquid at high temperatures and pressures and becomes a solid as the temperature lowers than the pour point. Wax appearance and pour point temperatures indicate the first appearance of waxes and the entire appearance of waxes in waxy crude oil, respectively, and because of these two temperatures, deposition of wax takes place on the pipe wall, which would result in a decrease in the operating internal pipe diameter, which may lead to sudden shutdown [375]. Rukhtong et al. (2015) reported that due to the higher influence of heat transfer near the pipe wall via conduction and convection from the surrounding weather condition, the temperature of crude oil near pipe wall is lower than the centre of the pipe [388]. Guozhong and Gang (2010) also agree with this, who observed the dropping of crude oil temperature towards the pipe wall and that the temperature around the pipe wall was almost similar with the surrounding temperature, making the viscosity higher near the pipe wall [389, 390].

## 15.2 Experimental

### 15.2.1 Materials and Instruments

Five crude oil samples were collected from five different wells of Upper Assam oilfields. The instruments were pour point apparatus and Grace viscometer (model no. M3600).

## 15.2.2 Methodology

### 15.2.2.1 Rheological properties

The pour point of crude oil samples was determined by ASTM D 97-17b, Standard Test Method, and the rheological test was done as per the manual of the instrument. The pour point apparatus was used to measure the pour point of the samples. Here ice was placed inside the apparatus. A crude oil sample was taken in the glass tube and a thermometer was placed at the top of it to check the temperature of the sample. The temperature was observed periodically to check at what temperature the flow ceased, that is, the pour point temperature of the sample.

### 15.2.2.2 Shear stress, shear rate and viscosity

The Grace viscometer was connected to the software of the computer, whereby choosing the test setup, the parameters for time, temperature and rotational speed were entered. After entering the rotational speed it automatically calculated the shear rate, which was displayed on the test setup of the screen. After the test steps were generated, the test was started and the results obtained in a Microsoft Excel sheet in which we obtained the dynamic viscosity in cp and shear stress in dyne/$cm^2$ against the parameters entered.

# 15.3 Results and Discussion

## 15.3.1 Pour Point

The pour point of the five samples S1, S2, S3, S4 and S5 was determined and is given in Table 15.1.

## 15.3.2 Effect of Shear Rate on Viscosity

The relationship between viscosity and shear rate for samples S1, S2, S3, S4 and S5 at different temperatures of 30°C, 40°C, 50°C, 60°C, 70°C and 80°C were plotted (Figs. 15.2–15.6). Upon analysing Figs. 15.2–15.6, it can be observed that viscosity changes significantly with shear rate. Figures 15.2–15.4 and 15.6 also show that viscosity

**Table 15.1** Pour point of the samples

| S. no. | Pour point (°C) |
|---|---|
| S1 | 9 |
| S2 | 12 |
| S3 | 24 |
| S4 | 3 |
| S5 | 21 |

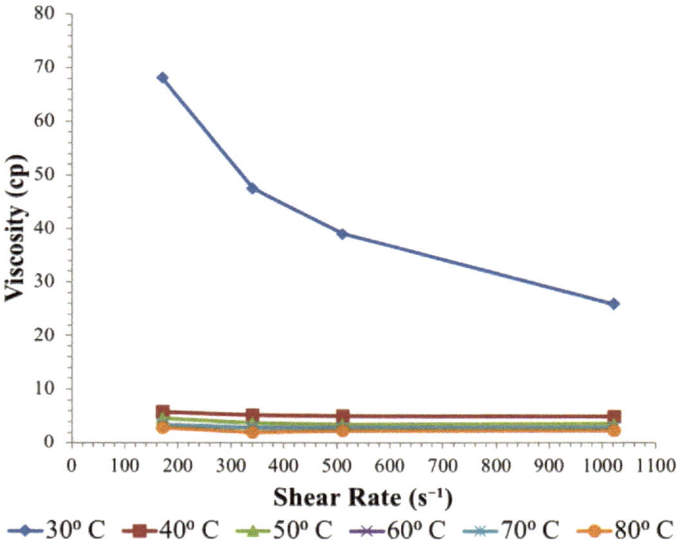


**Figure 15.2** Viscosity vs. shear rate for crude oil sample 1.

decreases with an increase in shear rate for samples S1, S2, S3 and S5, which possess shear thinning or pseudoplastic behaviour. For sample S4, as seen in Fig. 15.5, at 70°C and 80°C, it shows shear-thickening but at 30°C, 40°C, 50°C and 60°C it shows shear-thinning behaviour.

### 15.3.3 Effect of Temperature on Shear Stress

The relationship between shear stress and temperature at a constant shear rate for samples S1, S2, S3, S4 and S5 is plotted

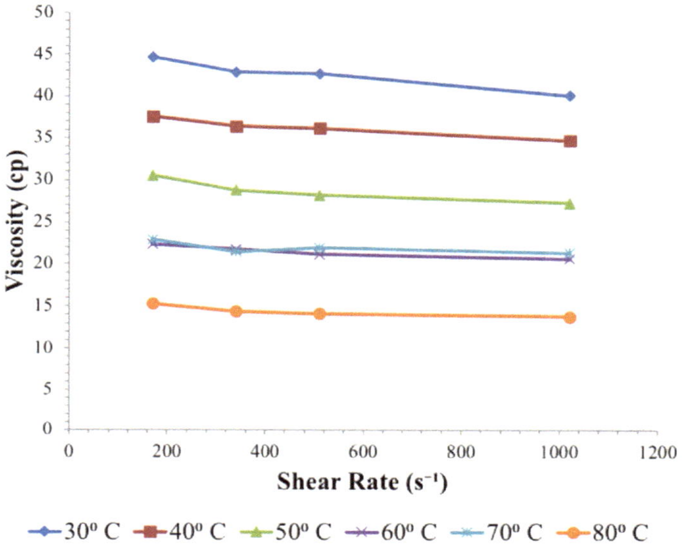


**Figure 15.3** Viscosity vs. shear rate for crude oil sample 2.

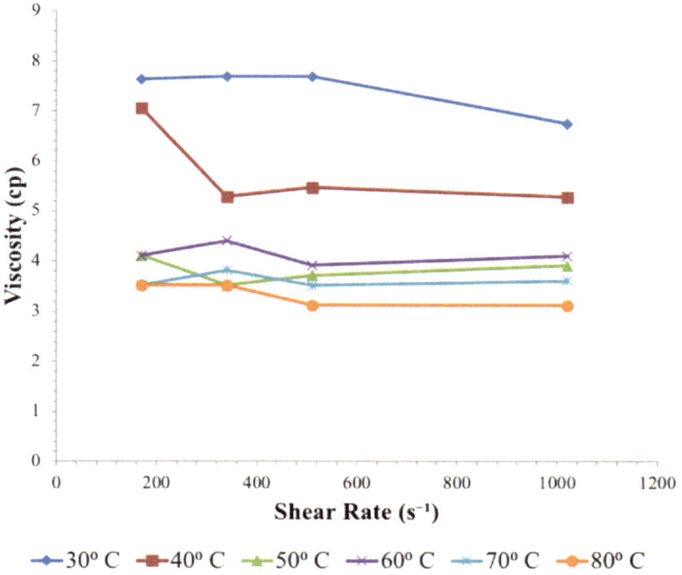


**Figure 15.4** Viscosity vs. shear rate for crude oil sample 3.

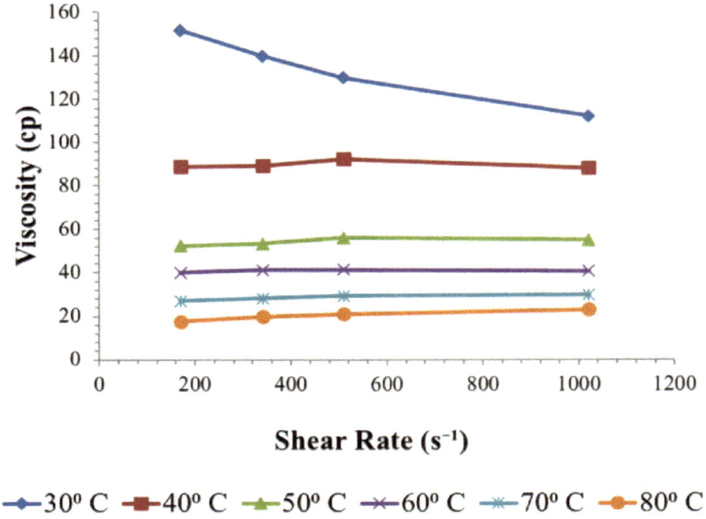


**Figure 15.5** Viscosity vs. shear rate for crude oil sample 4.

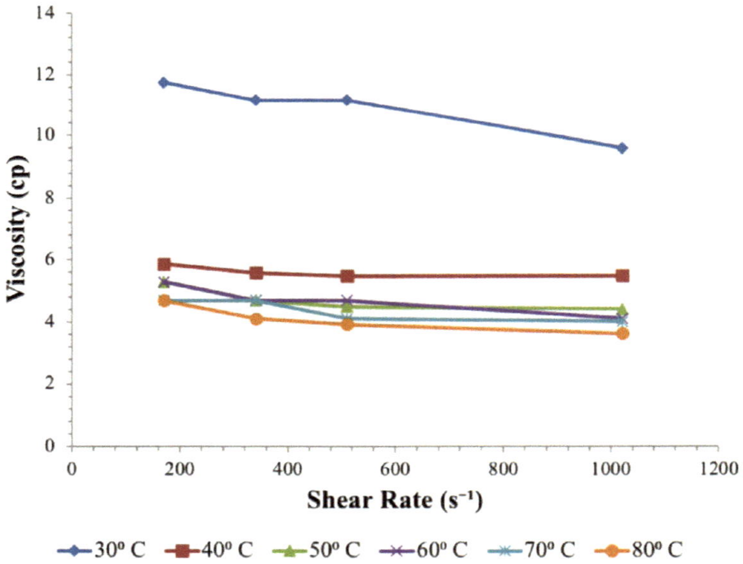


**Figure 15.6** Viscosity vs. shear rate for crude oil sample 5.

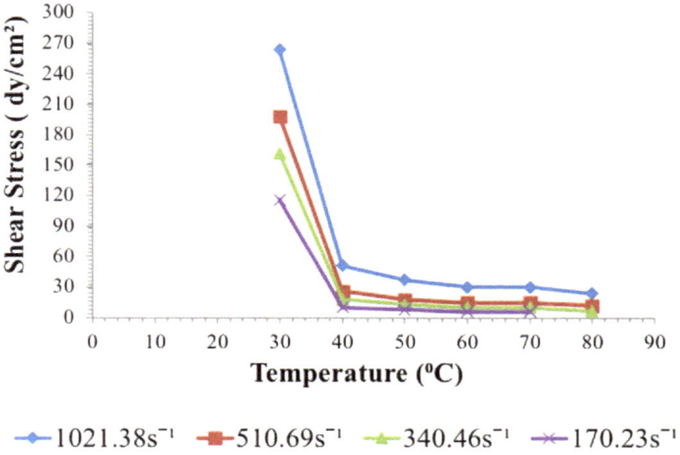


**Figure 15.7** Shear stress vs. temperature for crude oil sample 1.

in Figs. 15.7–15.11. Upon analysing Figs. 15.7–15.11, it is observed that as temperature increases from 30°C to 40°C, 50°C, 60°C, 70°C and 80°C, shear stress decreases rapidly and then stabilises after a certain value.

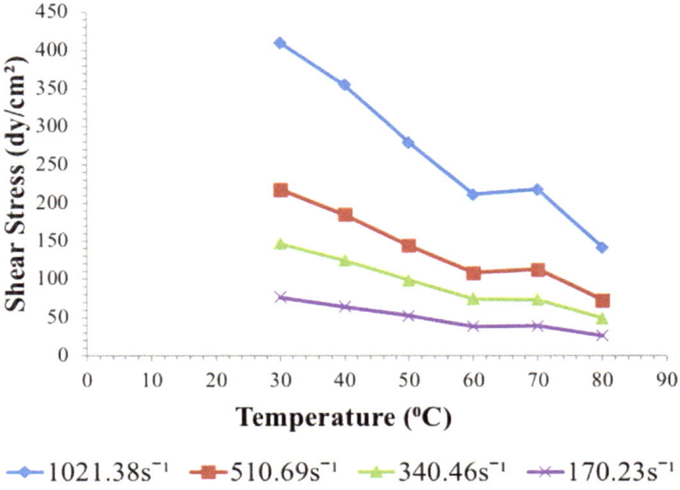


**Figure 15.8** Shear stress vs. temperature for crude oil sample 2.

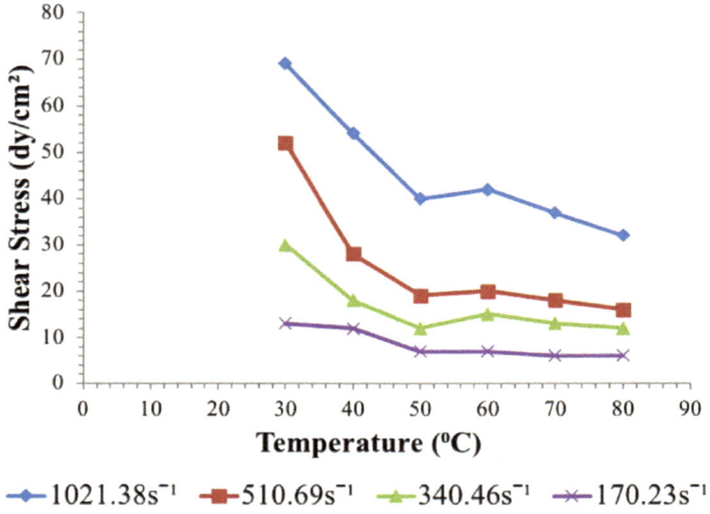


**Figure 15.9** Shear stress vs. temperature for crude oil sample 3.

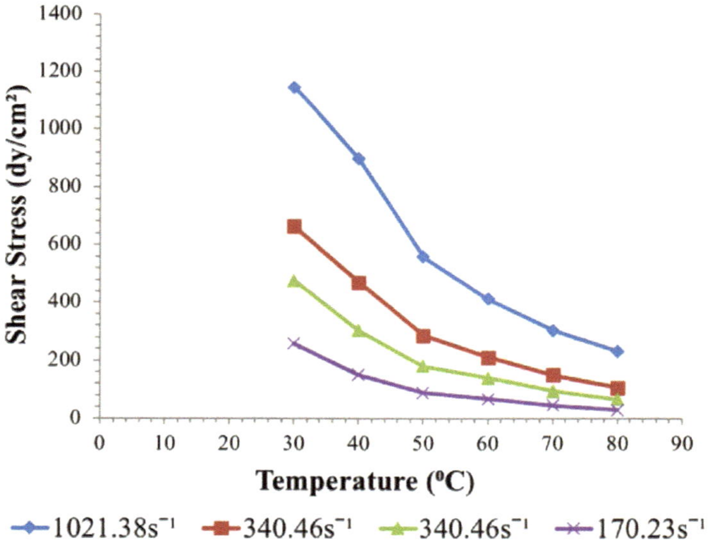


**Figure 15.10** Shear stress vs. temperature for crude oil sample 4.

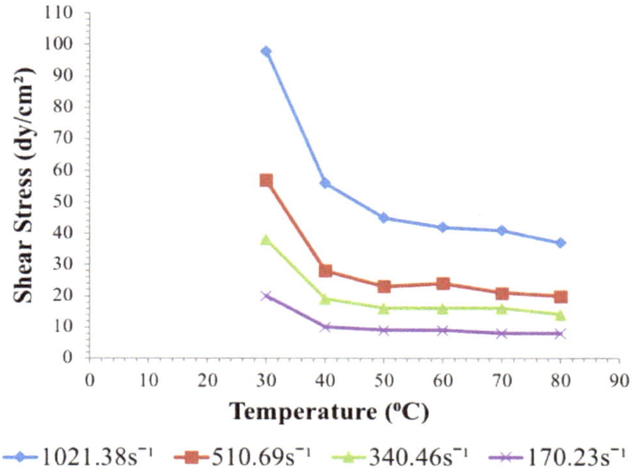


**Figure 15.11** Shear stress vs. temperature for crude oil sample 5.

### 15.3.4 Effect of Temperature on Viscosity

The relationship between viscosity and temperature at a constant shear rate for samples S1, S2, S3, S4 and S5 is plotted in Figs. 15.12–15.16. Upon analysing Figs. 15.12–15.16, it is observed that

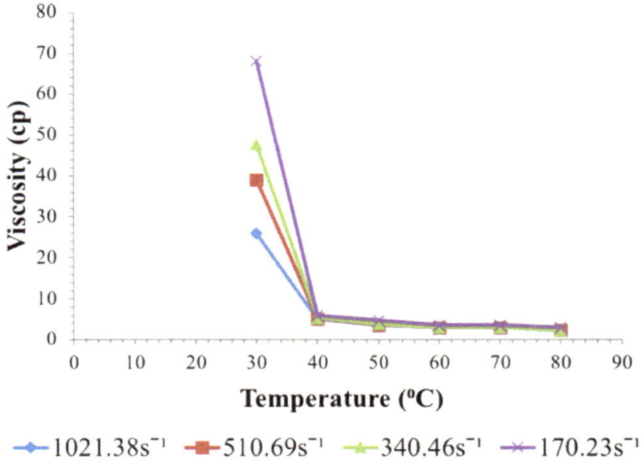


**Figure 15.12** Viscosity vs. temperature for crude oil sample 1.

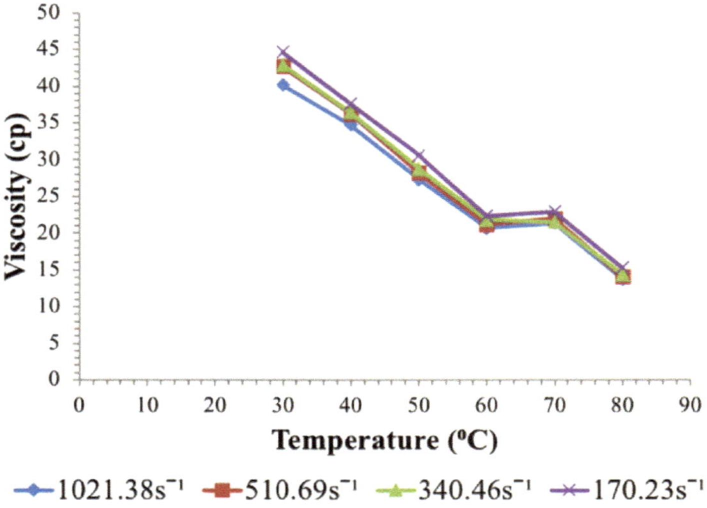


**Figure 15.13** Viscosity vs. temperature for crude oil sample 2.

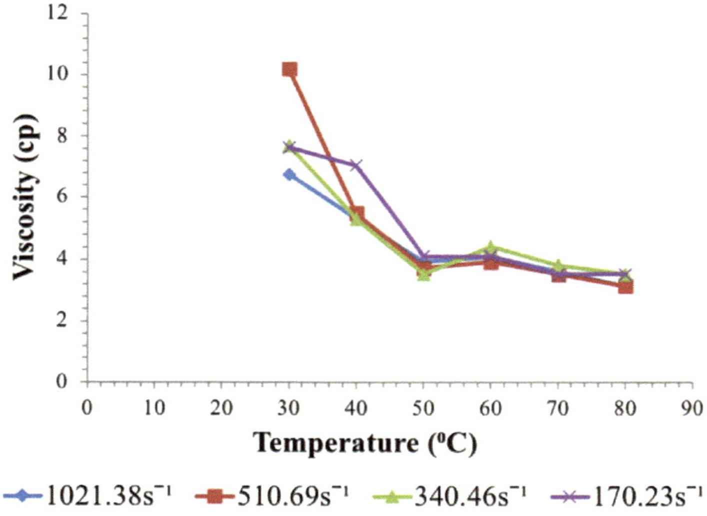


**Figure 15.14** Viscosity vs. temperature for crude oil sample 3.

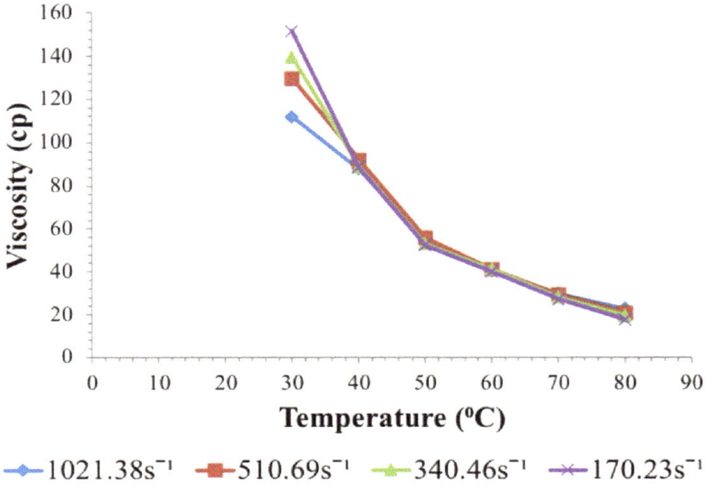


**Figure 15.15** Viscosity vs. temperature for crude oil sample 4.

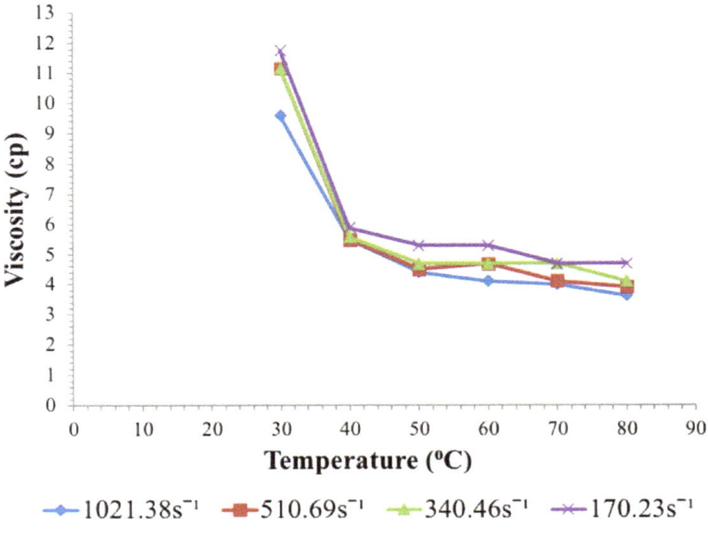


**Figure 15.16** Viscosity vs. temperature for crude oil sample 5.

as temperature increases from 30°C to 40°C, 50°C, 60°C, 70°C and 80°C viscosity decreases. It is also observed from the figures that viscosity is higher at a lower shear rate.

## 15.4 Conclusion

Viscosity behaviour of crude oil versus shear rate at different temperatures has been studied. Viscosities of the crude oil were examined over shear rate ranging from 0.1 to 1021.38 $s^{-1}$ and temperature ranging from 30°C to 40°C, 50°C, 60°C, 70°C and 80°C. The curves indicate that an increase in shear rate resulted in a decrease in viscosity. Increase in temperature of crude oil results in a significant decrease of viscosity. The viscosity reaches low values at a high shear rate because the flow encounters less resistance. At the moment of spindle rotations, the bonds of aggregated molecules start to be destroyed due to the energy exerted by shear being dissipated in the crude oil. As the shear rate increases, the molecules tend to break down into individuals; thus, resistance to spindle rotation is less and eventually reduced the viscosity. This phenomenon is also known as non-Newtonian shear-thinning behaviour. It is also proven from the graphics that the effects of viscosity differences are larger at a low shear rate than at a high shear rate. However, at a higher shear rate, viscosity decreases and exhibits a Newtonian profile. This occurs because with increasing shear rate, the molecules start to untangle from each other and start to align in the direction of flow. From the observation, viscosity is strongly affected by temperature.

Therefore, for smooth, continuous flow of crude oil through the pipeline, a low pour point is required. From the above discussions and observations, we can consider that we can easily transport samples having a low pour point and Newtonian flow behaviour.

## Acknowledgements

We would like to acknowledge the Department of Petroleum Technology and the Department of Petroleum Engineering of Dibrugarh University for providing laboratory facilities in undergoing this research work. We also sincerely acknowledge Oil India Limited (OIL), Duliajan, and Oil and Natural Gas Commission (ONGC), Nazira, for providing crude samples.

## Chapter 16

# Average Angle Method with Microsoft Excel for Directional Well Path Planning and Risk Monitoring

**Devtusti Rath and Prasun Banik**
*Department of Petroleum Engineering, DUIET, Dibrugarh University, Assam, India*
devtustirath@gmail.com, prasunbanik@dibru.ac.in

Directional drilling while has greatly increased the scope of oil and gas production from difficult and non-accessible areas, it has also greatly increased the chances of risks in extensive loss of investments, time and efforts, due to wrong drilling of the directional well. In an attempt to ensure accuracy in drilling a directional well, it is very essential to conduct surveys to ensure that the drill bit is moving in the right direction. The surveys along with well path calculation methodologies give the information as of whether the bit is moving in the correct direction or is drifting away from the required path.

Microsoft Excel, if employed to calculate the well path coordinates, can make the analyses of the wellbore trajectory easier and would bring access to such difficult computations from field experts to anyone who has basic knowledge on well path calculation methods and operation of Excel. Application of Excel at the field level

*Advances in Petroleum Technology*
Edited by Subrata Borgohain Gogoi

ISBN 978-981-4877-19-0 (Hardcover), 978-1-003-04993-7 (eBook)
www.jennystanford.com

can be close monitoring of the bit position at each survey station. The coordinates of the position of the drill bit at each directional survey station were found out by developing a user-friendly, easily modifiable and accurate method using Excel and Visual Basic for Excel.

## 16.1 Introduction

Directional drilling is the deviating of a wellbore along a planned course to a subsurface target whose location is at a given lateral distance and direction from the vertical (Fig. 16.1). Vertical wells may not be possibly constructed in areas which have hospitals and populated areas directly vertically above the target point. Directionally drilling of a well is the best way to exploit most of the pay zone, as a horizontally invaded pay zone gives maximum exposure to the reservoir in comparison to a vertical well. Also in offshore wells where the cost of drilling a well is very high and many platforms for drilling cannot be made, the directional approach to drilling is often considered. Reservoirs with very less vertical permeability in comparison to horizontal permeability can also be effectively exploited using directional or horizontal drilling.

$$\Delta\text{North} = \Delta\text{MD} \times \sin\left(\frac{I_1 + I_2}{2}\right) \times \cos\left(\frac{A_{z1} + A_{z2}}{2}\right)$$

$$\Delta\text{East} = \Delta\text{MD} \times \sin\left(\frac{I_1 + I_2}{2}\right) \times \sin\left(\frac{A_{z1} + A_{z2}}{2}\right)$$

$$\Delta\text{TVD} = \Delta\text{MD} \times \cos\left(\frac{I_1 + I_2}{2}\right)$$

where

MD = measured depth between surveys in feet
$I_1$ = inclination (angle) at upper survey in degrees
$I_2$ = inclination (angle) at lower survey in degrees
$A_{z1}$ = azimuth direction at upper survey
$A_{z2}$ = azimuth direction at lower survey

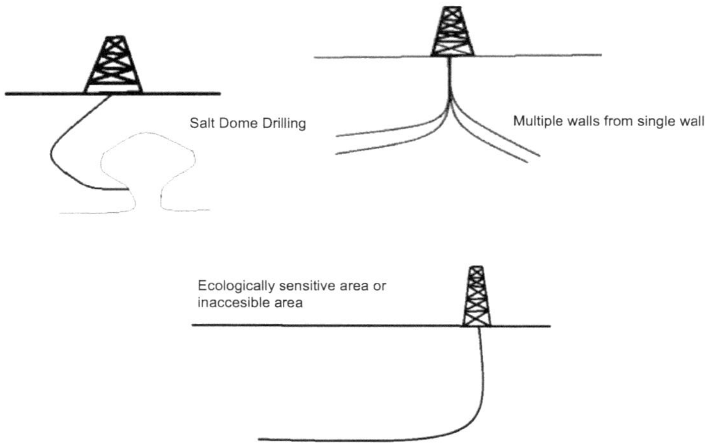


**Figure 16.1** Reasons for directional drilling.

### 16.1.1 Reasons for Directional Drilling

Directional drilling has been proven technically and economically feasible in a broad range of geologic settings, including tight gas, heavy oil and coal bed methane. This method is proven to substantially increase producible reserves of oil and gas. Some of the reasons for applications of directional drilling are as follows:

- Environmental issues
- To drill multiple wells from a single platform, mainly used in offshore
- Field development of offshore in deep waters or inaccessible locations
- To sidetrack any obstruction (fish) in the original wellbore
- To explore for additional producing horizons in adjacent sectors of the field
- To re-drill the well
- To put out fires resulting from blowout (relief wells)
- To drill to reservoirs avoiding inaccessible locations
- For salt dome drilling
- For fault controlling
- To get maximum productivity from the reservoir

### 16.1.2 Fundamental Concepts of Directional Drilling

In directional drilling some parameters are required in order to compute the position of the bit. A directional survey involves the measurement of such parameters. The depth, drift angle (inclination) and azimuth are measured [392]. Directional surveys involve employment of various tools that record such parameters downhole. Magnetic survey tools like magnetic single shot, magnetic multiple shot and Totco help in the recording of the parameters like azimuth and inclination, which can be later incorporated in directional well path calculations. Gyroscopic tools use a gyro instead of a magnetic compass, which helps eliminate the deflections in measurements due to magnetic noises. Another technique called measurement while drilling (MWD) helps in the recording of real-time bit movements inside the formations with the help of mud pulse telemetry [392].

#### 16.1.2.1 Directional surveying

The technique of measurement of the three components – depth, drift angle (inclination) and azimuth – is termed 'survey' [392].

*Depth measurement*: the course length which the drill bit has followed. A true vertical depth (VD) is the vertical distance between the surface point and the point the bit has reached.

*Inclination angle*: the angle between the vertical and a tangent to the well path at any point, or the point which the bit has reached. It is the angle measured in the horizontal plane between the direction of north and a point on the well path [393].

*Azimuth angle*: the angle measured in the horizontal plane between the direction of north and a point on the well path [393].

#### 16.1.2.2 Types of survey instruments

Basically, there are two types of survey instruments:

- Magnetic sensors can be run within the subsurface environment having less magnetic interference in order to obtain correct readings of a downhole survey [394]. There are two categories of magnetic sensors, mechanical compass and electronic compass.

- A mechanical compass uses a compass card, and a pendulum float device arrangement is employed in order to measure the inclination changes with respect to the magnetic north. In an electronic compass the bit position is measured by gravity accelerometers and fluxgate magnetometers [395].
- Gyroscopic sensors measure errors due to magnetic disturbances occurring due to the presence of magnetic elements downhole. Incorrect readings can be overcome by the use of gyroscopes. The principal element of the gyro is a rapidly spinning weighted wheel which maintains a present direction by resisting changes in direction, owing to its inertia [393].

#### 16.1.2.3 Methods of calculating wellbore trajectories

Wellbore surveys give the measurements of the parameters which can be later used to calculate the wellbore trajectory. This calculation of the coordinated position of the drill bit at each survey station can be done with the help of some wellbore trajectory calculation methods. Some of the methods are tangential, balanced tangential, mercury, angle averaging, radius of curvature and minimum radius of curvature [392, 396].

## 16.2 Average Angle Method for Directional Drilling Calculations

As the name implies, in the average angle method, inclination angles and direction angles are averaged and used for calculation of departure, total VD and easting and northing increments (Fig. 16.2). This method is simple and an accurate means of calculating wellbore surveys.

### 16.2.1 Calculations Involved

#### 16.2.1.1 Wellbore geometry using the average angle method

Figures 16.2 and 16.3 show the geometrical representation of a wellbore trajectory using the average angel method.

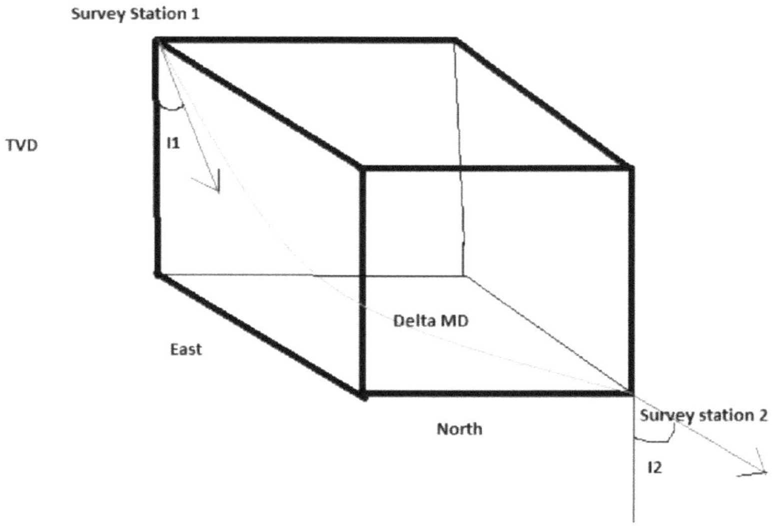


**Figure 16.2** Wellbore geometry using the average angle method.

## 16.3 Microsoft Excel Programme Development

### 16.3.1 Design

Excel (Fig. 16.4) is a very useful tool for decreasing the time and complexity of tedious and lengthy manual calculations. Data storage and manipulation can be easily done with the help of an Excel spreadsheet. Visual Basic for Excel (VBA) can be incorporated during programme development in order to construct user-friendly user forms which are less prone to errors due to mishandling and makes the Excel spreadsheet look more organised and interactive [392]. Development of an Excel spreadsheet employing the angle averaging method narrows the gap between an industry specialist and a layman trying to understand the bit positioning and directional drilling design. The developed Excel spreadsheet has two buttons opening two different user forms (Fig 16.4): one is 'enter your details' and the other is 'enter your plan'.

With the help of these two user forms, the VD, north coordinate and east coordinate of the position of the bit can be calculated

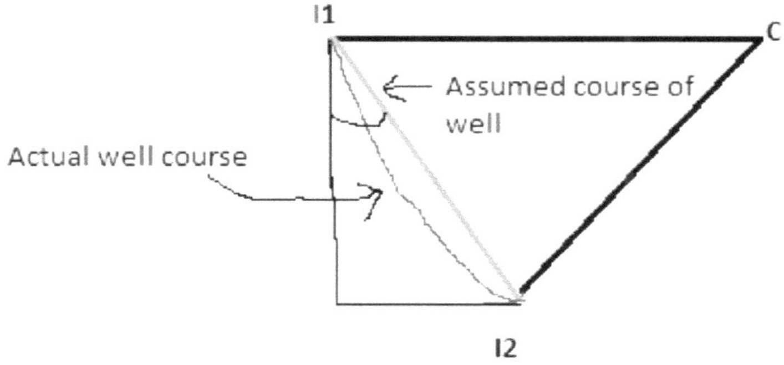


**Figure 16.3** Assumed course and actual coarse of wellbore using the average angle method.

at each survey station, which is then matched with the planned trajectory, which can give an idea of whether the bit is moving along the planned trajectory. Active X control buttons are designed in Excel which can be invoked to open the two user forms [397].

## 16.3.2 Introduction to Coding in Visual Basic

Visual Basic is a third-generation, event-driven programming language from Microsoft. Microsoft intended Visual Basic to be relatively easy to learn and use. Visual Basic was derived from BASIC and enables the rapid application development (RAD) of graphical user interface (GUI) applications; access to databases using Data Access Objects, Remote Data Objects, or ActiveX Data Objects; and creation of ActiveX controls and objects.

### 16.3.2.1 Field data interface

It allows the user to input the data (Fig. 16.5) about the field in which the survey was conducted. Record maintenance is very important to ensure segregation of the data generated for various different directional wells. This interface allows the user to enter the company name, field name, well no., reservoir engineer, reservoir fluid type and the date when the plan was implemented. The added data is displayed for reference on the right side of the screen.

**Figure 16.4** Developed Excel spreadsheet design.

### 16.3.3 Survey Input Data and Coordinate Generation Interface

This interface (Fig. 16.6) allows the user to input the values of VD, north coordinate and east coordinate at the kickoff point determined by the survey instrument. After taking measurements of inclination angles and azimuth angles at each station the user can add the values recorded. The programme calculates the changes and total VDs, changes and total east coordinates and changes and total north coordinates covered by the bit, using the average angle method. The computed coordinates can be matched with the coordinates of the required theoretical well path and also the coordinates of the target depth to ensure the correct drilling path and prevent unnecessary losses of resources and money.

**Figure 16.5** Developed Excel spreadsheet field data interface.

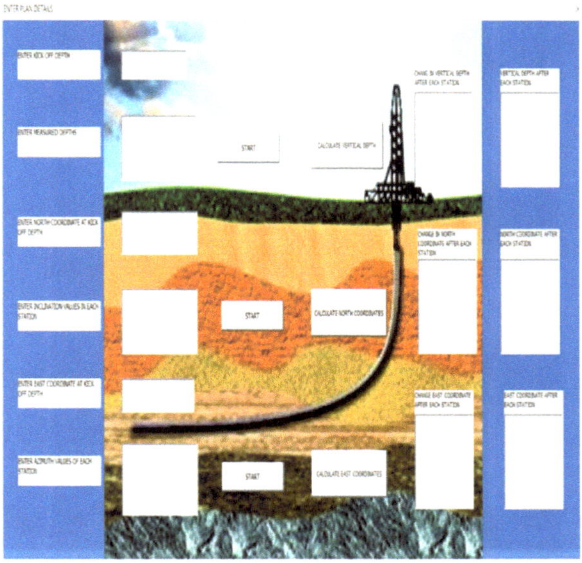


**Figure 16.6** Developed Excel spreadsheet and coordinate generation interface.

## 16.4 Results and Discussion

### 16.4.1 Input Data

The user form allows the user to enter data of the depth, north and east coordinates of the kickoff point (the point at which the well is deviated from the vertical), the measured depths and the inclination and azimuth values at each survey stations, as shown in Table 16.1.

**Table 16.1** Sample input data [393]

| | |
|---|---|
| North coordinates of kick-off point | 6241.6 feet |
| East coordinates of kick-off point | −8099.3 feet |
| Vertical depth of kick-off point | 9820.7 feet |
| Measured depth at station 1 | 15,000 feet |
| Measured depth at station 2 | 15,100 feet |
| Measured depth at station 3 | 15,200 feet |
| Measured depth at station 4 | 15,300 feet |
| Measured depth at station 5 | 15,400 feet |
| Hole inclination at station 1 | 0.9981 radian |
| Hole inclination at station 2 | 1.020 radian |
| Hole inclination at station 3 | 1.0225 radian |
| Hole inclination at station 4 | 1.0138 radian |
| Hole inclination at station 5 | 1.019 radian |
| Azimuth at station1 | 5.667 radian |
| Azimuth at station1 | 5.671 radian |
| Azimuth at station1 | 5.672 radian |
| Azimuth at station1 | 5.618 radian |
| Azimuth at station1 | 5.68 radian |

### 16.4.2 Programme-Generated Results

Tables 16.2–16.4 summarise the results obtained for coordinates (north and east) and VD using the developed Excel spreadsheet and their comparison with results obtained by calculation of the same, with the manual theoretical computational method and the resulting percentage of errors, for four survey stations.

**Table 16.2** Comparison of theoretically calculated value of vertical depth with programme-generated results for four survey stations

| S. no. | Theoretically vertical depth (ft.) | Programme vertical depth (ft.) | Error (%) |
|---|---|---|---|
| 1 | 9873.91 | 9873.966 | 0.0005 |
| 2. | 9926.085 | 9926.196 | 0.001 |
| 3. | 9978.557 | 9978.69 | 0.001 |
| 4. | 10031.03 | 10031.33 | 0.003 |

**Table 16.3** Comparison of theoretically calculated value of north coordinate with programme-generated results for four survey stations

| S. no. | Theoretically vertical depth (ft.) | Programme vertical depth (ft.) | Error (%) |
|---|---|---|---|
| 1. | 6310.8 | 6310.76 | 0.0006 |
| 2. | 6380.7 | 6450.39 | 1.09 |
| 3. | 6450.7 | 6655.47 | 3.1 |
| 4. | 6520.8 | 6929.43 | 6.26 |

**Table 16.4** Comparison of theoretically calculated value of east coordinate with programme-generated results for four survey stations

| S. no. | Theoretically east coordinate (ft.) | Programme east coordinate (ft.) | Error (%) |
|---|---|---|---|
| 1. | −8147.98 | −8148.0731 | 0.0011 |
| 2. | −8196.853 | −8196.8463 | 0.00008 |
| 3. | −8245.375 | −8245.619 | 0.002 |
| 4. | −8293.478 | −8294.39 | 0.010 |

## 16.5 Conclusion

The developed Excel spreadsheet was coded in Visual Basic which is accurate, simple, easy to understand and easy to modify. The user can click on the Start button and add all the measured depth values found in the survey station, and the programme will immediately calculate and display the change in the VD at each station, resulting in the concluding VD that the bit has reached. Similarly, the north and east coordinates of the position of the bit can be calculated using the developed spreadsheet. The results can be used to be compared to the planned course of the drilling path, and any anomaly observed can be prevented by changing the course of the bit, ensuring risk monitoring. The computed results were compared with manual theoretical calculations and showed substantial accuracy. The study revealed the possible use of the average angle method employed Excel for well path design by practising drilling engineers anywhere with the help of just a laptop or a desktop supporting Excel and a macro-enabled workbook. Future work can be done by incorporating field survey data for affirmation of Excel. Similar programmes can be developed by employing more accurate survey methods like the minimum curvature method for a more accurate wellbore trajectory calculation. With the advent of newer technologies like MWD and real-time wellbore monitoring, the uses of such geometrical commutations methods is getting lost. Development of such user-friendly Excel spreadsheets can help revive the essence of directional drilling without having to depend upon expensive software packages or monitoring systems and can ensure that every practising drilling engineer has convenient access to the aspects of directional wellbore design.

Chapter 17

# Using Digital Capabilities in Geoscience Software for Estimating the Resource Volume of a Reservoir Using the Map-Based Method: A Quick Evaluation Procedure

**Nitin Lahkar[a] and Shamina Sheena[b]**
[a] *Oil India Limited, Noida, Uttar Pradesh, India*
[b] *Schlumberger Asia Services Ltd., Gurugram, Haryana, India*
nitin_lahkar@oilindia.in

In petroleum geology and petroleum engineering work, various methods have been used for reserve estimation. Geoscience software has provided a major boost in these methods in bridging the gap between classical paperwork and calculations for static modelling to calculate the in-house estimates and to an extent recoverable reserve. But sometimes due to cut-throat competition and tight timelines for the potential investor, this job requires an in-between approach and a quick-look method for calculating the volumetrics without affecting the accuracy of calculation. Here, the concept of utilising the capabilities of digitisation and map-based volumetrics comes into play. In addition, often there may be a lack of

*Advances in Petroleum Technology*
Edited by Subrata Borgohain Gogoi

ISBN 978-981-4877-19-0 (Hardcover), 978-1-003-04993-7 (eBook)
www.jennystanford.com

sufficient data to create a proper static model. Even in this situation, the method described in this chapter can come into the forefront as an added bonus. In this chapter, the authors have focussed on solving the challenges mentioned above using the industry-standard Geoscience software. What was required for this work basically is a hard copy or picture of a structure map of the area/reservoir top/bottom surface and basic data of some wells drilled to date in that area (optional). The map was uploaded as an image in Geoscience software, and then it was digitised/contoured perfectly by the authors using the spectrum of digital capabilities of the software. The authors then used the map-based volume feature of Geoscience software to calculate the entire volumetrics of the reservoir. The entire process has been vividly illustrated in this paper. Later, it was observed that this method gave results exactly in line with those using a 3D static model, that too in much lesser time. If sparse well data is available, then property maps of the reservoir, that is, porosity map, saturation map, net-to-gross (NTG) map, etc., can be prepared in the software and input into the map-based volume calculation. Even if there is no well data available and only a hard copy of a map is available, even then a rough estimate of the volume can be obtained using fixed petrophysical parameters, as has been done in this case.

## 17.1 Introduction

Volume calculation of oil and gas reservoirs is one of the primary goals of interest when we are dealing with exploration or production of oil and gas in general. Today, volume estimation of geological structures, that is, hydrocarbon (HC) reservoir volume calculation, is generally done using various modelling software programmes. The reserves calculation method based on the high precision of 3D reservoir geological mode can be used to deal with the issues related to precisely capturing reservoir petrophysical parameters when estimating complex reservoir reserves in terms of the traditional volume method. With the rapid development of software technology, 3D modelling of petroleum reservoir has been the basic method of the petroleum industry making a reservoir description. Three-dimensional geological modelling of the reservoir can reflect the characters of distribution and regularities of variation in the reservoir geological space attribute, and this provides assistance in petroleum prospect development [398, 399].

## 17.2 Concept and Need of Screening Volumetrics in Geoscience Software for Quick Resource Evaluation by the Map-Based Method

A knowledge of the possible range of HC volumes in an exploration lead or prospect is indispensable to evaluation of the economics of a project; however, there is seldom sufficient data (or time) for a potential investor/partner/operator to perform detailed volumetric calculations, such as in a 3D geologic or simulation model. Simple screening methods may provide some information to take decisions, for example, to bid on a block, rank prospects for drilling or participate in a farmout. Other useful options in volumetric operations include creation of area or volume/depth functions and polygons around closure areas. The Geoscience software of today's era offers several methods to generate quick volumetric estimates using maps, the hard copies of which are generally available for a prospect – at the minimum, a depth surface and a contact. The method selected will depend on the amount of additional data available and the project's requirements for more or less detail and accuracy. In this scenario, as is evident, the map-based method can act as a boon. This method is significantly advanced. Besides calculating a variety of volumetric quantities, such as recoverable HCs for multiple zones, it can create property height maps and even perform rudimentary uncertainty analyses. This method is functionally similar to Petrel's 3D volumetric calculations and includes several enhancements compared to the simple methods. For each zone:

(1) Input of constant values or maps for contacts of different types (oil–water contact [OWC], gas–oil contact [GOC], etc.)

(2) Net-to-gross (NTG) and porosity maps or constants as input
(3) Oil, gas and water saturation maps or constants
(4) Surface conditions (e.g., Bo) and recovery factors
(5) Validation of surface geometry (checks for crossovers)
(6) Uncertainty workflows for contact depth, NTG and porosity (optional)

The output options are also enhanced, generating formatted spreadsheet reports, volume height maps and area/volume versus depth functions. Its uncertainty workflows can be used to evaluate the effects of variation in porosity, NTG and structure, or for a range of contact depths. In general, the Input panel is where the user can specify multiple zones, contacts and other properties. Each zone may have a different set of contacts and different properties. Properties are specified as either constants or maps – for example, porosity or NTG may have been mapped from log values. Properties including bulk, net or pore volume (PV), stock tank oil initially in place (STOIIP) or gas initially in place (GIIP) and recoverable oil or gas can be output as a formatted report, which can be cut and pasted into a spreadsheet. Results can be grouped by contact sets, zones or boundaries. Reporting units can also be selected, such as barrels or cubic meters. Output options include maps of selected volumes, which can be displayed and used as per requirement: This type of display can be extremely crucial to visualise 'sweet spots' in structural or stratigraphic plays. Another useful feature of map-based volumetric is its uncertainty workflow. By activating options to evaluate the effects of incremental contact depths or variability in NTG or porosity for each zone, multiple cases can be generated for further analyses, including histograms to display the range of potential outcomes [400].

## 17.3 Methodology and Work Done

The field for the study was located in the Volga–Urals Basin, Russia. Few details about the field are given in Table 17.1.

**Table 17.1** Relevant information of the field in study

| Parameter | Details |
|---|---|
| Basin | Volga–Urals Basin, Russia |
| Hydrocarbon type | Oil, gas |
| Discovery year | 1942 |
| Reservoir age | Devonian |
| Depositional environment | fluvial |
| Reservoir major unit | Pashiyskiy formation |
| Drive mechanism | Water drive |
| Average gross pay | 8.67 m |
| Average net pay | 4.26 m |
| Average porosity | 15% |
| Average permeability | 248 mD |
| Maximum reservoir temperature | 46°C |
| Trapping mechanism | Strati-structural |
| Source rock | Ardatovskiy formation |
| Source rock age | Devonian |
| Source rock lithology | Bituminous shale |

## 17.3.1 Map Loading

A hard copy of the structure contour map of the reservoir top is shown in Fig. 17.1. The map was first scanned using an office scanner and loaded into Geoscience software using the coordinates

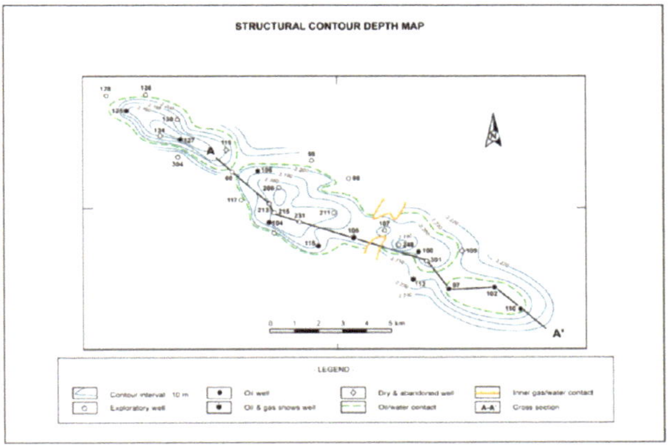


**Figure 17.1** Loading of structure contour map of reservoir top into Geoscience software.

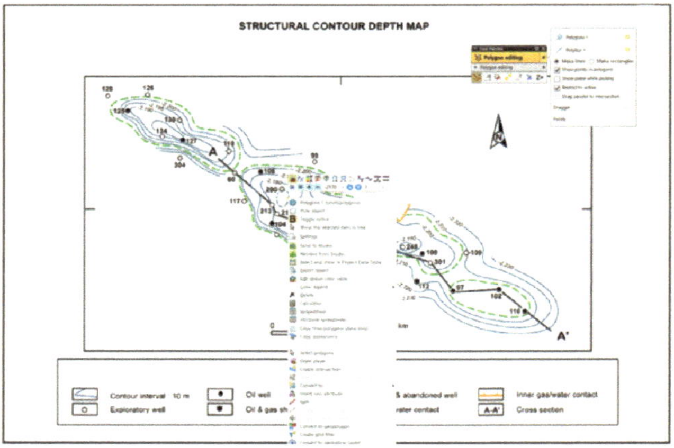


**Figure 17.2** Digitising of the scanned map.

given in the map. (Since a scanned map cannot be used directly in any geoscientific software, the authors needed to digitise the map for enabling it to be in a software-readable format.) All Geoscience software may not have a map-digitising feature. Hence, the operator or investor has to ensure that the purchased Geoscience software must have this feature for calculating map-based volumes.

## 17.3.2 Digitising the Map Using Geoscience Software

After loading the map in the software, the authors did a quality check to ensure the map was in right proportions with respect to the original map. After the quality check was complete the authors started digitising the contours using the polygon feature. Each contour was digitised as a polyline under a common polygon, as shown in Fig. 17.2.

It is to be noted that the authors set the depth of each contour after digitising it. Hence, the depth of each contour was also stored in the software parallel to the storage of contour points and contour lines in the software. Before making the surface, the authors did a quality check of the contour points along with ensuring that the depth range of the digitised contours was exactly the same as that of the hard copy of the map, as shown in Fig. 17.3.

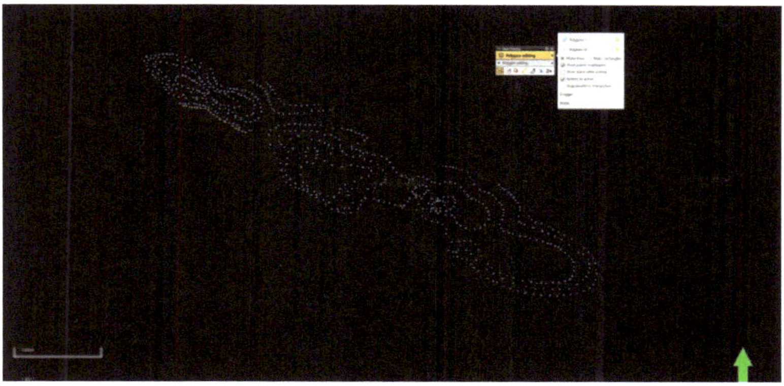

**Figure 17.3** Digitised contour points in a 2D window.

### 17.3.3 Making the Reservoir Top Surface

After the contour points were ready, the authors started making the reservoir top surface. The contour points were used as the major input for making the surface. Then an appropriate algorithm was applied to make the surface. The selection of an algorithm depends on a number of points, like availability of data, geometry of the trap and analogy. Some minor smoothening of the data was also conducted. Finally, after running the algorithm and smoothening process, the reservoir top surface was ready, as shown in Fig. 17.4.

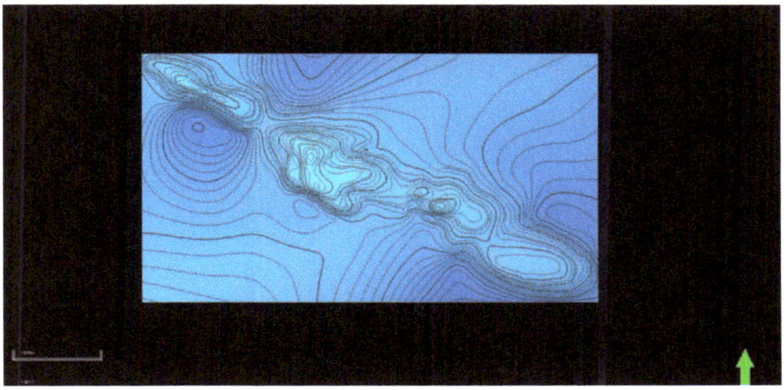

**Figure 17.4** Reservoir top surface.

The same surface was visualised by the authors in a 3D view for better, enhanced understanding of the structure, as shown in Fig. 17.5.

**Figure 17.5** 3D view of the structure in Geoscience software.

Then the authors did some further finishing touches and colour adjustments, as shown in Fig. 17.6, to arrive at the industry-accepted form of the reservoir top. The OWC depth was available to the authors. So, the OWC was also digitised as a separate contour.

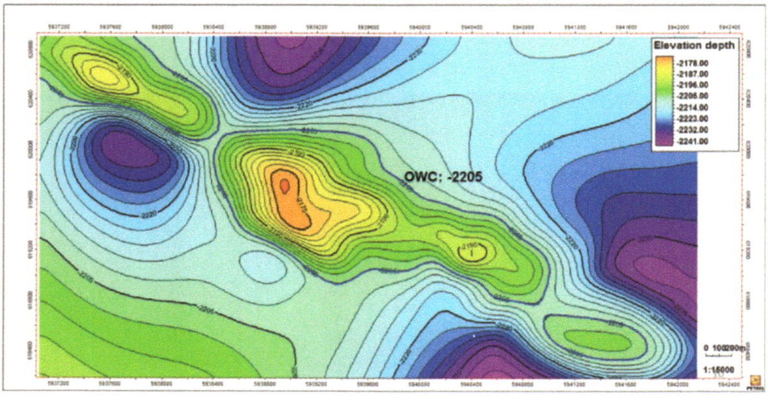


**Figure 17.6** Final reservoir top in a map window.

### 17.3.4 Prospect Differentiation

The three different prospects were also marked by the authors as separate polygons. This was done to enable volume calculation separately for the three prospects later. However, a polygon covering all the three prospects was also made and retained, as shown in Figs. 17.7 and 17.8.

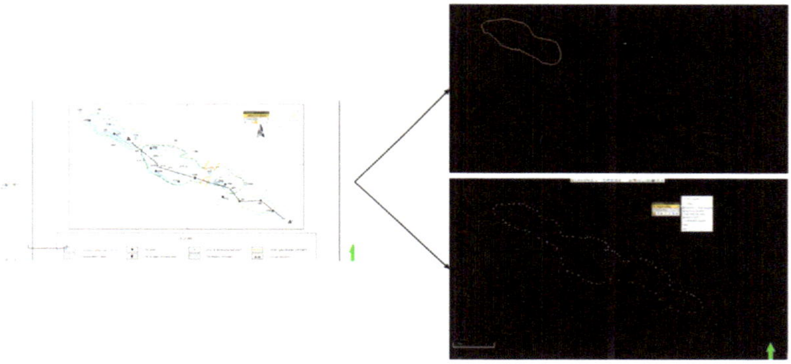

**Figure 17.7** Digitising and differentiating the three prospects.

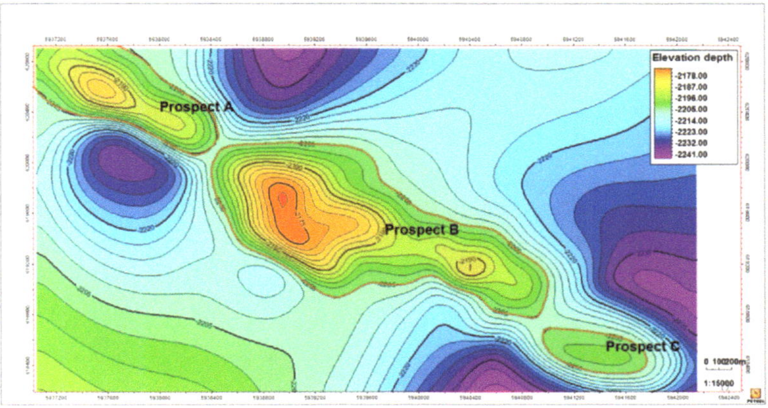


**Figure 17.8** Map showing the three prospects.

### 17.3.5 Volume Calculation

After finalising the reservoir top the authors started the process of volume calculation. The reservoir top surface was given as the main input. Both NTG and porosity options were being considered for volume calculation. Due to the absence of individual well data a fixed value of OWC, NTG, porosity, saturation of water ($S_W$), oil formation volume factor (Bo), etc., were given as inputs. It is to be noted that when some well data is available, it can be loaded in Geoscience software and property maps can be generated and put as input during volume calculation. The same has been done by authors on numerous occasions when such data was available. The required outputs as desired by the authors in the final output sheet were selected. Again, the requisite options as per the maps desired as output by the authors were selected. Figure 17.9 shows the input summary.

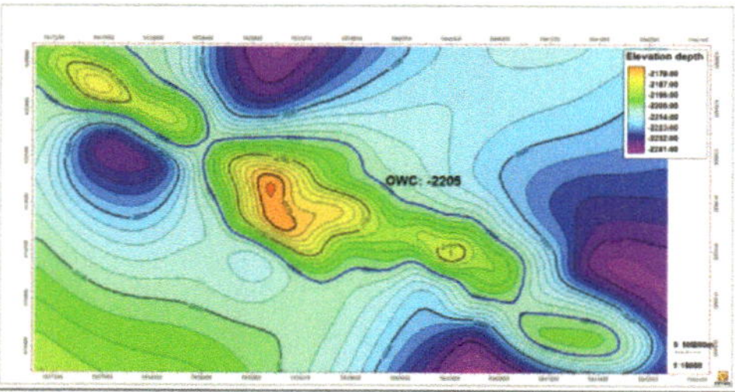


Input data

| Zone(s) | Top surface | Base surface/Isochore | Net-to-gross | Porosity | Contact Oil |
|---|---|---|---|---|---|
| | | -2205.00 | 0.4900 | 0.1600 | -2205.00 |

Properties in Oil interval

| Zone(s) | Sat. water | Sat. oil | Sat. gas | Bo [rm3/sm3] | REC | Rs [sm3/sm3] |
|---|---|---|---|---|---|---|
| | 0.20000 | 1-Sw-Sg | 0.00000 | 1.1340 | 0.4000 | 150.0000 |

**Figure 17.9** Showing input summary.

| Case | Bulk volume [*10^6 bbl] | HCPV oil [*10^6 rm3] | STOIIP [*10^6 bbl] | GIIP [*10^6 sm3] | Recoverable oil [*10^6 bbl] |
|---|---|---|---|---|---|
| Prospect_A | 38 | 0.38 | 2.11 | 50 | 0.84 |

| Case | Bulk volume [*10^6 bbl] | HCPV oil [*10^6 rm3] | STOIIP [*10^6 bbl] | GIIP [*10^6 sm3] | Recoverable oil [*10^6 bbl] |
|---|---|---|---|---|---|
| Prospect_B | 145 | 1.45 | 8.03 | 192 | 3.21 |

| Case | Bulk volume [*10^6 bbl] | HCPV oil [*10^6 rm3] | STOIIP [*10^6 bbl] | GIIP [*10^6 sm3] | Recoverable oil [*10^6 bbl] |
|---|---|---|---|---|---|
| Prospect_C | 3 | 0.03 | 0.15 | 4 | 0.06 |

| Case | Bulk volume [*10^6 bbl] | HCPV oil [*10^6 rm3] | STOIIP [*10^6 bbl] | GIIP [*10^6 sm3] | Recoverable oil [*10^6 bbl] |
|---|---|---|---|---|---|
| Totalprospect | 186 | 1.85 | 10.29 | 245 | 4.12 |

**Figure 17.10** Output summary.

After running the map-based volume calculation using the reservoir top surface made by the software and the inputs provided as above, the output sheet was generated with the requisite volumes, like STOIIP, hydrocarbon pore volume (HCPV), GIIP and recoverable oil. Figure 17.10 shows the output summary.

## 17.3.6 Uncertainty Analysis

Since there was uncertainty in the porosity and NTG values, an uncertainty analysis was done by the authors to arrive at a range of values for STOIIP and recoverable oil. Figures 17.11 and 17.12 show uncertainity analysis and calculations, respectively.

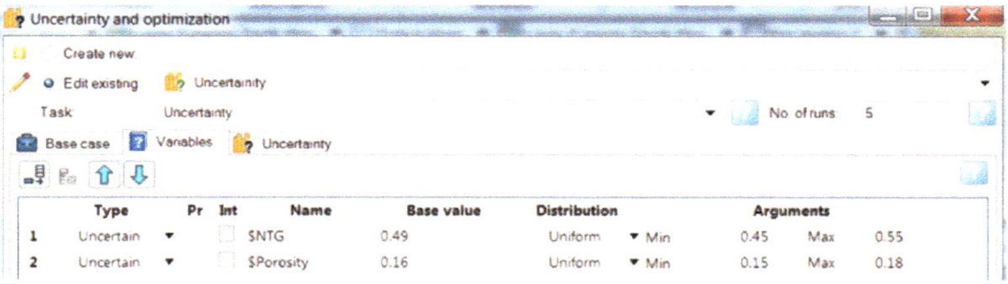


**Figure 17.11** Uncertainty analysis of volume calculations.

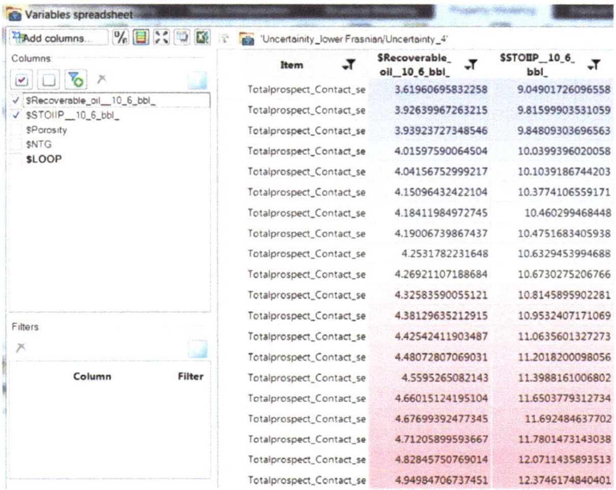


**Figure 17.12** Calculations.

## 17.4 Conclusion

The authors successfully completed map-based volume calculation utilising the digitisation capabilities of Geoscience software. It is believed that this method can be very useful for investors who are looking into entering new blocks, but data availability is very limited and the investor has to evaluate multiple blocks in very small window of time. This chapter will act as a ready reckoner for individuals who want to do map-based volume calculations in Geoscience software. The digitisation capabilities of Geoscience software have now made possible things which were not possible earlier and non-availability of digital maps meant that volume calculation using software was almost next to impossible. The scenario is different today.

## Acknowledgements

The authors acknowledge the technical guidance and support provided by Oil India Limited and Schlumberger, without which this research work would not have been possible.

## Chapter 18

# Chemical Consolidation as a Sand Control Method in Oilfields of the Upper Assam Basin

**Aditi Gogoi and Pradip Borgohain**
*Department of Petroleum Technology, Dibrugarh University, Assam, India*
rs_aditigogoi@dibru.ac.in

Sand production along with formation fluids due to the unconsolidated nature of the formation is a major problem in the oil and gas fields of Upper Assam. The excessive ingression of sand in producing wells leads to a decrease of crude oil production, damages downhole and surface tools, sand disposal issues, need of routine clean outs and stuck wells. It also decreases the economic-producing life of the well. The pressure at which a well will produce sand free from the reservoir is known as the threshold pressure. This threshold pressure is usually ignored by oil-producing companies to maximise the production rate from the reservoir. This results in increasing sand production, and so there is a need for improved sand control methods. The conventional methods of controlling the sand production is mostly through mechanical means such as gravel packs, stand-alone sand screens and hydraulic fracturing by providing a barrier in the wellbore and perforations.

*Advances in Petroleum Technology*
Edited by Subrata Borgohain Gogoi

ISBN 978-981-4877-19-0 (Hardcover), 978-1-003-04993-7 (eBook)
www.jennystanford.com

Most of the oilfields of the Upper Assam Basin are in the maturity phase of their producing life, and so there is an increase of water cut with high-pressure drawdown across the sand face. So these conventional methods are incapable of controlling fine sands of the formation. So, an alternative method of chemical consolidation treatment is necessary to minimise the sand production from the ageing oilfields of the Upper Assam Basin. The chemical sand consolidation technique is a sand control mechanism through which the sand is arrested within the reservoir in contrast to any mechanical sand control method in which sand is filtered at the formation wellbore.

## 18.1 Introduction

It is seen that 70% of the total oil and gas reserves in the world are located in poorly consolidated reservoirs [401, 402]. Poorly consolidated reservoirs have fines and sand migration problems. The fines movement can reduce the near wellbore fluid velocities. Hence, reduced fines migration can lead to better cumulative production. Recently, it was observed that the Tipam oilfield of the Upper Assam Basin was facing fines ingression problem. The conventional method of controlling the sand production by a gravel pack was not able to control the fines production, so the chemical consolidation method was taken into account. Various problems such as erosion of valves and pipelines, sand accumulation in the separators and plugging of the production liner are caused by sand influx into the wellbore. Cleaning and repair works related to sand production due to production rate restriction amounts to great costs incurred by the industry every year.

Since the early 1940s, resin consolidation has been used in the industry as a sand control method [403, 404]. To strengthen the formation in the immediate area of the wellbore with a chemical that will bind together the partially consolidated or unconsolidated grains at their point of contact is the main purpose of resin consolidation. The process involves mixing a binding agent, pore-retaining agent and curing (catalyst) agent in a uniformly specific ratio. The mixture is injected at the exterior of the casing and positioned at the sand-prone zone [134, 405]. The resin consolidation processes have been classified in various ways: minimum preparation time at the well site, low injection pressure, short cure time before restoring the well to production, high compressive strength of the resulting matrix, good resistance to deterioration from well fluids and commonly used treating fluids and high retained permeability are desirable characteristics for a consolidation process [406]. Several types of resins are presently used in the sand control art. Examples of hardenable organic resins which are suitable for use in accordance with this subject are epoxy resins, polyester resins, phenol–formaldehyde resins, urea–formaldehyde resins, furan resins, urethane resins and mixtures of such resins.

## 18.2 Experiments

### 18.2.1 Materials

A core sample and sand collected from the Tipam oilfield of the Upper Assam Basin were used. For the consolidation process of the produced sand, Tackifier was used as the consolidating agent.

### 18.2.2 Experimental Setup

Figure 18.1 shows a schematic diagram of the test setup used in this work. This model consisted of a core holder of length 28 cm and diameter 1.5 inch (3.81 cm) packed with sand grains by hammering to ensure uniform packing. A positive displacement pump was used for the injection of resins and saturating the sand pack with brine and crude oil.

### 18.2.3 Methodology

Sieve analysis of the sand sample was done by first weighing 100 g of the sand and then putting separately the coarsest sieve at the top and the finest at the bottom bearing ASTM nos. 25 (0.707 mm), 35

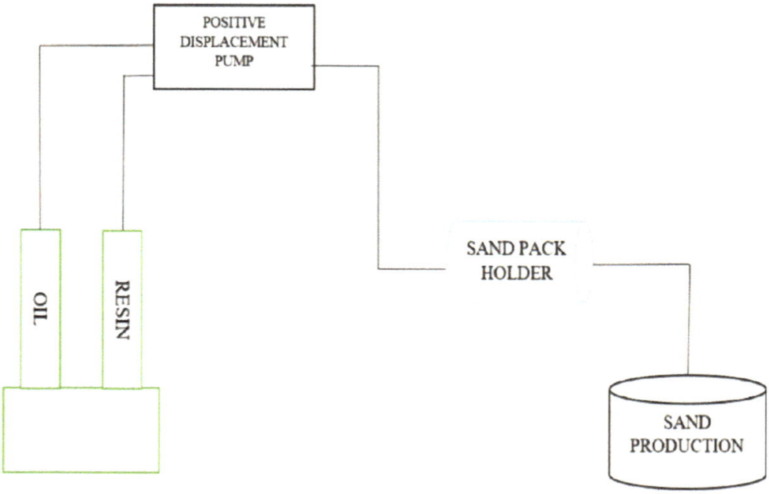


**Figure 18.1** Sand pack flooding apparatus.

(0.500 mm), 45 (0.354 mm), 60 (0.250 mm), 80 (0.177 mm), 120 (0.125 mm) and 170 (0.088 mm) arranged in descending sequence on the mechanical sieve shaker which provided vibration to the sieves. At the end of the analysis, the weights of the retained samples at the sieves were recorded to be used to generate sieve curves. X-ray diffraction (XRD) analysis of the core sample and produced sand sample was carried to find out clay minerals. Rock thin slide analysis of the core was performed using a Leica petrological microscope to study the clay mineralogy. A sand pack flooding experiment was done and the materials used for consolidation were Tackifier and isopropyl alcohol (IPA).

The following method was followed for the flooding experiment:

(i) A sand pack was prepared with dry sand into a core holder and hammered to ensure uniform packing.
(ii) 2% KCl solution was pumped to saturate the pack by applying a back pressure of 14 bar.
(iii) 200 ml of resin solution (Tackifier and IPA) was pumped through the apparatus.

(iv) Diesel was pumped after the solution at practically zero pressure.
(v) The core holder was kept in oven at 80°C for 24 h.
(vi) 2% KCl solution was pumped again through the core holder to saturate it.

Darcy's equation was used to calculate the permeability of the sand pack.

Before consolidation:

$$k_1 = \frac{\mu q L}{\Delta P A}$$

$$k_1 = 2.133 D$$

where $\mu_{\text{brine}} = 1.01$ cP, $q = 10$ cm$^3$/min, $L = 30$ cm, $A = 11.4$ cm$^2$ and $\Delta P = 3.053$ psi $= 0.207$ atm.

After consolidation:

$$k_2 = \frac{\mu q L}{\Delta P A}$$

$$k_2 = \frac{\mu q L}{\Delta P A}$$

$$k_2 = 1.628 D$$

where $\mu_{\text{brine}} = 1.01$ cP, $q = 10$ cm$^3$/min, $L = 30$ cm, $A = 11.4$ cm$^2$, $\Delta P = 4$ psi $= 0.272$ atm and increased permeability $= (k_1 - k_2)/k_2 = 31\%$.

## 18.3 Results and Discussion

### 18.3.1 Sieve Analysis

From Fig. 18.2, it can be concluded that the core sample was dominated mostly by the fines. Also, by looking at the curve trend it is evident that the fine grains were moderately sorted, whereas the coarse tail end was poorly sorted.

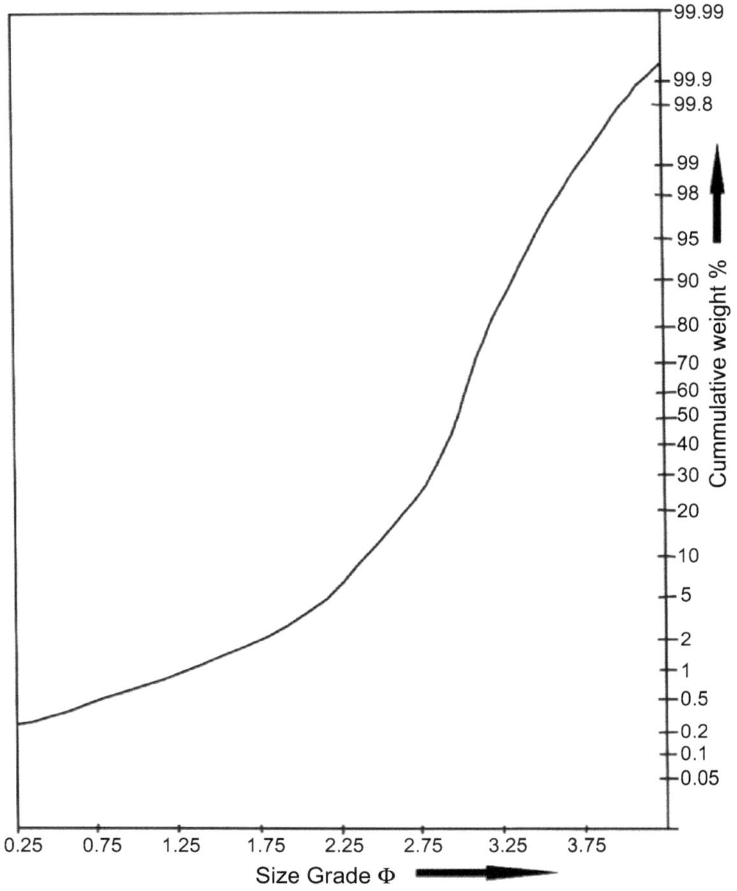


**Figure 18.2** Cumulative plot of the sand sample.

### 18.3.2 Rock Thin Slide Analysis

From Figs. 18.3 and 18.4, it was found that the sandstones in the core sample were composed mainly of quartz, feldspar and rock fragments with calcareous and siliceous cementing material. The matrix percentage was high with a low percentage of cements, which is vulnerable to the migration of sands with increased water saturation.

Figure 18.3 Photomicrographs of Tipam reservoir sandstones.

Figure 18.4 Photomicrographs of Tipam reservoir sandstones.

From Figs. 18.5 and 18.6, it is seen that both the core sample and the produced sand sample contained kaolinite, chlorite and illite as clay minerals. All these clays are migratory in nature, and with an increase in water saturation due to coning/ cusping or edge water encroachment, the clay matrix gets loosened and disintegrated, thereby making sand and clay mobile.

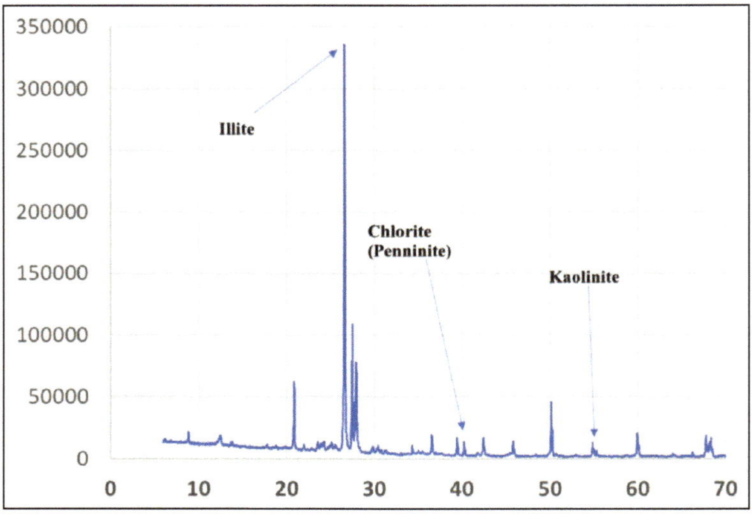


**Figure 18.5** XRD analysis of the core sample.

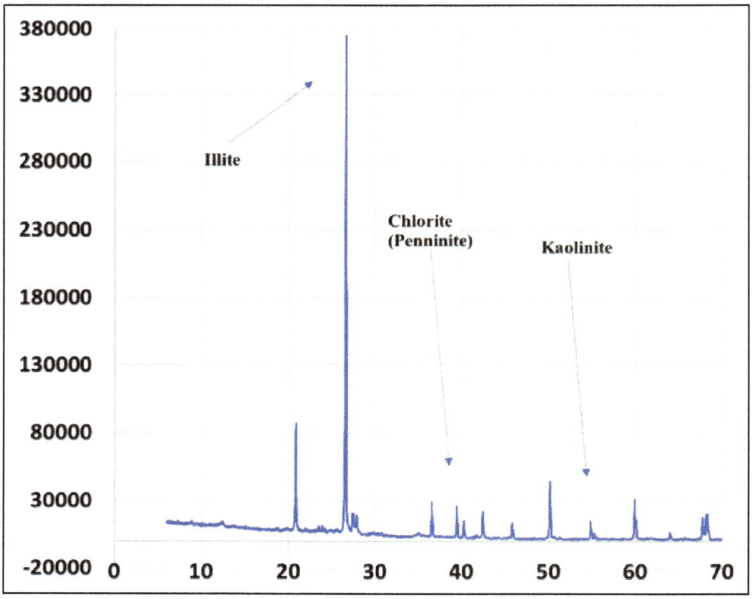


**Figure 18.6** XRD analysis of the produced sand sample.

## 18.4 Conclusion

Since fines are dominant in the core sample, sand production cannot be controlled by gravel pack and so chemical consolidation is necessary. The sandstones found in the core sample constitutes a low percentage of cementing agent which is mainly constituted of calcareous and siliceous matters and the clay minerals present are migratory in nature. An increase in water saturation due to coning or edge water encroachment loosens and disintegrates the clay matrix, which results in mobilisation of sand and clay particles. A high fluid withdrawal rate results in premature breakthrough of water due to coning, cusping and fast edge water encroachment, which results in an increased sand cut problem. So, an alternative method of chemical consolidation treatment is necessary. From Fig. 18.7, it was seen that after the resin solution was flooded through the sand pack, it showed an increase of 31% in permeability. Hence, Tackifier gave a good consolidation, enhancing the permeability of the sand pack.

**Figure 18.7** Sand pack after chemical consolidation.

## Acknowledgements

We would like express our sincere gratitude to the Institute of Oil and Gas Production Technology, Panvel; Central Sophisticated Instrumentation Centre, Dibrugarh University; and the Department of Petroleum Technology, Dibrugarh University, for providing us laboratory facilities.

## Chapter 19

# A Review of Microbial Degradation of Petroleum Hydrocarbons

**Amarjit Rajbongshi and Subrata Borgohain Gogoi**
*Department of Petroleum Technology, Dibrugarh University, India*
amarjit.rajbongshi@gmail.com

Petroleum is the main source of energy for industry and daily life. It is a liquid mixture of complex hydrocarbons (HCs). Some HC components have been known to belong to the family of carcinogens and neurotoxic organic pollutants. Accidental release or HCs during activities related to the petroleum industry is a potential threat to the environment. The petroleum industry also handles a huge volume of produced water (PW) during the exploration and production of crude oil which is contaminated with toxic HCs. PW is a complex mixture of organic and inorganic compounds. Nowadays, bioremediation is used extensively for the degradation of HCs, leading to their partial or complete mineralisation into carbon dioxide, water and inorganic compounds. The bioremediation process is a promising approach as it is cost-effective and has abundant availability of raw materials over mechanical and chemical methods. This chapter covers an overview of common HC pollutants, microbes known as HC degraders, biosurfactant-producing microorganisms, which helps in enhancing HC degradation.

*Advances in Petroleum Technology*
Edited by Subrata Borgohain Gogoi

ISBN 978-981-4877-19-0 (Hardcover), 978-1-003-04993-7 (eBook)
www.jennystanford.com

## 19.1 Introduction

Oil and gas are the major sources of energy for industry and daily life. Leaks and accidental spills occur regularly during the exploration, production, refining, transport and storage of petroleum, and their products lead to hydrocarbon (HC) contamination of the environment. One of the major environmental problems today is HC contamination resulting from the activities related to the oil and gas industry. Accidental releases of petroleum products are of particular concern in the environment, and these petroleum components have been known to belong to the family of carcinogens and neurotoxic organic pollutants. The technology commonly used for the soil remediation includes mechanical, burying, evaporation, dispersion and washing. However, these technologies are expensive and can lead to incomplete decomposition of contaminants. Microbial degradation is the major and ultimate natural mechanism by which one can clean up the petroleum HC pollutants from the environment [407, 408]. The recognition of biodegraded petroleum-derived aromatic HCs in marine sediments was reported by Jones et al. [409], who studied the extensive biodegradation of alkyl aromatics in marine sediments which occurred prior to detectable biodegradation of the n-alkane profile of the crude oil, and microorganisms, namely *Arthrobacter, Burkholderia, Mycobacterium, Pseudomonas, Sphingomonas* and *Rhodococcus* were found to be involved for alkylaromatic degradation. Microbial degradation of petroleum HCs in a polluted tropical stream in Lagos, Nigeria, was reported by Adebusoye et al. [410]. Nine bacterial strains, namely *Pseudomonas fluorescens, P. aeruginosa, Bacillus subtilis, Bacillus* sp., *Alcaligenes* sp., *Acinetobacter lwoffi, Flavobacterium* sp., *Micrococcus roseus* and *Corynebacterium* sp. were isolated from the polluted stream which could degrade crude oil [411]. Various industries and different daily life processes use petroleum-based products as the major sources of energy. Leaks and accidental spills occur frequently during transportation, production, exploration, refining and storage of petroleum and its derivatives. The amount of natural crude oil seepage was estimated to be 600,000 metric tons per year with an uncertainty of 200,000 metric tons per year [412a]. The success of

**Table 19.1** Major oil spills [412b]

| S. no. | Oil spill | Date | Site | Oil spilled (million gal.) |
|---|---|---|---|---|
| 1 | Arabian Gulf, Kuwait | January 19, 1991 | Persian Gulf, Kuwait | 382–520 |
| 2 | Gulf oil spill | April 22, 2010 | Gulf of Mexico | 206 |
| 3 | Ixtoc 1 oil spill | June 3, 1979 | Bay of Campeche, Mexico | 140 |
| 4 | *Atlantic Empress* oil spill | June 19, 1979 | Coast of Trinidad and Tobago | 90 |
| 3 | Kolva river oil spill | August 6, 1983 | Kolva river, Russia | 84 |
| 4 | Nowruz oilfield spill | February 10, 1983 | Persian Gulf of Iran | 80 |
| 5 | *Castillo de Bellven* oil spill | August 6, 1983 | Saldanha Bay, South Africa | 79 |
| 6 | *Moco Cadiz* oil spill | March 16, 1978 | Portsall, France | 69 |
| 7 | *ABT Summer* oil spill | May 28, 1991 | Coast of Angola | 51–81 |
| 8 | *MT Haven Tauker* oil spill | April 11, 1991 | Italy | 45 |
| 9 | *Odyssey* oil spill | November 10, 1988 | Coast of Nova Scotia, Canada | 40.7 |
| 10 | The *Sea Star* oil spill | December 19, 1972 | Gulf of Oman | 35.3 |
| 11 | The *Torrey Canyon* oil spill | March 18, 1967 | Scilly Isles, UK | 25–36 |

bioremediation efforts in the cleanup of the oil tanker *Exxon Valdez* oil spill of 1989 in Prince William Sound and the Gulf of Alaska created tremendous interest in the potential of biodegradation and bioremediation technology [413]. Some major oil spillages in history are shown in Table 19.1.

## 19.2 Identification of Microorganisms

There are many commercially growing microorganisms. The 16S rRNA sequencing generally identifies that bacteria rRNA is the least viable in all cells. Significantly, the similar portions of the rRNA sequence are remotely related to the organisms. It is easy to measure true differences by precisely aligned sequences with remotely related organisms. To determine the taxonomy, phylogeny and approximate species state of diverging rates between bacteria, the genes encoding rRNA have been extensively used. The advanced relationship of microorganism can be seen by a comparison of the 16s rRNA sequencing. It has been determined that the DNA sequence of the 16s rRNA gene can be used to find a large number of species. In fact, as in many species, there is no other gene as well-characterised. The National Center for Biotechnology Information (www.ncbi.nlm.nih.gov) provides a large number of sequences of clinical and environmental isolates on the internet. To compare new database sequences, such sites can provide search algorithms.

## 19.3 Microbial Degradation of Hydrocarbons

Microbial degradation of HCs is principally based on biodegradation. It is a process of complete removal of organic toxic pollutants into harmless or naturally occurring compounds like carbon dioxide, water and inorganic compounds which are safe for humans, animals, plants and aquatic life [414]. Bioremediation is a biological mechanism of recycling wastes into another form which can use and be reused by other organisms. There are several microorganisms which are involved in the biodegradation of a wide variety of organic compounds in the presence and absence of oxygen. In Table 19.2 the microorganisms involved in degradation of organic pollutants and in Table 19.3 anaerobic microorganisms isolated from various oil reservoirs and types of HCs used for degradation are shown.

## 19.4 Biosurfactants

Biosurfactants are surface-active compounds which can reduce the interfacial tension between two liquids or that between a liquid and a solid. They are nontoxic, easily biodegradable and eco-friendly and have high stability, and a wide variety of industrial applications

**Table 19.2** Microorganisms and pollutants [422]

| S. no. | Microorganisms | Compound |
|---|---|---|
| 1 | *Penicillium chrysogenum* | Monocyclic aromatic hydrocarbons, BTEX, phenol compounds |
| 2 | *Pseudomonas alcaligenes, P. mendocina, P. putida, P. veronii, Achromobacter, Flavobacterium, Acinetobacter* | Petrol, diesel, PAHs, toluene |
| 3 | *Pseudomonas putida* | Monocyclic aromatic hydrocarbons, e.g., benzene and xylene |
| 4 | *Phanerochaete chrysosporium* | Biphenyl and triphenylmethane |
| 5 | *Aspergillus niger, A. fumigatus, Fusarium solani, Penicillium funiculosum* | Hydrocarbons |
| 6 | *Coprinellus radians* | PAHs, methylnaphthalenes, and dibenzofurans |
| 7 | *Alcaligenes odorans, Bacillus subtilis, Corynebacterium propinquum, Pseudomonas aeruginosa* | Phenol |
| 8 | *Tyromyces palustris, Gloeophyllum trabeum, Trametes versicolour* | Hydrocarbon |
| 9 | *Candida viswanathii* | Phenanthrene, benzopyrene |
| 10 | *Cyanobacteria*, green algae, diatoms, *Bacillus licheniformis* | Naphthalene |

makes them a highly useful group of chemical compounds. Biosurfactants can be produced from a variety of microorganisms. Chemically synthesised surfactants are quite expensive, and they cause environmental pollution [415], but they have a wide range of applications in industries [416]. Biosurfactants become important over chemical surfactants in order to reduce the application of chemically synthesised surfactants in industries. Biosurfactants are environment-friendly products produced by microorganisms. They have unique properties over chemically synthesised surfactants, such as high surface activity, low toxicity, easy biodegradability [185], better environmental compatibility [186], greater foaming [417] and good antimicrobial activity at extreme temperatures, pH and salinity [230, 416]. The most effective biosurfactant can

**Table 19.3** Petroleum hydrocarbon–degrading anaerobic bacteria [323]

| S. no. | Organism | Hydrocarbon(s) used |
|---|---|---|
| 1 | *Anoxygenic photoheterotrophic* bacterium | Toluene |
| 2 | *Blastochloris sulphoviridis* ToP1 | |
| 3 | *Azoarcus* sp. strain EB1 | Ethylbenzene |
| 4 | *Azoarcus* sp. strain T | Toluene, *m*-xylene |
| 5 | *Azoarcus tolulyticus* Td15 | Toluene, m-xylene |
| 6 | *Azoarcus tolulyticus* To14 | Toluene |
| 7 | *Dechloromonas* sp. strain JJ | Benzene, toluene |
| 8 | *Dechloromonas* sp. strain RCB | Benzene, toluene |
| 9 | *Pseudomonas sp.* strain NAP-3 | Naphthalene |
| 10 | *Pseudomonas* strain EbN1 | Ethylbenzene, toluene |
| 11 | *Pseudomonas* strain HdN1 | $C_{14}$–$C_{20}$ alkanes |
| 12 | *Pseudomonas* strain HxN1 | $C_6$–$C_8$ alkanes |
| 13 | *Pseudomonas* strain M3 | Toluene, *m*-xylene |
| 14 | *Pseudomonas* strain mXyN1 | Toluene, *m*-xylene |
| 15 | *Pseudomonas* strain OcN1 | $C_8$–$C_{12}$ alkanes |
| 16 | *Pseudomonas* strain PbN1 | Ethylbenzene, propylbenzene |
| 17 | Strain pCyN1 | *p*-Cymene, toluene, *p*-ethyltoluene |
| 18 | Strain pCyN2 | *p*-Cymene |
| 19 | Strain T3 | Toluene |
| 20 | Strain ToN1 | Toluene |
| 21 | *Thauera aromatica* K172 | Toluene |
| 22 | *Thauera aromatica* T1 | Toluene |
| 23 | *Vibrio* sp. strain NAP-4 | Naphthalene |
| 24 | *Geobacter grbiciae* TACP-$2^T$ | Toluene |
| 25 | *Geobacter grbiciae* TACP-5 | Toluene |
| 26 | *Geobacter metallireducens* GS15 | Toluene |
| 27 | *Desulfobacula toluolica* To12 | Toluene |
| 28 | *Desulfobacterium cetonicum* | Toluene |
| 29 | Strain AK-01 | $C_{13}$–$C_{18}$ alkanes |
| 30 | Strain Hxd3 | $C_{12}$–$C_{20}$ alkanes, 1-hexadecene |
| 31 | Strain mXyS1 | Toluene, m-xylene, m-ethyltoluene, m-cymene |
| 32 | Strain NaphS2 | Naphthalene |
| 33 | Strain oXyS1 | Toluene *o*-xylene, *o*-ethyltoluene |
| 34 | Strain Pnd3 | $C_{14}$–$C_{17}$ alkanes, 1-hexadecene |
| 35 | Strain PRTOL1 | Toluene |
| 36 | Strain TD3 | $C_6$–$C_{16}$ alkanes |

reduce the surface tension of water from 72 mN/m to values in the range of 25–30 mN/m [418]. Therefore, the screening and evaluation of efficient biosurfactant-producing strains have wide biological and economic significance. Biosurfactants emulsify compounds, increase the water solubility and make the compounds more accessible for the microorganisms [419]. Biosurfactants are widely used in the petroleum industry, such as enhanced oil recovery (EOR), bioremediation of oil spills and transportation of crude oil [415]. Biodegradation is an effective method where microbes are utilised in the contaminated oil-spilled site. These microbes are involved in the breakdown process of the pollution components into low-molecular-weight compounds [420]. The breakdown and solubilisation of large masses of organic compounds are an important step in bioremediation [421]. The presence of surface-active compounds, such as biosurfactants, accelerates the biodegradation of organic compounds. Some types of biosurfactants and microorganisms responsible for their production are given in Table 19.4.

**Table 19.4** Biosurfactants and microorganisms involved [166, 213, 221, 419, 560–568]

| Biosurfactants | Microorganisms |
|---|---|
| Glycolipids | *Pseudomonas aeruginosa* |
| | *Aeromonas sp.* |
| | *Bacillus sp.* |
| Rhamnolipids | *Rhodococcus erithropolis* |
| | *Pseudomonas aeruginosa* |
| | *Pseudomonas fluorescens* |
| Trehalose lipids | *Arthrobacter paraffineus* |
| | *Rhodococcus erythropolis* |
| | *Mycobacterium spp.* |
| Sophorolipids | *Candida bombicola* |
| | *Candida apicola* |
| | *Candida lipolytica* |
| | *Torulopsis bombicola* |
| Mannosylerythritol lipids | *Candida antartica* |

(*Contd.*)

**Table 19.4** (*Continued*)

| Biosurfactants | Microorganisms |
|---|---|
| Lipopeptides | *Bacillus subtilis* |
| | *Pseudomonas fluorescens* |
| | *Bacillis pumilis* |
| | *Bacillus licheniformis* |
| Surfactin | *Pseudomonas fluorescens* |
| | *Bacillus subtili* |
| Viscosin | *Bacillus licheniformis* |
| | *Psudomonas fluorescens* |
| Lichenysin | *Serratia marcescens* |
| | *Bacillus licheniformis* |
| Polymeric surfactants | *Acinetobacter calcoaceticus* |
| Emulsan | *Acinetobacter radioresistens* |
| | *Acinetobacter calcoaceticus* |
| Alasan | *Candida lipolytica* |
| | *Acinetobacter radioresistens* |
| Liposan | *Candida tropicalis* |
| | *Candida lipolytica* |
| Lipomanan | *Candida tropicalis* |
| Polymixins | *Bacillus polymyxa* |
| Gramicidin S | *Bacillus brevis* |
| Phospholipids | *Acinetobacter spp.* |
| | *Thiobacillus thiooxidans* |
| Polyol lipids | *Rhodotorula glutinis* |
| | *Rhodotorula graminis* |
| Serrawettin | *Serratia marcescens* |
| Fatty acids (corynomycolic acids, spiculisporic acids) | *Corynebacterium lepus* |
| | *Arthrobacter parafineus* |
| | *Penicillium spiculisporum* |
| | *Talaromyces trachyspermus* |
| Sulfonylipids | *Capnocytophaga spp.* |
| Diglycosyl diglycerides | *Lactobacillus fermentii* |
| Biodispersan | *Acinetobacter calcoaceticus* |
| Mannan-lipoprotein | *Candida tropicalis* |
| Food emulsifier | *Candida utilis* |
| Insecticide emulsifier | *Pseudomonas tralucida* |
| Sulfated polysaccharid | *Halomonas eurihalina* |
| Acetyl heteropolysaccharide | *Sphingomonas paucimobilis* |
| Microbactan glycolipoptide | *Microbacterium* |

## 19.5 Factors Influencing the Degradation of Hydrocarbons

The activity of microbes can be affected by the following factors: temperature, oxygen, pH and nutrients.

### 19.5.1 Temperature

The physical and chemical compositions of HCs are affected by temperature during biodegradation. At low temperature, the degradation rate decreases and reaches the maximum level at high temperatures ranging from 30°C to 40°C.

### 19.5.2 pH

The highest degradation rates are generally observed at neutral pH. Some HC-degrading microorganisms have been isolated from historically contaminated sites, even at a pH range of 2–3.

### 19.5.3 Activity of Water

The biodegradation rates of HCs in terrestrial ecosystems may be reduced because of the water available for metabolism and growth of microbes. The availability of water directly impacts the movement and growth of microorganisms.

### 19.5.4 Salinity

A higher concentration of salt inhibits microbial degradation. HC degradation in a saline environment depends upon the types of microorganisms involved.

### 19.5.5 Oxygen Availability

HC degradation can take place both in the presence and in the absence of oxygen. Degradation rates in aerobic conditions are higher than anaerobic conditions due to the availability of the primary enzyme oxygenase.

### 19.5.6 Nutrient Availability

Nutrients are important components for successful contaminant biodegradation, including nitrogen, iron and phosphorus in some cases, which enhance the degradation process.

## 19.6 Conclusion

Microbial degradation helps eliminate oil contamination from the environment after large quantities of oil are removed through various physical and chemical methods. The increase in our knowledge of the HC-degrading microbial communities and the mechanisms by which petroleum biodegradation occurs will prove helpful for predicting the environmental fate of these compounds and for developing practical HC bioremediation strategies in the future. Also, it will provide a better understanding of the mechanism of biosurfactants in biodegradation from indigenous microorganisms to transform or mineralise organic contaminants. On the basis of the chapter, it may be concluded that microbial degradation of HCs can be considered as a key component in the cleanup strategy for petroleum HC remediation.

## Acknowledgements

We sincerely acknowledge the DBT's Twinning programme no. NE–ENV\2013\222 entitled 'Remediation of Produced Water from NE Oilfields by Microbial and Membrane Intervention with Modelling of an Effluent Unit for the Recovery of Usable Water' for funding this research work.

## Chapter 20

# Evaluation of the Corrosion and Scaling Potential of Oilfield Produced Water of the Upper Assam Basin

**Debasish Konwar, Subrata Borgohain Gogoil and Tapan Jyoti Gogoi**

*Department of Petroleum Technology, Dibrugarh University, Assam, India*
konwar.debasish110@gmail.com

Oilfield produced water (OFPW) is a complex mixture of organic and inorganic compounds and the largest volume of by-product generated during oil and gas recovery operations. Large volumes of water are produced from various oilfields of the Upper Assam Basin. OFPW contains different toxic compounds which adversely affect the environment; such petroleum contaminants and impurities are oil, naturally occurring radioactive minerals (NORM), oil and grease (O&G), hydrocarbons (HCs), production chemicals and various metals. The scaling and corrosion potential is one of the most important indices in OFPW quality evaluation. OFPW separated from oil and gas is either injected into the reservoir for pressure maintenance or discharged to the environment after preliminary treatment at the well site. Therefore, evaluation of OFPW and its characterisation are very important and essential

*Advances in Petroleum Technology*
Edited by Subrata Borgohain Gogoi

ISBN 978-981-4877-19-0 (Hardcover), 978-1-003-04993-7 (eBook)
www.jennystanford.com

for environment and reservoir management. This chapter deals with the characterisation of OFPW generated from different depths and horizons of Upper Assam oilfields. It includes the analysis of pH, conductivity, salinity, total dissolved solids (TDS), turbidity, O&G, hardness, alkalinity and inorganics. The corrosion and scaling potential of OFPW is evaluated on the basis of the aggressive index (AI), Langelier saturation index (LSI), Ryznar stability index (RSI) and Puckorius scaling index (PSI). These indices indicate whether the water will form scales and/or corrosives or be in equilibrium with calcium carbonate ($CaCO_3$).

## 20.1 Introduction

Currently, it is well known that the oil industry has to handle more oilfield produced water (OFPW) than oil, which makes the oil industry look more like a 'water industry' [423]. OFPW is the largest waste stream of oil and gas production facilities, which needs to be treated properly in order to prevent bioaccumulation in aqueous organisms and the transfer of harmful chemicals to the food chain [424–426]. The United States produces about 7–10 barrels of water for each barrel of oil [427]. OFPW consists of formation water (FW), injected water and condensed water. Out of various options for the management of OFPW, the most adopted options by hydrocarbon exploration industries are underground injection for increasing oil recovery in the deep reservoir zone and the injection to nonpay zones in porous rocks for mere disposal purpose [428, 429]. OFPW contains low- and high-molecular-weight petroleum hydrocarbons (HCs), dispersed and dissolved oil, dissolved compounds, suspended solids, naturally occurring radioactive minerals (NORM), dissolved gases, salts and heavy metals [430, 431]. The major compounds found in OFPW are aliphatic HCs, less soluble aromatic HCs, organic acids and phenols. Volatile aromatics and napthalenes are the most toxic compounds in OFPW and are sometimes discharged to the surrounding environment or may be used for water flooding or reservoir pressure maintenance [427]. The quality of OFPW varies considerably depending on the geographic location of the reservoir oilfield, the geochemistry of the producing formation, the type of HC product being produced and the characteristics and type of the producing well [425, 432].

Northeast India OFPW is a by-product generated during various operations in hydrocarbon exploration industries and is a major source of pollution in the area. These pollutants find their way and traverse to various environmental receptors, causing potential danger of polluting the same. The water bodies, both surface and underground, are more susceptible to such pollution in oilfields. Management of OFPW is a key issue for the oil and gas producers because of its large volume of generation and high handling cost. The ageing oilfields of the Upper Assam Basin belonging to Oil India Limited (OIL), Duliajan and Oil and Natural Gas Corporation Limited (ONGCL), Nazira, produce crude oil with very high water cuts of more than 90%. Production of produced water (PW) from these ageing oilfields is also in an increasing trend. This OFPW is highly saline, very rich in dissolved minerals and has a high temperature than the normal atmospheric water temperature. The salinities of OFPW from these oilfields range from 1000 to 3000 mg/l [429, 432], that is, 1000–3000 ppm. The oilfield in the study area has disposed in an average 3000 $m^3$ of OFPW on a daily basis from its various oilfields, which are an environmental concern since the injected OFPW has a potential to migrate and contaminate the ground water of the area. The deterioration of OFPW quality is dependent on various water quality parameters. There exists a strong correlation amongst different parameters of OFPW, and the combined effect of their interrelatedness indicates the quality of OFPW. The regression equations for the parameters having significant correlation coefficients can be used to estimate the concentrations of the other constituents. A study of correlation coefficients helps to access the overall water quality and to quantify relative concentrations of various pollutants in water and provide necessary cues for water management initiatives [425].

The most common indicators of water quality assessment are scaling and corrosion [433]. The physical and chemical interaction between a metal and its environment has an electrochemical nature which results in a change in metal properties. This phenomenon is known as corrosion [434]. This imports some harmful by-products such as lead and arsenic to the OFPW, which may have serious environmental problems and will also lead to serious health problems and reduction of the lifetime of pipes and fittings, whereas scaling can aid corrosion resistance in the water distribution system by creating a barrier between the conductive water and the surface of the pipe [435]. Scaling is a multiphased process in which ions (such as $Ca^{2+}$ and $Mg^{2+}$) react with existing substances in water to form a film deposited on the inner surface of the tubes. Temperature, hardness, pH, acidity, alkalinity, chlorine, total dissolved solids (TDS), gases, salts and microorganisms in the water are the several factors which influence the corrosion and scaling processes. Among all the factors, biological factors are of secondary importance [436]. Langelier saturation index (LSI), Ryznar stability index (RSI), Puckorius scaling index (PSI) and aggressive index (AI) are important indicators to explain the state of water corrosion or scaling [423, 425, 430]. The most widely used indicator of scale formation potential in OFPW is LSI, which is a pure equilibrium index which incorporates the thermodynamic driving forces on the formation and growth of $CaCO_3$ scaling properties [437]. PSI incorporates an estimate of buffering capacity of water to investigate the relationship between the saturation state and scale formation [438]. RSI correlates an empirical database of scale thickness observed in OFPW to water chemistry. Like LSI, it uses the concept of saturation level. AI is derived from pH, $Ca^{2+}$ hardness and total alkalinity and is used in the studies of corrosion quality of water resources [439].

In this chapter the physiochemical characterisation of OFPW is identified and this data is compared with the World Health Organization (WHO) specification for its safe disposal. A correlation and regression study was done to quantify the association between the parameters, and it showed a linear relationship among different pairs of parameters. Water quality indices LSI, RSI, PSI and AI studies were done to compare the scaling and corrosion potentials of the water.

## 20.2 Materials and Methods

### 20.2.1 Materials and Equipment

Ten samples of OFPW were collected from oilfields of the Upper Assam Basin. These were obtained from well heads along with OFPW, as shown in Table 20.1. The equipment is in Table 20.2.

### 20.2.2 Physical and Chemical Characterisation

#### 20.2.2.1 Water analyser

The physical properties of OFPW like pH, salinity, electrical conductivity (EC), TDS, turbidity and dissolved oxygen (DO) are determined in a water analyser.

**Table 20.1** Materials used in the experiments

| S. no. | Materials | Specification | Source |
|---|---|---|---|
| 1 | OFPW samples | Barail formation | OIL, Duliajan |
| 2 | EDTA | 0.01M | |
| 3 | Erichrome black indicator | 0.5 g in 100 ml | |
| 4 | NaCl | M.W. 58.44 g/mol | RFCL Limited |
| 5 | HCl | M.W. 36.46 g/mol, specific gravity at 25°C is 1.18 | |
| 6 | Phenolphthalein | pH range 8.2–10 (colourless to pink) | |
| 7 | pH Buffer | pH 10 | WTW |
| 8 | KCl | M.W. 74.56 g/mol | Avantor Performance Materials India Limited |
| 9 | $CaCO_3$ | M.W. 100.09 g/mol | Merck Specialities Pvt. Ltd. |
| 10 | $Li_2CO_3$ | M.W. 73.89 g/mol | Spectrochem Pt. Ltd. Mumbai India |
| 11 | Petroleum ether | Wt. per ml at 20°C, 0.630–0.645 g | Fisher Scientific Human Diagnostics |
| 12 | Methyl orange | — | and Surgichem |

**Table 20.2** Equipment used in the experiments

| S. no. | Parameter | Instrument | Chemicals used |
|---|---|---|---|
| 1 | pH | | |
| 2 | salinity | | |
| 3 | EC | Model: Water Analyzer 371 | |
| 4 | TDS | Make: Systonics | |
| 5 | Turbidity | | |
| 6 | DO | | |
| 7 | Na | Model: Flame photometer 128 | NaCl |
| 8 | K | Make: Systronics | KCl |
| 9 | Ca | Photometer 128 | $CaCO_3$ |
| 10 | Li | Compressor 126 | $Li_2CO_3$ |
| 11 | Mn | Model: Atomic Absorption | |
| 12 | Fe | Spectrophotometer Analyst 200 | |
| 13 | Cr | Make: Perkin Elmer | |
| 14 | Cu | | |

### 20.2.2.2 Titration

Titration was employed for the determination of alkalinity and hardness as per ASTM D974-64 (IP Standards for Petroleum and Its Products, 31st Edition, Part I, The Institute of Petroleum, 61, New Cavendish Street, London).

### 20.2.2.3 Alkalinity

Here, 0.05 N HCl was introduced in a burette, which was titrated with 50 ml of OFPW. To the OFPW, 2–3 drops of methyl orange were added by means of a pipette as an indicator till the colour turned orange. Then, 0.05 N HCl solution was prepared by diluting 10 ml of 1 N HCl solution (1 N HCl = 36.5 gms of HCl in 1 l of distilled water [DW]) by applying Eq. 20.1.

We have,

$$N_1 V_1 = N_2 V_2, \tag{20.1}$$

where $N_1 = 1\ N$, $V_1 = 10$ ml and $N_2 = 0.05\ N$.

Therefore,

$$V_2 = 200 \text{ ml}.$$

Hence by diluting 10 ml of 1 M HCl with 190 ml of DW, 0.05 N HCl was obtained. This 0.05N HCl was added from the burette to the solution containing OFPW and methyl orange till the colour changed to pink, indicating the end point of titration [440].

$HCO_3^-$ and alkalinity were calculated by applying Eq. 20.2 and Eq. 20.3, respectively:

$$HCO_3^- \text{ in g/[(mol)(l)]} = \frac{T \text{ ml} \times 0.05^{-}l \times 61\,\text{g/mol}}{\text{Vol. of OFPW ml}} \quad (20.2)$$

$$\text{Total alkalinity of OFPW in mEq/l} = HCO_3^- \times 0.0.82 \quad (20.3)$$

The following are applicable to Eq. 20.1 and Eq. 20.2:

$T$ = 50 ml of OFPW + methyl orange titrate with 0.05 N HCl
Normality = $N$ = 0.05 $^{-}l$
Molecular weight of $HCO_3$ = 61 g/mol
Volume of OFPW in ml
0.82 is the conversion factor from ppm to mEq or mmol/l.

#### 20.2.2.4 Total hardness

Here, 0.01 M edetate disodium salt (EDSS) with a molecular weight of 372.24 g/mol and a chemical formula of $C_{10}H_{14}N_2Na_2O_8$ was introduced in the burette, which was titrated with 50 ml of OFPW. To the OFPW, 2–3 drops of Erichrome black was added by means of a pipette as an indicator till the colour turned greyish-black. Then, 0.01 M EDSS solution was prepared by dissolving 3.7224 g of EDSS in 1 l of DW in a volumetric flask. This 0.01 M EDSS was added to the burette with OFPW and Erichrome black till it attained an ink colour, indicating the end point of titration. Calculations were done by Eq. 20.4.

$$\text{Total hardness in mg/l} = \frac{\text{EDTA} \times 1000}{\text{Vol. of OFPW}}, \quad (20.4)$$

where $W$ = weight of EDSS in mg.

#### 20.2.2.5 O&G

The separation of oil and grease (O&G) from OFPW was done by using the gravity separation method (ASTM D 3921). Equal volumes of the OFPW sample and petroleum ether were taken

(1:1 = 20:20 ml) in a separating funnel. The mixture was thoroughly mixed manually for 30 min. The appearance of two layers was observed, where the lower layer was separated out in a beaker and dried in an oven and weighed. The calculations were done by Eq. 20.5.

$$\text{O\&G in mg/l} = \frac{(y - x) \times 10^6}{\text{Vol. of OFPW}} \tag{20.5}$$

#### 20.2.2.6 Flame photometer

A flame photometer was used to determine Na, K, Li and Ca present in the OFPW. Preparation of stock standard solution

Na: A standard solution of 1000 ppm was prepared by dissolving 2.5416 g of NaCl in 1 l of DW.

K: A standard solution of 1000 ppm was prepared by dissolving 1.9070 g of KCl or 2.5869 g of $KNO_3$ in 1 l of DW.

Ca: A standard solution of 1000 ppm was prepared by dissolving 2.497 g of $CaCO_3$ in approx. 300 ml of DW and adding 10 ml of conc. HCl in 1 l of DW.

Li: A standard solution of 2000 ppm was prepared by dissolving 4.945 g of $Li_2CO_3$ in approx. 300 ml of DW and adding 15 ml of conc. HCl in 1 l of DW.

#### 20.2.2.7 Atomic absorption spectrophotometer

Determination of Fe, Cu, Mn and Cr in OFPW was done using an atomic absorption spectrophotometer (AAS). Prior to the determination of the amount present, the AAS was calibrated and analysed using a blank solution of 2% $HNO_3$ and standard 1 ppm, 2 ppm, 3 ppm and 4 ppm of Fe solutions for the Fe test. Similarly for Cr, Mn and Cu, 1 ppm, 2 ppm, 3 ppm and 4 ppm of Cr, Mn and Cu standard samples were made, respectively. After calibration, the samples were aspirated and the results were obtained.

### 20.2.3 Correlation Index

The correlation index (CI) is the relationship and association between variables. Relationship and association refer to the extent to which one variable changes in quantity or quality in response to a change in another variable [433].

Let $x$ and $y$ be the two variables. The correlation coefficient (Pearson) ($r$) between the variables $x$ and $y$ is given by Eq. 20.6:

$$r = \frac{N\sum xy - (\sum x)(\sum y)}{\sqrt{\left[N\sum x^2 - (\sum x)^2\right]\left[N\sum y^2 - (\sum y)^2\right]}}, \tag{20.6}$$

where $N$ = number of pairs of parameters, $\sum xy$ = sum of the product of the paired parameters, $\sum x$ = sum of $x$ parameters, $\sum y$ = sum of $y$ parameters, $\sum x^2$ = sum of squared $x$ parameters and $\sum y^2$ = sum of squared $y$ parameters.

The quantity $r$ measures the strength and direction of a linear relationship between two variables. The linear correlation coefficient is sometimes referred to as the Pearson product moment correlation coefficient in honour of its developer Karl Pearson [441]. The value of $r$ can be expressed as a value between $+1$ and $-1$. A coefficient of $+1$ indicates a perfect positive correlation: a change in the value of one variable will predict a change in the same direction in the second variable. A coefficient of $-1$ indicates a perfect negative correlation: a change in the value of one variable predicts a change in the opposite direction in the second variable. Lesser degrees of correlation are expressed as nonzero decimals. A coefficient of 0 indicates there is no discernible relationship between fluctuations of the variables [442]. If the values of the correlation coefficient $r$ between two variables $x$ and $y$ are fairly large, it implies that these two variables are highly correlated. In such cases it is feasible to try a linear relation in the form as in Eq. 20.7:

$$y = a + bx \tag{20.7}$$

The values of empirical parameters $a$ and $b$ are calculated with the help of Eq. 20.8 and Eq. 20.9, respectively:

$$a = \frac{\sum y \sum x^2 - \sum x \sum x}{n\sum x^2 - (\sum x)^2} \tag{20.8}$$

$$b = \frac{n(\sum x) - \sum x \sum y}{n(\sum x)^2 - (\sum x)^2} \tag{20.9}$$

### 20.2.4 Water Stability Indices

In this study, LSI, RSI, PSI and AI were used to determine the corrosion and scaling tendency of OFPW samples. These indices are briefly discussed next.

#### 20.2.4.1 LSI

LSI indicates the degree of saturation of $CaCO_3$ in water. It is calculated by using the pH, alkalinity, $Ca^{2+}$ concentration, TDS and water temperature of a water sample collected at a tap. If:

(1) LSI is negative, then the water is undersaturated with respect to $CaCO_3$. Undersaturated water has a tendency to remove existing $CaCO_3$ protective coatings in pipelines and equipment.
(2) LSI is positive, then the water is supersaturated with respect to $CaCO_3$ and scale forming may occur.
(3) LSI is zero, then water is considered to be neutral, which is neither scale forming nor scale removing.

LSI is calculated by Eq. 20.10:

$$\mathrm{pHs} - \mathrm{pH} = \mathrm{LSI}, \tag{20.10}$$

where pHs is the saturation pH and pH is the pH value of OFPW.

#### 20.2.4.2 RSI

RSI uses a database of scale thickness measurements to predict the effect of water chemistry. RSI is calculated by Eq. 20.11:

$$2\mathrm{pHs} - \mathrm{pH} = \mathrm{RSI} \tag{20.11}$$

If:

(1) $\mathrm{RSI} < 5.5$, then heavy scales will form.
(2) $5.5 < \mathrm{RSI} < 6.2$, then scales will form.
(3) $6.2 < \mathrm{RSI} < 6.8$, then there will be no difficulties.
(4) $6.8 < \mathrm{RSI} < 8.5$, then water is aggressive.
(5) $\mathrm{RSI} > 8.5$, then water is very aggressive.

#### 20.2.4.3 PSI

OFPW with high $Ca^{2+}$ content but low alkalinity and buffering capacity can have a high calcite saturation level. This high $Ca^{2+}$ level leads to an increase in ion activity product. A plot of ion activity product versus precipitate for the water indicates a rapid decrease in pH as $Ca^{2+}$ precipitates due to the low buffering capacity. Even a minute decease in carbonate OFPW would drastically decrease the ion activity product due to the small quantity present prior to the initiation of precipitation. It can be calculated by Eq. 20.12:

$$\text{PSI} = 2(\text{pHs}) - \text{pHeq}, \qquad (20.12)$$

where pHs is the pH at saturation in calcite or $CaCO_3$, pHeq is 1.465 $\times$ log10[alkalinity] + 4.54 and $H$ is Ca hardness (mg/l as $CaCO_3$).

If:

(1) PSI > 6, scaling is unlikely to occur.
(2) PSI < 7, OFPW is likely to dissolve scales.

#### 20.2.4.4 AI

AI is an indicator of the corrosivity of water. AI is calculated from the actual pH, $Ca^{2+}$ hardness and total alkalinity. AI is simpler and more convenient than LSI as AI does not include the effects of temperature or dissolved solids. It is calculated by Eq. 20.13:

$$\text{AI} = \log[(\text{H} \times \text{A})] + \text{pH}, \qquad (20.13)$$

where $H$ is Ca hardness (mg/l as $CaCO_3$) and $A$ is total alkalinity.

If:

(1) AI > 12, water is non-aggressive.
(2) AI = 10 − 11.9, water is moderately aggressive.
(3) AI < 10, water is very aggressive.

The summary of indices equations, results and their indication are shown in Table 20.3.

**Table 20.3** Summary of water indices

| Index | Equation | Index | Indication |
|---|---|---|---|
| | | LSI $\leq$0 | Water is undersaturated w.r.t. $CaCO_3$. Undersaturated water has a tendency to remove existing $CaCO_3$ protective coatings in pipelines and equipment. |
| LSI | pHs-pH=LSI | LSI = 0 | Water is considered to be neutral, neither scale-forming nor scale removing. |
| | | LSI$\geq$0 | Water is supersaturated w.r.t. $CaCO_3$ and scale forming may occur. |
| | | RSI$\leq$5.5 | Heavy scale will form. |
| | | 5.5$\leq$RSI$\leq$6.2 | Scale will form. |
| RSI | 2pHs-pH=RSI | 6.2$\leq$RSI $\leq$ 6.8 | No difficulties. |
| RSI | | 6.8$\leq$ RSI$\leq$8.5 | Water is aggressive. |
| | | RSI$\geq$8.5 | Water is very aggressive. |
| RSI | PSI=2(pHs)-pHeq | PSI$\leq$6 | Scaling is unlikely to occur. |
| | | PSI$\geq$7 | Likely to dissolve scale. |
| | | AI$\geq$12 | Water is non-aggressive. |
| AI | AI=log[(H**)(A***)]+pH | AI = 10 − 11.9 | Water is moderately aggressive. |
| | | AI$\leq$10 | Water is very aggressive. |

## 20.3 Results and Discussion

### 20.3.1 Physical and Chemical Characterisation

The physical and chemical characterisation of OFPW is in Table 20.4, which was compared with WHO specifications for OFPW disposal, and all the values exceeded the limit, except pH and TDS. Ten samples were collected from the well heads from various oilfields of the Upper Assam Basin. The maximum and minimum values among the 10 samples were determined and the standard deviation

**Table 20.4** Characterisation of OFPW and its comparison with WHO values

| Measure parameters | Max. | Min. | Mean | Standard deviation | WHO specifications |
|---|---|---|---|---|---|
| pH | 7.99 | 7.5 | 7.69 | 0.15584893 | 6.0–9.0 |
| EC (mS) | 6.5 | 1.43 | 3.242 | 1.50144967 | 0.25–0.75 |
| TDS (ppt) | 3.89 | 1.24 | 1.89 | 0.825536862 | 24.59 |
| DO (ppt) | 4.7 | 4 | 4.29 | 0.19692074 | 0.007 |
| Salinity (ppt) | 4.5 | 1.5 | 2.12 | 0.914986339 | 0.2–0.4 |
| Turbidity (NTU) | 117 | 1.5 | 58.079 | 42.9891 | 5.0–10 |
| Alkalinity (ppm) | 725.29 | 28.7 | 343.23 | 278.0459228 | 200 |
| Hardness (ppm) | 70 | 40 | 52.4 | 12.20382454 | 0.3 |
| O&G (ppm) | 3500 | 1000 | 1920.76 | 791.6304219 | 250 |
| $Fe^{2+}$ (ppm) | 2.47 | 0.024 | 0.7792 | 0.905377245 | 0.003 |
| $Cu^{2+}$ (ppm) | 0 | 0 | 0 | 0 | 0.003 |
| $Cr^{2+}$ (ppm) | 0 | 0 | 0 | 0 | 0.002 |
| $Mn^{2+}$ (ppm) | 0.075 | 0.031 | 0.0524 | 0.01359902 | 0.05 |
| $Na^{+}$ (ppm) | 613.2 | 323 | 430.52 | 81.37202358 | 0.002 |
| $K^{+}$ (ppm) | 44.2 | 7.1 | 19.37 | 12.12793012 | 250 |
| $Ca^{2+}$ (ppm) | 982 | 21 | 329.64 | 299.7951901 | 0.075 |
| $Li^{+}$ (ppm) | 5.3 | 2.2 | 3.879 | 1.090040264 | 2 |

calculated. The pH of all the samples was neutral and was within the permissible limit of WHO. EC is a measure of the capacity of a solution to carry an electric current, and the permissible value of EC for water to discharge into the environment is 0.25–0.75 mS/cm. OFPW is highly conductive, and so its experimental values are crossing the limits of WHO. The high value (0.55 mS) of EC above pure water can be attributed to the fact that PW has ions [429]. OFPW may contain inorganic and organic substances in molecular, ionised or microgranular suspended form. These substances are known as TDS. The range of values of TDS of OFPW is within the acceptable limit of WHO for its disposal. Dissolved oxygen (DO) levels in surface water bodies indicate the ability to support aquatic life in a marine ecosystem. The amount of DO retained in OFPW is also influenced by temperature. Oxygen ($O_2$) is less soluble in warm water. Warm water also enhances decomposer activity. Therefore, increasing the temperature of a water body increases the rate at which $O_2$ is depleted from OFPW. The DO values in the 10 samples are below the permissible limits, which can mean that

**Table 20.5** Nonparametric Spearman Rank Order Correlation correlation factor

| | pH | EC | TDS | DO | Sal. | Td | Ak | Hd | O&G | Fe | Mn | Na | K | Ca | Li |
|---|---|---|---|---|---|---|---|---|---|---|---|---|---|---|---|
| **pH** | 1 | | | | | | | | | | | | | | |
| **EC** | −0.25 | 1 | | | | | | | | | | | | | |
| **TDS** | 0.132 | 0.409 | 1 | | | | | | | | | | | | |
| **DO** | 0.560 | 0.032 | 0.633 | 1 | | | | | | | | | | | |
| **Sal** | −0.34 | 0.421 | 0.463 | 0.282 | 1 | | | | | | | | | | |
| **Td** | 0.535 | 0.310 | 0.053 | 0.194 | 0.268 | 1 | | | | | | | | | |
| **Ak** | 0.704 | −0.05 | 0.488 | 0.578 | 0.184 | 0.329 | 1 | | | | | | | | |
| **Hd** | 0.435 | 0.354 | 0.687 | 0.290 | 0.121 | 0.553 | 0.645 | 1 | | | | | | | |
| **O&G** | 0.785 | 0.181 | 0.356 | 0.336 | −0.05 | 0.643 | 0.813 | 0.770 | 1 | | | | | | |
| **$Fe^{2+}$** | −0.68 | −0.23 | −0.233 | −0.64 | −0.07 | −0.80 | −0.43 | −0.37 | −0.56 | 1 | | | | | |
| **$Mn^{2+}$** | −0.34 | 0.320 | 0.638 | 0.275 | 0.525 | −0.21 | −0.11 | 0.156 | −0.22 | −0.009 | 1 | | | | |
| **$Na^{+}$** | −0.53 | 0.564 | −0.107 | −0.35 | 0.329 | −0.21 | −0.16 | −0.25 | −0.14 | 0.249 | 0.011 | 1 | | | |
| **$K^{+}$** | −0.21 | 0.109 | 0.593 | 0.222 | 0.355 | −0.29 | −0.06 | 0.146 | −0.12 | 0.079 | 0.925 | −0.08 | 1 | | |
| **$Ca^{2+}$** | −0.16 | 0.380 | 0.759 | 0.209 | 0.636 | −0.11 | 0.069 | 0.434 | 0.111 | 0.006 | 0.876 | 0.034 | 0.92 | 1 | |
| **$Li^{+}$** | 0.23 | −0.26 | −0.529 | −0.57 | −0.34 | 0.372 | 0.149 | 0.190 | 0.358 | 0.064 | −0.53 | −0.12 | −0.44 | −0.40 | 1 |

EC, electrical conductivity (mS); TDS, total dissolved solids (ppt); DO, dissolved oxygen (ppt); Sal, salinity (ppt); Td, turbidity (NTU); Ak, alkalinity (ppm); Hd, hardness) (ppm).

there was less amount of $O_2$ available dissolved in OFPW for marine organisms. The salinity of OFPW is well above the permissible limits of WHO. This highly saline OFPW affects the agricultural and aquatic habitats. Various suspended and dissolved solids are present in OFPW. This leads to various problems for the stream system as higher turbidity levels are often associated with higher levels of viruses, parasites and some bacteria because they can sometimes attach themselves to the dirt in the water [425]. The experimental values of OFPW samples were higher than the permissible limits. Alkalinity is the capacity of water to resist changes in pH which would make the water more acidic. It is the capacity of OFPW to neutralise an acid. It showed values higher than the normal values of discharge regulations. Hardness, a physiochemical property of water, is generally a measure of $Ca^{2+}$ and $Mg^{2+}$ ions in water [443]. High values of hardness are detrimental for human health. The experimental values of hardness are above the permissible limit of WHO and also the concentration of Ca ions in the OFPW samples. $Na^+$, $K^+$, $Li^+$, $Fe^{2+}$ and $Mn^{2+}$ ions were present in excessive amounts in OFPW samples. As the OFPW samples were highly saline, so their $Na^+$ ion concentration was very high. O&G in OFPW is due to the presence of suspended oil particles in the water samples.

## 20.3.2 Spearman Rank Order Correlation Coefficient

A correlation coefficient is a statistical measure of the degree up to which a change in the value of one variable predicts a change in the value of another. A positive correlation factor indicates that a change in the value of one variable will predict a change in the value of another variable in the same direction, whereas a negative correlation factor indicates that an increase in the value of one variable will lead to a decrease in the value of the other variable. The correlation coefficient ranges from $+1$ to $-1$. $Fe^{2+}$, $Mn^{2+}$, $Na^+$, $K^+$ and $Ca^{2+}$ showed a negative correlation factor with alkalinity. Similarly there was significant correlation between the other parameters of OFPW (Table 20.5). Korfali (2012) studied the impact of corrosion and scaling in the water distribution system on the water quality in a coastal region of Beirut, Lebanon [444]. The water was found to be oversaturated in terms of $CaCO_3$ and was

**Table 20.6** LSI values

| Sample | pHs | LSI | Indication based on Langelier (1936) | Indication based on improved Langelier by Carrier (1965) |
|---|---|---|---|---|
| PW-1 | 6.4 | 1.3 | Water is supersaturated with $CaCO_3$ and scale forming may occur. | Scale forming but noncorrosive |
| PW-2 | 6.9 | 0.60 | | |
| PW-3 | 8.5 | −0.85 | Water is undersaturated with $CaCO_3$. Undersaturated water has a tendency to remove existing $Ca^{2+}$ carbonate protective coatings in pipelines and equipment. | Serious corrosion |
| PW-4 | 7.5 | 0.31 | Water is supersaturated with respect to $CaCO_3$ and scale forming may occur. | Slightly scale forming and corrosive |
| PW-5 | 7.2 | 0.41 | | |
| PW-6 | 8.2 | −0.70 | Water is undersaturated with $CaCO_3$. Undersaturated water has a tendency to remove existing $CaCO_3$ protective coatings in pipelines and equipment. | Serious corrosion |
| PW-7 | 7.7 | −0.027 | | Slightly corrosive but non-scale forming |
| PW-8 | 8.0 | −0.042 | | |
| PW-9 | 7.3 | 0.58 | Water is supersaturated with $CaCO_3$ and scale forming may occur. | Scale forming but noncorrosive |
| PW-10 | 7.7 | −0.15 | Water is undersaturated with respect to $CaCO_3$. Undersaturated water has a tendency to remove existing $CaCO_3$ protective coatings in pipelines and equipment. | Slightly corrosive but non-scale forming |

expected to delay metallic corrosion. From the result of his study, it can be inferred that there is an inverse relation of $Cl^-$ with metal scale formation, except for lead [444].

### 20.3.3 Water Stability Indices

#### 20.3.3.1 LSI

LSI values were determined for OFPW and are presented in Table 20.6. This index determines the scaling potential of the water. Table 20.1 shows the corrosion indications of the OFPW samples. If the LSI value is positive and water is being supersaturated with $CaCO_3$, scaling can be formed. With increasing of LSI the scaling potential increases. Water temperature has a positive effect on LSI, that is, with water temperature increasing LSI becomes more positive. In this study the LSI value ranged from –0.85 to 1.3. OFPW samples 3 and 6 showed a high value of corrosion according to Carrier (2002) [445]. Also these two samples were undersaturated with respect to $CaCO_3$. Undersaturated water has a tendency to remove existing $CaCO_3$ protective coatings in pipelines and equipment according to Langelier (1938) [446]. The positive value of LSI indicates the samples to be saturated with respect to $CaCO_3$. These supersaturated waters tend to form scaling on the pipelines, thus restricting free flow. According to Asano et al. (2002), OFPW sample 9 indicated that the water is noncorrosive [445]. According to DeMartini (1938), these samples were supersaturated with respect to $CaCO_3$ and scale forming may occur [446].

#### 20.3.3.2 RSI

The RSI values of the OFPW samples ranged from 3.6 to 8 (Table 20.7 and Fig. 20.1). RSI less than 5.5 indicated the formation of heavy scale, whereas RSI greater than 8.5 indicated the water to be aggressive, that is, such water actively seeks to gain minerals and metals which results in the corrosion of iron and copper surfaces and dissolving of concrete and stone surfaces. RSI in the range of 6.2 to 6.8 indicated the water to be neutral. From Fig. 20.1, the highest value of RSI was in OFPW sample 3 [443]. This water was aggressive according to the RSI value and also OFPW was found to form heavy scale, according to the improved RSI by Asano et al. (2002). OFPW samples 7, 8 and 10 had no difficulties, according to Ryznar [443], and there will be formation of less corrosion, according to the

**Table 20.7** RSI values

| Sample | pHs | LSI | Indication based on Ryznar [443] | Indication based on improved Ryznar index [445] |
|---|---|---|---|---|
| PW-1 | 5.6 | 3.6 | Heavy scale will form. | Corrosion intolerable |
| PW-2 | 6.1 | 4.6 | | Heavy scale |
| PW-3 | 7.8 | 8.0 | Water is aggressive. | Heavy corrosion |
| PW-4 | 6.8 | 5.8 | Scale will form. | Light scale |
| PW-5 | 6.6 | 5.6 | | |
| PW-6 | 7.3 | 7.1 | Water is aggressive. | Corrosion significant |
| PW-7 | 7.2 | 6.6 | No difficulties. | Little scale or corrosion |
| PW-8 | 7.4 | 6.8 | | |
| PW-9 | 6.5 | 5.1 | Heavy scale will form. | Light scale |
| PW-10 | 7.0 | 6.4 | No difficulties. | Little scale or corrosion |

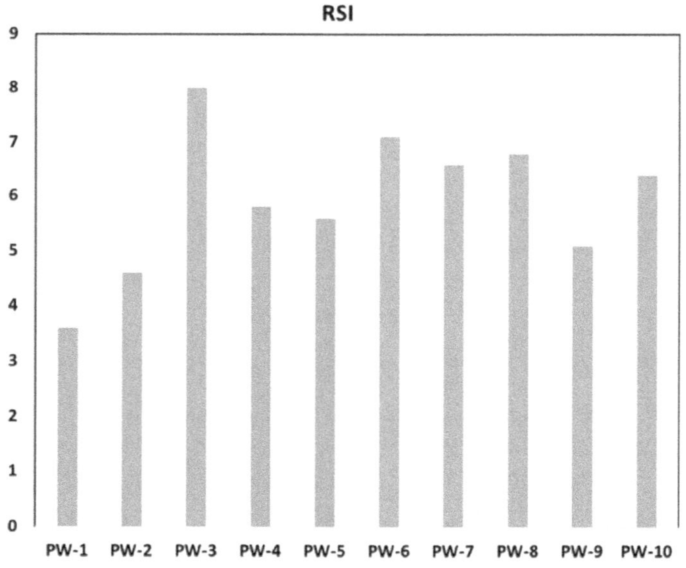


**Figure 20.1** Value of RSI of OFPW samples.

improved Ryznar index by Asano et al. (2002) [445]. PSI values of OFPW ranged from 6.31 to 8.40 (Table 20.8 and Fig. 20.2). The scaling of OFPW is unlikely to occur if the value of PSI is less than 6. For PSI greater than 7, OFPW was likely to dissolve scale. OFPW

Table 20.8 PSI values

| Sample | pHs | pHeq | PSI | Indication |
|---|---|---|---|---|
| PW-1 | 7.7 | 8.73 | 6.66 | Scaling unlikely to occur |
| PW-2 | 7.5 | 8.68 | 6.31 | |
| PW-3 | 7.62 | 7.36 | 7.87 | |
| PW-4 | 7.77 | 7.53 | 8.00 | |
| PW-5 | 7.65 | 7.68 | 7.61 | |
| PW-6 | 7.54 | 6.67 | 8.40 | Likely to dissolve scale |
| PW-7 | 7.68 | 7.94 | 7.41 | |
| PW-8 | 7.99 | 8.5 | 7.41 | |
| PW-9 | 7.89 | 8.67 | 7.10 | |
| PW-10 | 7.56 | 8.01 | 7.10 | |

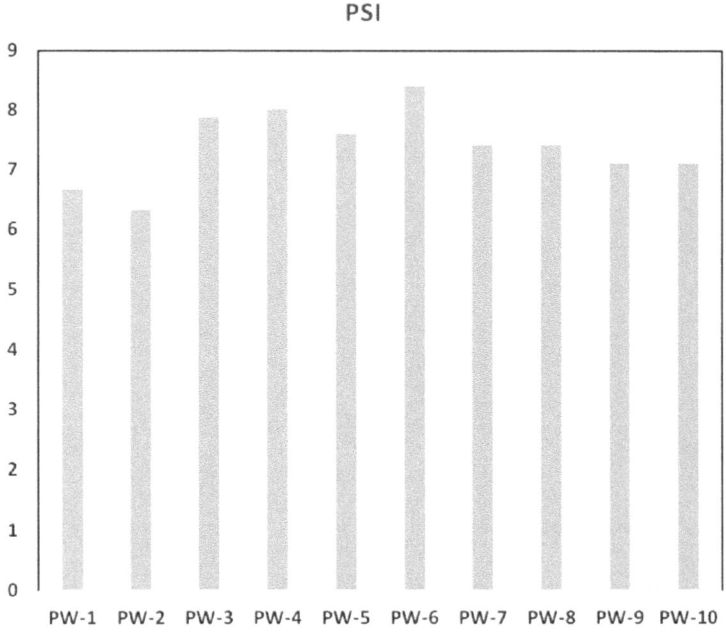


**Figure 20.2** Value of PSI of OFPW samples.

sample 6 had the highest value of PSI and is likely to dissolve scale formation. OFPW samples 1 and 2 had PSI values less than 7 and so scaling was unlikely to occur.

**Table 20.9** AI values

| Sample | pHs | pHeq | PSI | Indication |
|---|---|---|---|---|
| PW-1 | 725.29 | 982 | 13.55 | Water is non-aggressive. |
| PW-2 | 675.27 | 379 | 12.90 | |
| PW-3 | 5.034 | 56.6 | 11.30 | Water is moderately aggressive. |
| PW-4 | 10.044 | 446.9 | 12.46 | Water is non-aggressive. |
| PW-5 | 140.056 | 513.4 | 12.50 | |
| PW-6 | 28.7 | 522 | 11.71 | Water is moderately aggressive. |
| PW-7 | 210.084 | 98 | 11.99 | |
| PW-8 | 560.224 | 21 | 12.06 | |
| PW-9 | 662.36 | 141.9 | 12.86 | Water is non-aggressive. |
| PW-10 | 235.26 | 135.6 1 | 2.06 | |

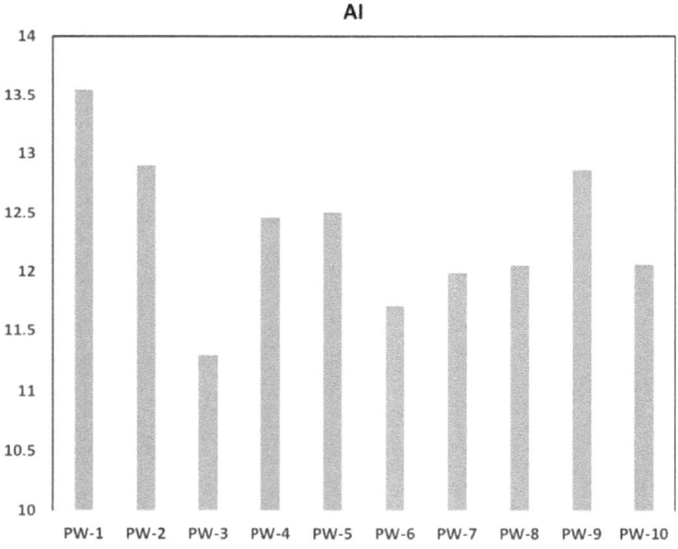


**Figure 20.3** Value of AI of OFPW samples.

The values of AI of OFPW samples ranged from 11.30 to 12.90, as in Table 20.9 and Fig. 20.3. AI values greater than 12 indicated OFPW to be non-aggressive. The values of AI between 10 and 11.9 indicated the water to be moderately aggressive, whereas values less than 10 indicated water to be very aggressive. Most of the samples were indicating a moderately aggressive or non-aggressive nature of OFPW. With the increase in the aggressive potential of OFPW, it tries to gain minerals and metals, which results in the corrosion of iron and copper surfaces and dissolving of concrete and stone surfaces.

## 20.4 Conclusion

OFPW collected from the oilfields of the Upper Assam Basin was characterised for its physical and chemical properties. Its characterised values were compared with WHO specifications for disposal. From the comparison study, it was inferred that some of its parameters have higher values than the safe limits, which can adversely affect the environment. EC, turbidity, hardness, $Na^+$ and $Ca^{2+}$ showed values much higher than the safer limits for its disposal, whereas pH, salinity, DO and TDS gave satisfactory results. The high Na concentration can be related to the fact that all the samples have high salinity. The correlation study is a useful mechanical tool for monitoring of the parameters of the water samples. The results showed that all the parameters are correlated. Scaling and corrosion are important indicators which must be considered for the transportation of OFPW and also form economic and health hazards. In fact all the samples showed a high corrosion tendency in all the indices. OFPW must be stabilised before its transportation and also must be treated before disposal to overcome health hazards.

## Acknowledgements

We would like to sincerely acknowledge the following:

(1) Indo-Tunisia Project No. DST/INT/TUNISIA/P-02/2017 entitled 'Characterisation of Oilfield Water to Mitigate a Major Environmental Hazard with Recovery of Usable Water'
(2) Indo-US Project F. No. 194-1/2009(IC) dated February 2015, entitled 'Foam-Assisted $CO_2$ Flooding for the Depleted Reservoirs of Upper Assam Basin and in Candidate Reservoirs in

Louisiana' of the 21st Century Knowledge Initiative programme under University Grants Commission
(3) DBT Twinning Project No. BT/485/NE/TBP/2013 entitled 'Remediation of Produced Water from NE Oilfield by Microbial and Membrane Intervention with Modelling of an Effluent Unit for the Recovery of Usable Water'
(4) Dibrugarh University for laboratory facilities

## Chapter 21

# A Review on Treatment and Management of Oilfield Produced Water

**Tapan Jyoti Gogoi, Subrata Borgohain Gogoi and Pranab Boral**
*Department of Petroleum Technology, Dibrugarh University, Assam, India*
pf_tapanjyoti@dibru.ac.in

During exploration and production of crude oil and natural gas, a large amount of oilfield produced water (OFPW) is generated, which contains both organic and inorganic pollutants. It is the largest waste stream generated in the oil and gas industry. Due to the increasing volume of OFPW all over the world in the current ongoing state, the outcome and effect of discharging OFPW on the environment has lately become a significant issue of environmental concern. It is conventionally treated through different physical, chemical and biological methods, but the current technologies cannot remove suspended oil particles and dissolved elements to a large extent. Besides, many conventional treatment methods produces hazardous sludge. Therefore, it is convenient to treat OFPW with membrane treatment methods. The submerged membrane filtration concept is well-established for low-pressure microfiltration (MF) and ultrafiltration (UF) membrane applications in the oil and gas industry. But using of nanofiltration (NF) and reverse osmosis (RO) methods is not feasible for the industry

*Advances in Petroleum Technology*
Edited by Subrata Borgohain Gogoi

ISBN 978-981-4877-19-0 (Hardcover), 978-1-003-04993-7 (eBook)
www.jennystanford.com

due to their high cost. The high salt concentration and variations of water characteristics have a direct influence on the treatment of OFPW through the membrane separation process; therefore it is appropriate to incorporate a conventional treatment method initially.

There are a large number of studies surrounding the treatment of OFPW for the removal of both organic and inorganic pollutants. And major research efforts in the future could focus on the optimisation of current technologies and the use of combined conventional and membrane treatment methods for treatment of OFPW in order to comply with reuse and discharge limits. Hence, there are some increasing interests for identifying a novel treatment method for disposal of OFPW under permissible limits. Additionally, the characteristics, opportunities and challenges of OFPW were studied. This chapter provides an important insight into using different novel treatment methods for management of OFPW.

## 21.1 Introduction

The basis of the hydrocarbon (HC) industry is to carry out exploration and exploitation and to produce oil and gas from subsurface formations. While producing oil and gas, a large amount of oilfield produced water (OFPW) is produced where the water-to-oil ratio (WOR) is nearly 7:1 [447–450]. When HCs are produced by water drive reservoirs, OFPW production increases with the ageing of oilfields. OFPW contains both organic and inorganic compounds, dissolved and suspended solids, production chemicals (which were used during production activities) and solid particles due to leaching of rocks and corrosion of pipelines [447, 451]. If OFPW is directly disposed into the environment without treatment it may affect the salinity, biochemical oxygen demand (BOD) and chemical oxygen demand (COD) of the water sources which contain contaminants and are lethal to human, animal and plant life [122, 452].

OFPW has to be treated before disposal into the environment as the usable water resources are decreasing rapidly. A variety of OFPW treatment techniques have been proposed and applied during past decades, which can be subdivided into physical, chemical,

**Table 21.1** Various sources of oilfield water

| Explora-tion | Drilling | Production | Transportation |
|---|---|---|---|
| Natural seeps | Drilling fluids disposal | OFPW | Tankers |
| | Pits | Work over fluid | Pipeline |
| | Blowout | Blowout and wreckage | Terminal |
| | | Production testing | |
| | | Operation failure and accidents | |
| | | EOR applications | |

biological and membrane treatment processes [447, 453, 454]. The conventional methods to handle OFPW after treatment are direct discharge, re-injection into the disposal well or use for secondary reservoir pressure maintenance. The later method is expensive as the treated OFPW has to be transported back to the injection well often by a pipeline. Sometimes the HC industry sets up OFPW treatment systems on the field site, so the cost of transportation can be cut down, but the cost of bringing it to the surface is constantly present [425, 448, 455–457]. The various sources of oilfield waters are given in Table 21.1.

## 21.2 Characteristics

The physical and chemical properties of OFPW depend on the geographic field location, properties of the geologic formation at the subsurface where the water was present and the type of HC being produced [454, 458]. OFPW discharged from gas wells is more toxic than OFPW produced from oil wells, but its production is very less in comparison to oil wells [122, 459, 460].

### 21.2.1 Constituents

#### 21.2.1.1 Dispersed oil

OFPW contains contaminants of small discrete oil droplets suspended in water as dispersed oil, which affects the aquatic or marine ecosystem by altering the BOD. The amount of dispersed oil in OFPW will vary depending on several factors such as the density of the oil, the shear history of the droplet, the amount of oil precipitation and the interfacial tension (IFT) between oil and water [427, 461–463]. On average, the size of the droplet varies from 4 to 6 μm whose recovery requires a complex process.

#### 21.2.1.2 Dissolved organic compounds

A study of dissolved organic compounds is important because they are related to the origin and migration of oil accumulation, as well as to the disintegration or degradation of oil accumulation [456, 464–466]. The dissolved organic compounds include volatile HCs such as polycyclic aromatic hydrocarbons (PAHs), organic acids and phenols. The concentrations of these compounds are usually higher in OFPW of gas-producing platforms than in OFPW of oil-producing platforms. In general, the more alkaline the water is, the more likely it will contain higher concentrations of dissolved organic constituents [428, 429, 467, 468]. The solubility of some dissolved organic compounds increases with temperature and pressure if pressure is maintained on the system. The solubility decreases with increasing water salinity.

#### 21.2.1.3 Treatment chemicals

Biocides, reverse emulsion breakers and corrosion inhibitors are used during the treatment of OFPW, which are highly toxic even at low concentrations. The corrosion engineer needs to know how much of the inhibitor added at the well remains in OFPW so that corrosion can be controlled [432, 469, 470]. Organic scale inhibitor compounds are monitored routinely to verify that their concentration exceeds the minimum effective dose.

#### 21.2.1.4 Produced solids

The precipitated solids, clay, sand, silt, carbonates and corrosion products are produced from the subsurface formation and wellbore operations. The composition of produced solids in OFPW depends on several factors such as composition of the water in the depositional environment of the sedimentary rock, subsequent changes by rock–water interaction during sediment compaction, changes by rock–water interaction during water migration and changes by mixing with other waters, including infiltrating meteoric waters [471–473].

#### 21.2.1.5 Metals

The elements zinc (Zn), lead (Pb), manganese (Mn), iron (Fe) and barium (Ba) are usually present in OFPW. Their presence in low concentration is less toxic, but they may form undesired solids which adversely affect the treatment process [474–476].

### 21.2.2 Theory of Oil–Water Emulsion

Environmental studies on OFPW deal with the investigation of oil-in-water emulsions, and dissolved oils will be present as microdroplets in water. While studying oil-in-water concepts, we have to understand the theoretical and empirical concepts of the interfacial phenomena underlying emulsion behaviours [477–479]. Due to free-energy considerations, an emulsion should be inherently unstable because the dispersed oil could lower its interfacial free energy per unit mass by coalescing into progressively larger droplets. Two concepts may be considered to explain why emulsions can remain dispersed for long time periods. Firstly, droplets must be able to come into reasonable proximity before aggregation and finally coalescence can take place [480–482]. Secondly, even if droplets succeed in coming into reasonable proximity, coalescence may not be favoured because of properties existing at the interface.

### 21.2.3 Particle Size Analyses

The distribution of droplet sizes as a function of time is a fundamental indicator of emulsion stability. Furthermore, the degree of spread of particle sizes will also have an effect on stability. Given the inherent instability of emulsions and the tendency towards minimisation of interfacial energy, larger droplets should be favoured to grow at the expense of smaller ones. Hence, dispersion with a wide range of particle sizes will exhibit a continuous shift in particle size distribution towards a predominance of larger particles [483–485]. Dispersion with a narrower range of particle size will exhibit greater stability.

Apart from stability considerations, the functional behaviour of oil-in-water emulsions with regard to oil removal is inevitably related to particle size distribution because of the influence of that factor on the performance of processes such as sedimentation/flotation, filtration and adsorption. As well, one would intuitively expect the biological activity in either toxicity or biological accumulation studies to be strongly influenced by particle size distribution [457, 486, 487]. The techniques for measurement of particle size distribution may be grouped into three categories:

(1) Counting based on electrical properties of a solution as modified by particles
(2) Counting based on light transmission/scattering properties of particles
(3) Counting based on microscopic observation

## 21.3 Treatment Methods

### 21.3.1 Physical Treatment

#### 21.3.1.1 Sand filter

The sand filter treatment is the most effective and economical for treating OFPW. Filtration through granular media is designed to remove the residual flocs in the settled water. This generally removes metals, which requires a series of pretreatment steps such as pH adjustment, an aeration unit and a solid separation unit. Filter performance depends on the type, depth and size of filter media and the hydraulic loading rate on the filter. Performance is measured by filtered water quality and by the length of filter runs between basic washes [488–491].

#### 21.3.1.2 Hydrocyclones

Hydrocyclones can remove dispersed oil from OFPW by injecting a high-velocity stream tangentially into the conically shaped hydrocyclones, which creates a vortex. The radial acceleration created in the hydrocyclone forces the more dense water to the outer edge of the hydrocyclone and the less dense oil to the centre. The oil is then produced out of one end of the hydrocyclone and the water out of the other. The effectiveness of hydrocyclones in separating oil and water depends on a large number of parameters, including oil droplet size and oil–water density difference, inlet water velocity, solution gas, solids and system geometry [457, 488]. The major drawback of using a hydrocyclone is its low efficiency and inability to remove dissolved components.

#### 21.3.1.3 Evaporators

Evaporators use the evaporation principle for treatment of OFPW which does not require any chemicals and eliminates the risk of secondary sludge handling and highly skilled labour. On the other hand, the requirement of energy is very high, which increases the operating cost. The energy consumption could be brought down by reusing hot vapour to heat the fresh feed [447, 492].

#### 21.3.1.4 Dissolved air precipitation

OFPW is saturated with air in a packed column separator, where the pressure is released for the formation of micro air bubbles. It induces the flotation of aliphatic and aromatic HCs and removes the aliphatic compounds more efficiently than aromatic compounds. The reduction of bubble size and increase of air flow rate would lead to higher removal efficiency [457, 486].

#### 21.3.1.5 Freeze-thaw process

OFPW can be treated by freezing, which is a crystallisation process and uses the principle of solubility dependency on temperature. When the solution is cooled below the freezing point of the solvent but not below the depressed freezing point of the solution, relatively pure crystals of solvent and unfrozen concentrated solutions are obtained. The process is capable of removing huge amounts of HCs at low cost. But the requirements of subzero ambient temperatures and large land surface are its disadvantages [447, 457, 486].

#### 21.3.1.6 Physical adsorption

OFPW can be treated by activated carbon, organo-clay, copolymers, zeolite and resins. The combination of activated carbon and organo-clay proved to be more efficient in removing HCs. Copolymers reduce the oil content and zeolites are efficient in removing volatile organic compounds (VOCs) [453, 479, 493]. Adsorption is the most favourable method for pollutants removal from OFPW because of its low cost, ease to operate and simple design. The activated carbon is in granular powdered form, where the stronger residual forces on the surface adsorb the oil when they come in contact but the bonds between adsorbent and adsorbate are very weak and can be easily desorbed. The adsorption capacity of carbon can be measured by determining the adsorption isotherm experimentally where data can be obtained by treating a fixed volume of OFPW with a series of known weights of carbon. Activated carbon has been utilised for treatment of OFPW, but the focus has been shifted to developing low-cost agroindustrial waste (AIW) adsorbents such as fruits, rice husk, straw, coconut wastes, coffee wastes, vegetable peels and sludge [453, 494].

### 21.3.2 Chemical Treatment

#### 21.3.2.1 De-emulsifiers

De-emulsifiers are used to break OFPW emulsions into two clear immiscible phases, oil and water phases, by varying concentrations and temperatures. Most of the de-emulsifiers are polymers and surfactants which have a complex structure and reduce the oil–water IFT. The presence of solids like silts, iron sulphide and paraffin in OFPW complicates the process of separation. The separation rate of OFPW emulsions depends on the type of de-emulsifier used, the stability of the emulsion, temperature of de-emulsification, concentration of the de-emulsifier, residence time and the agitation energy of the de-emulsification process [495, 496].

#### 21.3.2.2 Precipitation

The suspended solids and colloidal particles can be removed by coagulation and flocculation. The most widely used coagulants is lime ($CaOH_2$) or sodium hydroxide (NaOH) to increase the pH of water. At high pH, dissolved solids, including heavy metals, tend to precipitate as a $OH^-$ sludge. Lime plus $Na_2CO_3$ can also be used to enhance the precipitation of $CaCO_3$. The main disadvantage of this process is its ineffectiveness for high concentration of metals in the sludge formed [479, 497, 498].

#### 21.3.2.3 Oxidation

It uses a combination of strong oxidants such as ozone ($O_3$) and hydrogen peroxide ($H_2O_2$), along with irradiation such as ultraviolet (UV) radiation and a catalyst such as a photocatalyst. It oxidises the organic components to their highest stable oxidation states. When OFPW contains highly toxic compounds in the form of aromatic fractions, then this method is effective to degrade these compounds [460, 479]. In photoelectron catalysis, the COD removal efficiency and toxicity reduction were found to be higher. The COD and dispersed oil content can also be reduced by combining flocculation with the Fenton oxidation adsorption process [493, 499].

#### 21.3.2.4 Electrochemical process

The electrochemical process consists of mostly one active metal and graphite as an anode and iron as a cathode, with the intermediate space filled with an electrolyte. During the process, $Mn^{2+}$ is formed, which oxidises and coagulates the organic contaminants [500, 501]. The advantages of this process are simple operational management, robustness and short-term adaption to OFPW fluctuations by simply switching the power on and off or adjusting the current density. The process can eliminate both organic and inorganic contaminants. The process can also degrade the organic contaminant and contains the potential for complete, residue-free mineralisation as a solution against water scarcity. It also has increased application for substances which are impossible to biodegrade [468, 479, 492].

#### 21.3.2.5 $O_3$

Ozone is an excellent disinfectant of OFPW through oxidation. Mostly, $O_3$ is added for pre-oxidation, before a sand filter as the filter can remove the remaining organic matter. It removes micro-pollutants such as pesticides, enhances the coagulation process and eliminates odour and taste through oxidation [432, 466, 473]. The organic matter is partly oxidised and becomes more easily biodegradable later in subsequent filters. The process is very efficient for preparing drinking and usable water, but it requires a high initial operating cost [473, 481].

### 21.3.3 Biological Process

The objective of biological treatment is to reduce the concentration of organic and inorganic compounds. The process involves oxidation and stabilisation of pollutants by microbes such as bacteria, protozoa, algae and fungi in the process of their growth and metabolism [464, 487, 499]. They use pollutants as nutrients and break them down into harmless products. Biological treatment processes combined with physical and chemical methods make a treatment system which is widely used in domestic and industrial waste treatment. The earliest used method of natural dilution was based on the principle that with time the low concentration of pollutants in the diluted waste will be destroyed by the natural flora of the receiving body. The domestic septic tank is a simple example of the biological treatment process. The basic concept in biological treatment is to cultivate a mixed culture of actively growing microbes as minifactories in which the organics serve as food [425, 465, 485]. Since the process is based on the bulk growth of living entities it is controlled by factors such as type of nutrients, temperature, hydrogen ion concentration of the medium and presence of toxic substances in OFPW. Some microbes require oxygen, while others grow in the absence of oxygen. Different groups of microbes prefer to grow on different nutrients. Therefore, selection of a suitable biological process depends on the pollutants to be removed, the type of microbes to be employed, convenience of the operation and the ultimate use of the treated water [478, 493, 501].

The biodegradation of petroleum and other HCs in the environment is a complex process, whose quantitative and qualitative aspects depend on the nature and amount of the oil or HCs present, the ambient and seasonal environmental conditions and the composition of the autochthonous microbial community [449, 466, 498]. Microbial degradation of oil has been shown to occur by attack on aliphatic or light aromatic fractions of the oil, with high-molecular-weight aromatics, resin and asphaltenes considered to be exhibiting only low rates of biodegradation, although some studies have reported their removal at high rates under optimal conditions. In aquatic ecosystems, dispersion and emulsification of oil slicks appear to be pre-requisites for rapid biodegradation; large masses of mousse, tar balls or high concentrations of oil in quiescent environments tend to persist because of low surface areas available for microbial activity. Petroleum spilled on or applied to soil is largely adsorbed to particulate matter, decreasing its toxicity but possibly also contributing to its persistence [427, 486, 491]. Biodegradation rates generally increase with increasing temperature; ecosystems exposed to extremely low temperatures degrade HCs very slowly. The microbial degradation of petroleum in aquatic environments is limited primarily by nutrients such as nitrogen and phosphorus; salinity and pressure may be important in estuarine and deep sea regions, respectively. Oxygen, nutrient concentrations, moisture and pH are predominant factors in determining biodegradation rates in soil [451, 468, 471].

The principal aim of biological treatment is to reduce the BOD load of effluents and to remove toxic substances from OFPW. The biological treatment results, in part, in the complete oxidation of the organics to mineral end products and, to a large extent, by conversion of dissolved organic compounds to settle able microbial biomass and reduction in heavy metal concentrations [463, 476, 477]. The need for oxygenation in HC biodegradation, along with the great sensitivity of anaerobic treatment systems to chemical toxicants, effectively restricts biological treatment systems for

OFPW to aerobic ones. Oxidation ponds, activated sludgefacilities, trickling filters and rotating biological contractors are utilised for the biological treatment of OFPW [459, 466].

### 21.3.4 Membrane Separation

The conventional methods of treating OFPW are capable of removing suspended particles with a particle size of 5 μm or above and could not meet the stringent disposal and re-injection regulations which are being updated periodically. Therefore, new technologies such as the membrane separation should be implemented to remove fine particles, colloids, high-molecular-weight organics and dissolved components [489, 502]. The membrane separation process includes MF, UF, NF and RO, which treat and generate water with high standards to meet regulations. The treated OFPW, which is also known as permeate, passes through various pore sizes of the membrane. The membranes are categorised as porous or dense by the type of materials and the mechanism of separations [429, 460, 475, 493]. The separation mechanism of dense membranes is based on the physicochemical interaction of the permeate and the membrane material, whereas separation of porous membranes is based on the mechanical separation by size of the permeate and pore sizes of the membrane. The mechanical resistance of the membrane depends on the membrane thickness, and therefore it should have a thin layer of material, narrow pore sizes, and high porosity [462, 475, 496]. On the basis of pore size, the membrane processes could be classified as MF, UF and NF. The membranes are also classified as organic, inorganic and composite membranes. The membranes are also classified as flat sheet membrane (FSM) and hollow fibre membrane (HFM), the latter being highly recommendable as it has higher packing density, ability to induce movement by mechanisms such as bubbling and the feasibility of backwashing. An ideal membrane should be mechanically resistant, have a high permeate flow rate and have high selectivity for a specific component [462, 501, 502].

The most important advantages of using membrane separation are the ease of operation and little or no requirement of chemicals. It also has other advantages such as sludge reduction, high quality of permeate, smaller space needed, minimal impact on permeate quality with variation in feed water quality, recycling of waste streams, possibility of having an automated plant, ability to be combined easily with other separation processes, low energy consumption and continuous separation [455, 466, 475, 489]. The primary disadvantage of using membranes is fouling. Irreversible and reversible fouling occurs while treating OFPW. The usage of an appropriate pretreatment process reduces membrane fouling to a greater extent. The other disadvantage is that it has certain high moderate capital and operational recurring costs [464, 485, 492].

During the treatment of OFPW using membranes, the permeate flux declines gradually. This decline may be due to concentration polarisation or fouling—a complex mixture comprising of colloidal, bacterial floes and dissolved macromolecules [448, 466, 473]. It may be due to biofouling, scaling, organic fouling and colloidal foulingṪhe microbial contamination of water may produce a biofilm on the surface of the membrane, which will decrease water permeation through the membrane. The salts may precipitate on the membrane surface, which will cause scaling. The organic fouling leads to coating of the surface or plugging of the pores in the porous support layer by HCs [459, 470, 498]. The accumulation of clay and silica on the surface of the membrane will cause colloidal fouling. The increase of suspended and extractable extracellular polymeric substance (EPS) concentration leads to a decrease in filterability. Polysaccharides often form a significant component of EPS and have higher molecular weights and a wider molecular weight distribution than proteins. Due to higher transmembrane pressure (TMP), fouling is more severe and reduction of membrane pore size occurs, resulting in higher final rejection [429, 452]. At the beginning of filtration, the rejection profile was low, indicating that the maximum water went through the membrane. But with time, the permeate decreases due to reduction of the overall size of the membrane pores and increase of the rejection profile [474, 497].

There are two types of membrane fouling, reversible and irreversible. In reversible fouling, the colloidal particles are deposited on the membrane surface and membrane pores to decrease the flux, which can be abolished by pure water backwashing. Irreversible fouling occurs due to strong chemical or physical sorption of particles on the membrane surface and in the membrane pores [495, 499, 501]. The method for recovering the flux declination of irreversible fouling is by washing the membrane with acid or alkali solutions. However, aggressive cleaning methods may not be able to recover the initial permeability of irreversibly fouled membranes. In general, fouling is a widespread and costly problem that affects membrane performance and is a complex function of the feed characteristics, membrane properties and operating conditions [456, 461, 491].

## 21.4 Management Practices

The main focus of the oilfield industry is on economic efficiency and productivity. Simultaneously, environmental regulation adds a new factor to petroleum engineering economics, the cost of working within the constraints of an environmental issue. The industry is absorbing this cost, while revising oilfield technology to reduce the environmental impact. The technical approach to the environmental issue is undertaken after the nontechnical means become redundant or uneconomical [466, 468, 495]. The waste of OFPW has to be managed efficiently because there is no other way to protect the environment. Oilfield operations are waste-intensive and have a lot of environmental implications. With the increasing public awareness on pollution and environmental protection, the governments of the countries world over have come out with stricter regulatory measures in line with the policy of sustainable development [429, 453, 481].

India is one of the foremost countries to make environmental protection as the constitutional commitment of the nation and protection of wildlife and quality of water bodies as the duty of every Indian citizen. The developments in science and changes in technology throw up new challenges in environment management by way of newer regulations by the government. In addition to this, India's commitments in the international forum on the environment do add to our responsibility in managing the environmental impacts related to operations [463, 464, 481].

The increased volume of OFPW handled in both onshore and offshore petroleum production operations is becoming a major concern, especially with the possibility of further reduction in the oil content allowed in the discharged water as well as the fact that OFPW contains a number of undesirable toxic components [450, 481, 493]. Handling this volume of water is of prime concern to all oil companies wherever they operate.

### 21.4.1 Reuse

Oil and gas operators routinely inject the treated OFPW into petroleum formations for secondary reservoir pressure maintenance which displaces oil to production wells. The life of the reservoir goes on decreasing due to production of oil and gas, which eventually decreases the subsurface reservoir pressure [473, 481, 492]. Therefore, the reservoir pressure of the matured field has to be maintained to continuously produce oil through the water-conning or water-flooding process. The production wells of the matured field generate a huge amount of water, where certain wells have a WOR value of 9:1. The reuse of OFPW in oil and gas operations also includes the use of water during well drilling and hydraulic fracturing, which require large water volumes [466, 480, 495]. Use of freshwater supplies in these practices can be minimised by treating and recycling of OFPW resources. The treated OFPW for reuse in oil and gas operations is highly recommended for the industry, which is cost-beneficial as it cuts the transportation and disposal costs.

### 21.4.2 Injection into Disposal Wells

OFPW has to be injected into a porous and permeable non-HC-bearing zone which should not be an aquifer [474, 487]. A previously productive oil and gas zone well which is now depleted and is porous and permeable is also suitable for disposal of treated OFPW after the parameters meet provided government regulations. The cost of transportation of the treated OFPW to the disposal wells has to be considered [474, 492].

### 21.4.3 Direct Discharge

The direct discharge of OFPW is a relatively inexpensive method of OFPW disposal. However, this is only an option for relatively high-quality OFPW which can meet the regulation norms set by the pollution control board [471, 492]. If the water is of poor quality, then the surrounding soil, water and vegetation will be contaminated, and the regulatory guidelines do not permit the oil industry to contaminate the environment. The direct discharge of OFPW in the water bodies will also affect marine flora and fauna, which will later affect human organisms consuming them [495, 501].

### 21.4.4 Consumption for Beneficial Use

In areas where disposal opportunities are limited or disposal capacity of a well is less than plentiful, it may be advantageous to utilise some form of treated water for beneficial use in the industry for cleaning purposes or in agriculture and household cleaning purposes [463, 496]. It can also be used for farm needs such as cooling of facilities for the animals and products, dairy sanitation and wash-down of facilities and animal waste disposal systems. Water quality is an important consideration for consumption of water for beneficial use. If discharge of the treated water does not meet the government regulations for domestic use and the available water sources are abundant, then an efficient system of treating OFPW will be uneconomical for the industry [428, 495]. Then the industry should either go for reuse in oil and gas operations or inject it in disposal wells.

## 21.5 Opportunities and Challenges

The rapid growth of the world population poses a great challenge to our drinking water supply. Agriculture and energy production draw a huge amount of fresh water, which is depleting the already scarce freshwater resources. OFPW streams which were once treated as waste are now considered as a valuable resource [463, 492, 494].

The reuse of OFPW faces challenges which include high treatment cost, potential chronic toxicity of the treated OFPW and public acceptance. Also, the amount and properties of OFPW change over time, which makes the job more difficult. The treatment cost of OFPW strongly depends on the physical and chemical characteristics of OFPW, which can vary widely among different production fields and the regulatory environment [458, 475, 477]. The PW from gas production typically has lower oil and grease (O&G) and total dissolved solids (TDS) content than that from OFPW. Therefore technology solutions for treatment and reuse of water would need to be customised according to the properties of the water and the amount to be treated.

The challenges of the membrane treatment process include concentration, polarisation or membrane fouling, low selectivity or low flux and low membrane lifetime. Membrane fouling is a function of feed properties, membrane properties and hydrodynamics [462, 494]. One of the most important parameters in controlling the flux is the crossflow velocity, because as pressure increases, flux increases linearly. The effect of fouling can be mitigated through different strategies. For example, oil, solids and gels can be reduced by using some pretreatment processes of coagulation, granular activated carbon, and low-pressure membrane process. The backwashing of the membrane, operating at higher temperature, and addition of disinfectants and antiscaling agents may control fouling. The most possible foulant for NF and RO membranes after UF treatment could be $Ca^{2+}$ bound to inorganic materials and silica bound to organic materials [450, 496]. Among the cleaning methods, backwashing has been used widely but is not recommended, as the membrane usually experiences a reduction in permeate flux after each backwash. Another problem with backwashing is that the operation must be stopped to backwash the membrane. Washing intervals depends on the O&G concentration in the feed, permeate flow rate, success of washing and the period of membrane degradation [462, 488, 501].

The RO process is most likely used for potable reuse applications due to desalination and removal of a large number of organic compounds. Although RO is capable of removing many organic compounds at high efficiency, the combined chronic toxicity of the organic compounds existing in the permeate needs to be carefully evaluated before direct reuse is implemented [449, 455]. The current drinking water standards were established on the basis of the human health risk associated with individual contaminants. A method for establishing a maximum safe contaminant concentration on the basis of the combined impact of the contaminant mixture is needed. The total organic carbon (TOC) concentration of the RO-treated water is low, but the presence of iodide and bromide can potentially form disinfection by-products (DBPs) which are more toxic than chlorinated DBP [477, 487, 498].

Finally, the greatest barrier to domestic wastewater reuse is probably public acceptance. However, the existing technologies have been demonstrated to meet current water standards, and there are concerns on unknown toxic effects or unknown toxic compounds. Until the human health impact of chemical mixtures is better understood, toxicity assays are necessary to address concerns on the potential synergy between toxic compounds and the possibility of undetected toxic compounds in the treated OFPW [458, 466].

## 21.6 Conclusion

OFPW has been treated using different methods of operation. The criteria used for comparing the technologies are reliability, mobility, robustness, flexibility, modularity, cost, chemical and energy demand, and residual disposal requirements. While selecting an OFPW treatment process, many processes and water quality–specific factors should be taken into account. The temperature of the feed water determines which type of desalination treatment process should be employed, since many technologies work more efficiently at high temperatures, while others use feed streams at low temperature. It is also important to consider the type of ions which need to be removed through different ion removal treatment processes. The treatment of OFPW through membrane processes removes divalent ions to a greater extent than monovalent ions, which may make the sodium adsorption ratio higher and render the water less suitable for beneficial use as irrigation water or surface discharge. Fouling of membranes is one of the major drawbacks of treating OFPW, which depends on permeate flux and its stability on time but can be minimised by using hydrophobic membranes. The modelling of permeate flux decline during treatment of OFPW is important from both economic and technological points of view. The empirical models need more experimentation for making accurate predictions. The theoretical models are useful for the prediction of permeate flux under different operating conditions without the need for time-consuming experiments, but there is no theoretical model which can accurately describe the whole process. Therefore, it is required to work on the improvement of the theoretical models to make more accurate predictions and to better understand the fouling mechanisms. The membrane life cycle can be increased through some pretreatment methods. Each method has some advantages as well as some disadvantages, and a combination of methods may be more useful for treatment of OFPW to meet different water quality requirements. Utilisation of a certain type of process or a combination of processes is highly dependent on the characteristics of OFPW. The developments of AIW as adsorbents which are effective in pollutant removal needs to be modified by some physical and chemical processes, including drying, milling and sieving grafting co-polymerisation, in order to increase the adsorption capacities.

The design of a treatment process wholly based on the literature is not possible as the characteristics of OFPW differ from well to well. The unpredictable and rapid onset of upsets in the OFPW treatment process often results in unplanned maintenance and production losses. Ceramic membranes may be a viable treatment technology for many OFPW applications. The capital cost of ceramic membranes is presently higher than polymeric membranes, but ceramic membranes are advantageous over polymeric membranes because of increased chemical, mechanical and thermal stability and, therefore, higher lifespan and higher productivity. The international standards demand more efficient separation systems than those now in common use. More research and development is required in membrane development, which is considered to be at the development stage, although many new products show promises. OFPW, which was once considered as waste, has to be treated properly to consider it a valuable resource, but unknown toxic effects and public acceptance are important barriers which have to be eliminated.

## Acknowledgements

We would like to acknowledge the following:

(1) Indo-Tunisia Project No. DST/INT/TUNISIA/P-02/2017 entitled 'Characterisation of Oilfield Water to Mitigate a Major Environmental Hazard with Recovery of Usable Water'
(2) DBT Twinning Project No. BT/485/NE/TBP/2013 entitled 'Remediation of Produced Water from NE Oilfield by Microbial and Membrane Intervention with Modelling of an Effluent Unit for the Recovery of Usable Water'

Chapter 22

# Experimental Work to Evaluate the Effects of Disposed Produced Water on the Surrounding Environment: A Case Study in Libya

**Jeevarupini Balakrishnan,[a] Juhairi Aris Muhamad Shuhili,[a] Tarek Arbi Omar Ganat,[a,b] Subrata Borgohain Gogoil,[b] Mohamad Nasir Mohamad Ibrahim[c] and Raoof Gholami[a]**

[a] *Department of Petroleum Engineering, Universiti Teknologi PETRONAS, P.O. Box 32610 Seri Iskandar, Perak Darul Ridzuan, Malaysia*
[b] *Department of Petroleum Technology, Dibrugarh University, Assam, India*
[c] *School of Chemical Sciences, University of Sains, Malaysia*
tarekarbi.ganat@utp.edu.my

Numerous studies showed that surrounding environments located nearby oil or gas fields are often contaminated by the disposal of produced water (PW), which often contains residual hydrocarbons (HCs) and some other contaminants if the effluent is improperly treated. This problem becomes more critical once the areas of PW disposal are located near residential or agricultural areas, which is the main objective of investigation in this study. The aim of the study was to analyse the impact of PW disposal from oil and gas activities towards a nearby area which is inhabited and surrounded by many farms. The oilfield is located less than 40 km

*Advances in Petroleum Technology*
Edited by Subrata Borgohain Gogoi

ISBN 978-981-4877-19-0 (Hardcover), 978-1-003-04993-7 (eBook)
www.jennystanford.com

away from the residential area and farmland (FL). There are six existing freshwater wells in the area within the vicinity of the oilfields and two freshwater wells in the FL used for daily activities and irrigation of the plantation. Six samples were collected from freshwater wells around the oilfield and two freshwater samples from the FL area. Three samples were collected from the three wells in the PW disposal pit (DP). Mainly, tests were conducted to measure the contents of the water samples for impurities such as $K^+$, $Na^+$, $Ca^{2+}$, $Mg^{2+}$, $Cl^-$, $HCO^{3-}$, $SO_4{}^{2-}$ and total dissolved solids (TDS). A variety of experimental methods were performed to measure different impurity concentrations. The results indicated that pH values are not governed by any concentration of anions. Well *H* showed the highest concentration of overall contaminants of 73,500 mg/l. The concentration of contaminants showed a decreasing trend in overall contents of TDS as the samples were taken from wells away from the PW DP, and the findings of the study show an alarming situation for areas surrounding the oilfield.

## 22.1 Introduction

Typically, the produced water (PW) composition during oil and gas production activities is different than the common ground water composition. PW usually registers a higher reading of impurities, such as traces of hydrocarbons (HCs), solid particles and chemicals used during the operation [427]. An endless debate on how to handle the produced solids, liquids and chemicals left over is occurring since the discussion often ends up in an impasse between cost optimisation and safety precautions [503–505]. Demulsification of PW and HCs is first carried out in a separator, which is affected by pH and temperature. The reduction of pH of the oil–water emulsion promotes the dissolution rate of silica ($SiO_2$) in the effluent, which facilitates the separation of oil and water. Temperature has a significant role to play when it comes to separation of oil and water since the interfacial tension (IFT) between two immiscible phases is too governed by temperature [506]. Emphasis shall be given on adjusting pH for better separation of oil and water, which ensures the quality of water disposed.

This is the reason why pH is always an important measurement during water treatment. In the demulsion process, the IFT must be increased to facilitate the separation of oil and PW [507].

## 22.2 Experimental

### 22.2.1 Sample Collection

Various PW samples were collected from disposal pits (DPs) of wells G, H and I, as shown in Figs. 22.1–22.5.

### 22.2.2 Methodology

#### 22.2.2.1 Spectrophotometry

Different methods were carried out in determination of the dissolved ions in the water samples collected from different wells. The experiment used to estimate the concentration of potassium

**Figure 22.1** Produced water at the disposal pit.

Figure 22.2 Disposal pit.

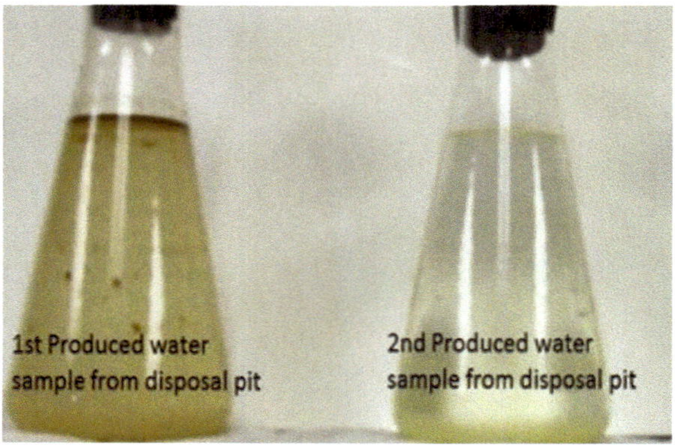


Figure 22.3 Disposal pit samples.

and sodium ion was spectrophotometry. The light emitted by the light source was reflected or absorbed by the potassium ($K^+$) and sodium ($Na^+$) ions in the sample solution. A higher amount of light absorbed by the photocell indicates a low concentration of sodium and potassium. The concentration of divalent cations such as

**Figure 22.4** Freshwater well sample.

**Figure 22.5** Farmland sample.

**Table 22.1** Composition analysis in mg/l of freshwater, disposal pit (DP) and farmland (FL) samples from all wells (W)

| W | SS | pH | $K^+$ | $Na^+$ | $Ca^{2+}$ | $Mg^{2+}$ | $Cl^-$ | $HCO_3^{1-}$ | $SO_4^{2-}$ | TDS |
|---|---|---|---|---|---|---|---|---|---|---|
| A | | 6.36 | 74.6 | 1036.7 | 389.5 | 146.1 | 1825 | 766.2 | 701.5 | 4939.6 |
| B | | 6.92 | 97.4 | 1555.0 | 665.5 | 118.7 | 2730 | 751.5 | 1139.5 | 7057.6 |
| C | Fresh water | 6.60 | 75.1 | 548.6 | 422.4 | 126.4 | 1700 | 778.4 | 719.9 | 4370.8 |
| D | | 6.81 | 108.5 | 2162.7 | 596.0 | 220.0 | 4260 | 490.5 | 792.7 | 8630.4 |
| E | | 6.16 | 114.8 | 1396.7 | 667.5 | 187.9 | 2875 | 424.7 | 1191.0 | 6857.6 |
| F | | 6.42 | 75.8 | 1015.0 | 409.3 | 128.7 | 1677 | 823.2 | 786.8 | 4915.8 |
| G | | 7.31 | 309.0 | 12150.0 | 1609.3 | 363.2 | 20971 | 742.2 | 1587.7 | 37732.4 |
| H | DP | 6.37 | 450.6 | 3710.3 | 24096.7 | 601.9 | 45286 | 277.5 | 652.3 | 75075.3 |
| I | | 6.81 | 211.0 | 1379.5 | 15081.3 | 408.1 | 19625 | 453.9 | 968.8 | 38127.6 |
| J | FL | 6.70 | 23.0 | 365.2 | 201.0 | 110.2 | 900 | 115.2 | 475.3 | 2189.9 |
| K | | 6.70 | 22.0 | 400.4 | 207.1 | 108.7 | 950 | 110.9 | 444.9 | 2244.0 |

$Mg^{2+}$ and $Ca^{2+}$ was measured using complexometric titration. The filtered sample was titrated with cyclohexanediaminetetraacetic acid (CDTA) solution until a blue endpoint appeared where the colour changed from red to blue.

#### 22.2.2.2 Mohr's method

The $Cl^{-}$ ion concentration was determined by Mohr's method, which was done by titrating with silver nitrate in the presence of $K_2CrO_4$ solution. The titration process changed the colour of the solution from lemon-yellow to red.

#### 22.2.2.3 Gravimetry and titration

The $SO_4{}^{2-}$ ion concentration was determined using the gravimetric method, which requires precipitation and heating of a $SO_4$ compound. The $HCO_3{}^{1-}$ ion concentration was measured using common titration with HCl.

## 22.3 Results and Discussion

### 22.3.1 Sulphate Concentration

As shown in Table 22.1, well G recorded the highest pH value due to a high $SO_4^{2-}$ ion concentration. $SO_4^{2-}$ ions are basic in nature, whose presence leads to a rise in pH. This relationship can be clearly construed for well G, which had the highest $SO_4^{2-}$ ions contents of 1587 mg/l compared to other wells and showed the highest pH reading of 7.31. Wells located in the DP showed the highest concentration of total contamination and the highest among the three wells located in the DP was well H, which had a TDS of 73,500 mg/l.

### 22.3.2 Bicarbonate Concentration

The main constituent which contributes to the acidity of the water sample is $HCO_3{}^{1-}$ ions. $HCO_3{}^{1-}$ ions are the ions in carbonic acid.

However, carbonic acid is considered as a weak acid and thus may not have much governance over the pH. On the contrary, well H, which had less concentration of $HCO_3^{1-}$ ions measured, showed a lower pH. In fact, no consistency can be inferred between the concentration of $HCO_3^{1-}$ ions and pH. The concentration of $HCO_3^{1-}$ ions was similar for all the water samples except for well H. The origin of $HCO_3^{1-}$ ions can be divided into $HCO_3^{-1}$ ions used in drilling fluid and $HCO_3^{1-}$ ions produced with formation water (FW) rich in $CO_4^{2-}$ and $Ca^{2+}$. However, the previous theory is more evident as the bicarbonate ($HCO_3^{-1}$) ion concentration was higher in the vicinity of the PW disposal site and showed minimal concentration at the FL. The filtration loss from drilling fluid should only affect the area around the field. Well H plausibly contained less $HCO_3^{1-}$ ions due to the severe reduction in the pressure and temperature of $CO_2$ at the well, which initiated precipitation of $CaCO_3$ from the PW [508].

## 22.3.3 Calcium and Sodium Concentration

The highest contaminant concentration was chloride ions, followed by $Au^{2+}$ and $Na^+$ ions. This finding did not concur with the common PW contaminant concentration, whereby the most abundant cation is $Na^+$. The $Ca^{2+}$ ion concentration is usually kept in check due to insolubility in water or brine due to some chemical compound formed by $Ca^{2+}$, such as $CaSO_4$. Unlike $Ca^{2+}$, a $Na^+$-based ionic compound has 100% solubility in water or brine. The insolubility of the compound induces precipitation in which the compound formed will be deposited as scale rather than being transported or dissolved. There are two different sources which can contribute to a higher concentration of $Ca^{2+}$ than $Na^+$. Drilling fluid used in drilling usually contains $CaCO_3$ and $Ca(OH)_2$ for density and pH regulation, respectively. Thus, the filtrate loss during the drilling operation contains these two minerals, which may be reproduced with PW from the reservoir. Another suggestion is that reservoir rock is $CO_4$, which dissolves in FW and is produced with PW. This theory is supported since reservoir rock of the Sirt Basin in Libya is predominantly carbonate reservoir rock [509].

## 22.4 Conclusion

The trend of contamination concentration decreases over the distance from the DP. The pertinent issue here was whether the contamination in the DP would be transported to the FL. The contaminate concentration from the water samples from the DP did not meet the European Union Water Quality Standard, which was used as a benchmark in this study. Supposedly, with the premise that the contamination is contained in the DP, there should not be any variation between the freshwater samples taken from the FL and oilfields. However, in this case, a significant difference can be noticed between the two samples, where the contamination level is higher in oilfield samples. This indicated the contaminants were being transported from the DP to the surrounding area. Despite this fact, another important element to take into account is the nature of rock permeability. Oilfields which bear HCs have the property of conducting fluid flow. Thus, the spread of contaminants is significant within the oilfield but limited inside the field. Nevertheless, the root cause of this problem which began from improper treatment of water in the DP shall be addressed at the utmost.

## Acknowledgements

The author would like to extend her utmost gratitude to her supervisor, Mr. Juhairi Aris, who supervised her throughout the analysis. A deep appreciation is also conveyed to Assoc. Prof. Tarek Arabi Omar Ganat, who collected the samples, ran preliminary analyses and shared the data with the author. Thanks also to the Libyan Petroleum Research Institute (LPRI), the laboratory where screening and preliminary runs were conducted. Thanks to personnel of the Petroleum Engineering Undergraduate Lab of Universiti Teknologi PETRONAS, who assisted in some of the experiments which were conducted for the sake of concept-proving in the analysis. Thanks to Prof. Subrata Gogoi from Dibrugarh University and Assoc. Prof. Mohamad Nasir, who shared the expertise in PW treatment from oilfields.

## Chapter 23

# Role of Sand Particle Size in the Retention of Total Petroleum Hydrocarbons and Heavy Metals

**Hassan ElGharbi,[a] Fatma Aloulou,[a] Wajdi Haj Ali,[b] Subrata Borgohain Gogoi[c] and Monem Kallel[a]**

[a] *Laboratoire de Génie de l'Environnement et Echo-Technologie GEET, Ecole Nationale d'Ingénieurs de Sfax, Université de Sfax, Tunisie*

[b] *La compagnie Franco-Tunisienne des Pétroles CFTP, QHSE, Sfax, Tunisie*

[c] *Department of Petroleum Technology, Dibrugarh University, Assam, India*

hassan.gharbi@enis.tn, monemkallel@gmail.com

Produced water (PW) is a mixture of formation water (FW) and injected water which is produced at surface facilities with oil and gas. In the Sidi Litayem oilfield, large volumes of PW are produced with oil-pumping operations. PW is generated on the surface as hot oilfield PW containing dissolved and dispersed oil, high concentrations of salt, heavy metals, polycyclic aromatic hydrocarbons (PAHs) and sometimes even radioactive materials. The present work focuses on the treatment of the Sidi Litayem oilfield PW with a sand filtration process and assesses its quality in comparison with Tunisian standards. One of the possible methods for getting rid of this production waste consists of re-injecting it into the geological structure. Another method consists of storing it near

*Advances in Petroleum Technology*
Edited by Subrata Borgohain Gogoi

ISBN 978-981-4877-19-0 (Hardcover), 978-1-003-04993-7 (eBook)
www.jennystanford.com

the drill sumps in special basins designed purposefully to separate hydrocarbons from the PW. Therefore, the environment assessment of PW has a big interest in order to preserve ecosystems in an area of exceptional biodiversity and to protect underground aquifers from such pollution. Five filtration tests in relation with the particle size of the sand were carried out under the same conditions. The yield of separation between water and oil was about 99% for the finest particle size with a very slight increase in comparison with Tunisian standards. For heavy metals, levels were decreased with a retention percentage of 84%, 93% and 82% for nickel (Ni), manganese (Mn) and chromium (Cr), respectively. Only Cr showed levels exceeding Tunisian standards.

## 23.1 Introduction

Produced water (PW) is a mixture of organic and inorganic pollutants produced on the surface during oil extraction. Annually, large volumes of PW are produced by the oil and gas industry. In 2009, PW volumes were estimated to be more than 70 billion barrels, of which 20 billion were produced by the United States only [510]. Several oil companies are using gravity separators for the removal of free oil as a first step of treatment, or 'primary treatment'. Emulsified oil is treated with different technologies such as dissolved air flotation (DAF), electrocoagulation-flotation and membrane separation as secondary treatment. However, these methods present several disadvantages, especially sludge production and/or high energy requirements. One of the possible methods of secondary treatment steps is sorption technology, which is a low-cost method which enables the recovery of emulsified oil with a possibility of regeneration. Different materials have been tested for the treatment of PW, such as zeolite [511], organoclay [512], bentonite [513] and activated carbon [514], demonstrating variable rates of adsorption. Sand was tested in the retention of heavy metals and organic compounds, showing a great impact on the retention of organic pollutants and metals [515, 516]. In this context, we have focused on the study of sand adsorption for the treatment of oilfield PW containing variable concentrations of emulsified oil or total petroleum hydrocarbons (TPH) and heavy metals. The study was kept as a function of the particle size of the sand.

## 23.2 Materials and Methods

### 23.2.1 Materials

Water samples are the PW of the Sidi Litayem oilfield. They were taken from the exit of the SIT1 station separators. They were stored in glass bottles and transported to the laboratory, where they were stored at a room temperature of 4°C. The used sand was purchased from the Gargouri Company, which makes simple preliminary treatments not affecting mineralogical composition. The analysis of the used sand was done as follows:

**Particle size analysis (PSA)**: The sand PSA was done on a set of AFNOR series sieves.

**X-ray diffraction (XRD)**: Powder samples were investigated using a Philips XRD with CuKa radiation, 45 kV and 35 mA, and scanning between 2 and 40 2q.

Chemical solutions were hydrochloric acid (HCl), sodium hydroxide (NaOH) and nitric acid ($HNO_3$).

### 23.2.2 Methodology

#### 23.2.2.1 Experimental

Separation experiments for the emulsion were carried out under the condition of a gravity-driven column filtration device of 40 mm diameter. A mass of 40 g of sand with a thickness of 25 mm was fixed at the bottom of the column. The emulsions were filtrated through this layer at a constant water head of 63 + 0.5 mm. Filtration of the samples was performed with an initial volume of 80 ml. Five particle sizes were tested, as follows:

- Test 1: Between 1 and 0.80 mm
- Test 2: Between 0.63 and 0.50 mm
- Test 3: Between 0.40 and 0.31 mm
- Test 4: Between 0.25 and 0.20 mm

- Test 5: Between 0.20 and 0.10 mm

Before each test, the sand column was purified with distilled water to eliminate turbidity. The process of sand filtration was performed at room temperature (25°C). Characterisation of the samples: The analysis of crude effluents and the filtrate of the five tests covered the following parameters: TPH, pH, electrical conductivity (EC), turbidity and suspended matter (SM).

(1) TPH: Approximately 250 ml of the sample to be analysed was extracted by a decanter with a volume separating funnel of 250 $cm^3$ of dichloromethane. This technique allowed the extraction of all organic compounds through the contact of the effluent with the solvent. After evaporation in a rotary evaporator, extracted hydrocarbons (HCs) were recovered and dried in ambient air. The dry residue weighed corresponded to the TPH. The extracted HCs were analysed through gas chromatography coupled with mass spectrometry (GC-MS) to identify and/or quantify accurately many substances.
(2) Physicochemical parameters (PCPs): The pH and EC were measured using a pH meter and conductivity meter, respectively. Turbidity was measured for samples filtered using the portable Turbidimeter Hach 2100 p. SM weight was determined by weighing the sample before and after drying overnight at 105°C. Heavy metals were determined by atomic absorption (Fisher Scientific Ice 3000).

## 23.3 Results and Discussion

### 23.3.1 Characterisation of PW

Characterisation of different parameters of the Sidi Litayem oilfield PW is given in Table 23.1. The PW of the Sidi Litayem oilfield had parameters exceeding largely the Tunisian standards for SM, TPH and chromium (Cr) levels. The pH was slightly below the standards, and the levels of nickel (Ni) and manganese (Mn) were under the required values. For copper (Cu), lead (Pb), zinc (Zn) and cadmium (Cd), levels were in the standard range.

**Table 23.1** Characterisation of the Sidi Litayem oilfield PW

| S. no. | Parameter | Value | Standard |
|---|---|---|---|
| 1 | pH | 6.16 | 6.5–8.5 |
| 2 | EC | 144.5 | – |
| 3 | SM (mg/l) | 337 | 30 |
| 4 | Turbidity | 202 | – |
| 5 | TPH (mg/l) | 1024 | 11 |
| 6 | Cu (mg/l) | 0.236 | 0.500 |
| 7 | Mn (mg/l) | 1.079 | 0.500 |
| 8 | Pb (mg/l) | 0.048 | 0.100 |
| 9 | Zn (mg/l) | 0.226 | 0.500 |
| 10 | Ni (mg/l) | 0.271 | 0.200 |
| 11 | Cd (mg/l) | QL* | 0.005 |
| 12 | Cr (mg/l) | 37.267 | 0.500 |

*Quantification limit of the used setup.

### 23.3.2 Filtration

The results of the TPH are given in Table 23.2. It can be observed that there was a decrease in the levels of TPH as a function of a decrease of the particle size of the sand. This decrease can be easily observed in the GC-MS spectra of the fives tests in Figs. 23.1–23.6. The GC-MS spectrum of the HCs extracted from the crude effluent was characterised by a unimodal appearance with abundance of low-molecular-weight n-alkanes. Alkanes began to appear from the carbon $C_{11}$ with a significant increase up to the $C_{14}$ carbon which showed the greatest abundance. The rate of alkanes then decreased with an increase in the molecular weight of n-alkanes.

**Table 23.2** TPH levels of the filtration tests

| S. no. | Sample | TPH content (mg/l) |
|---|---|---|
| 1 | Raw effluent | 1024 |
| 2 | Test 1 | 59.2 |
| 3 | Test 2 | 36.4 |
| 4 | Test 3 | 23 |
| 5 | Test 4 | 17 |
| 6 | Test 5 | 11 |

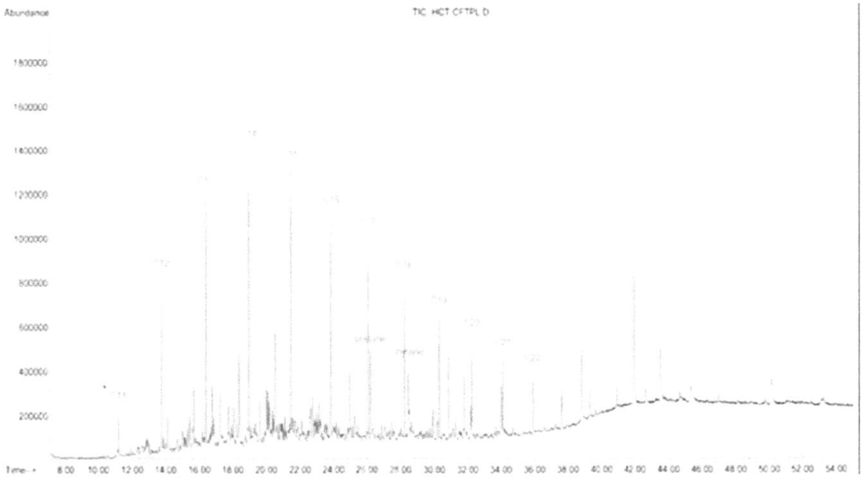


**Figure 23.1** GC-MS spectrum of the crude effluent.

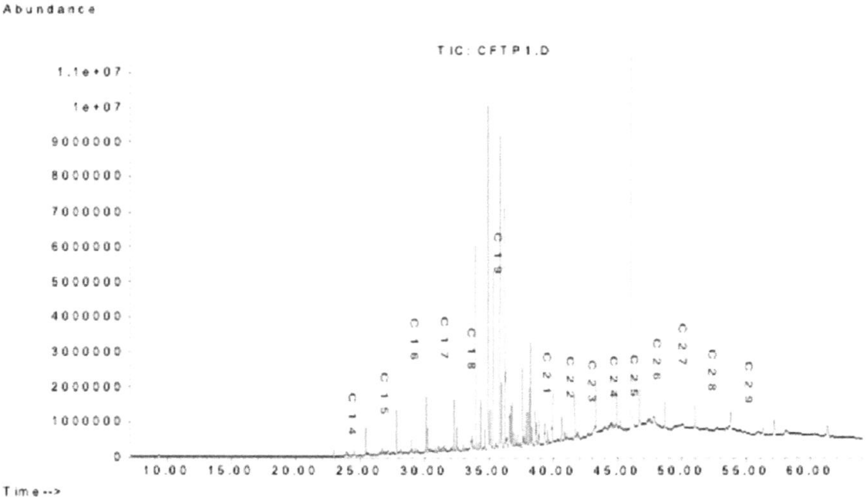


**Figure 23.2** GC-MS spectrum of the filtrate of test 1.

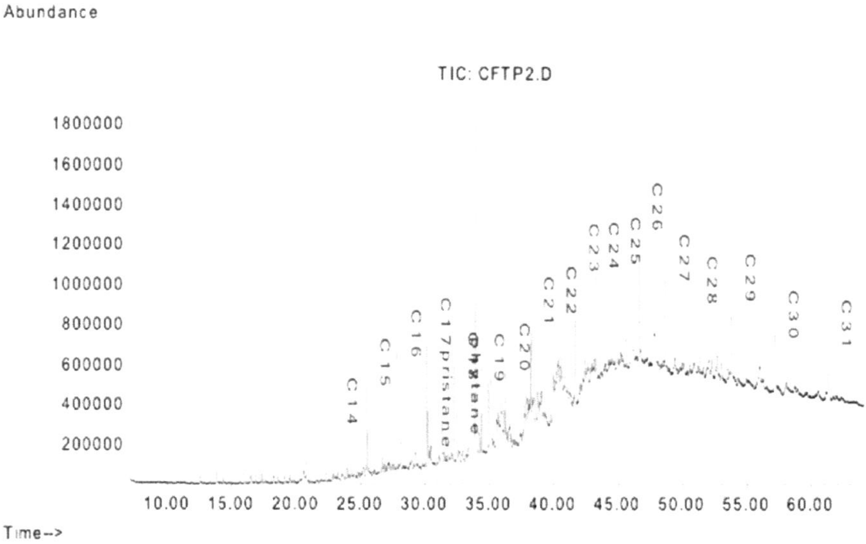


**Figure 23.3** GC-MS spectrum of the filtrate of test 2.

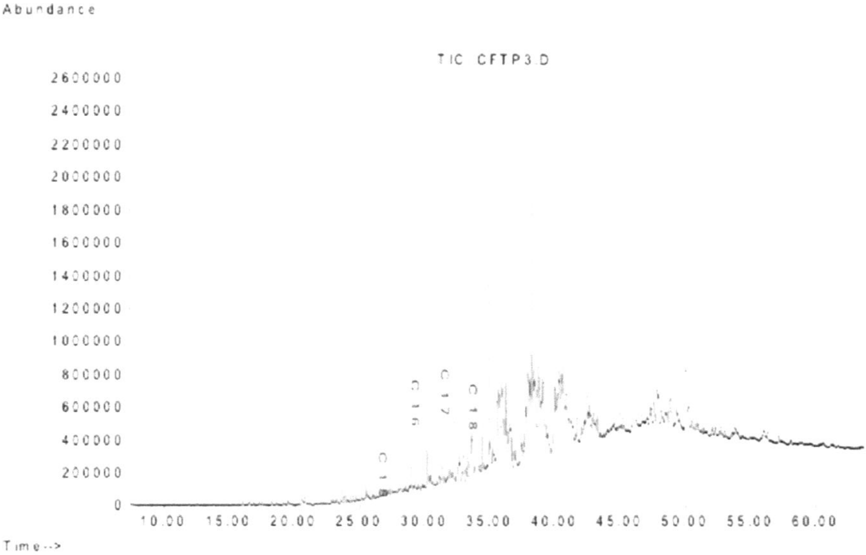


**Figure 23.4** GC-MS spectrum of the filtrate of test 3.

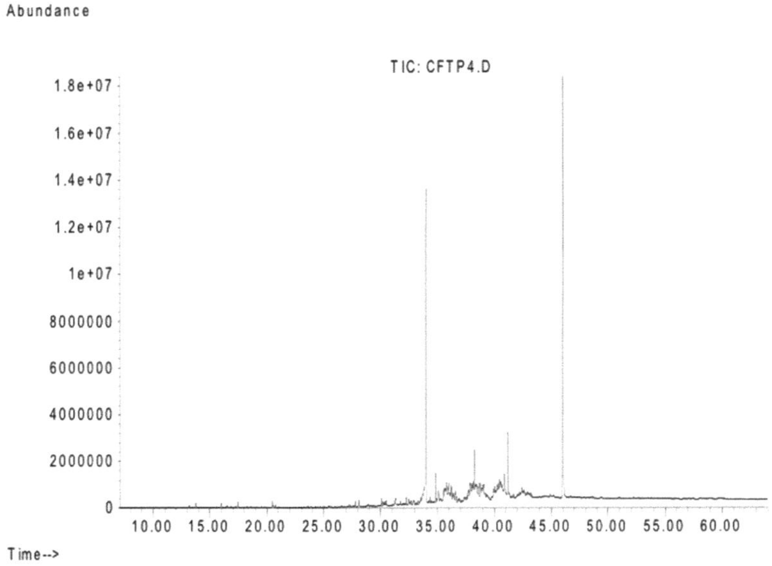


**Figure 23.5** GC-MS spectrum of the filtrate of test 4.

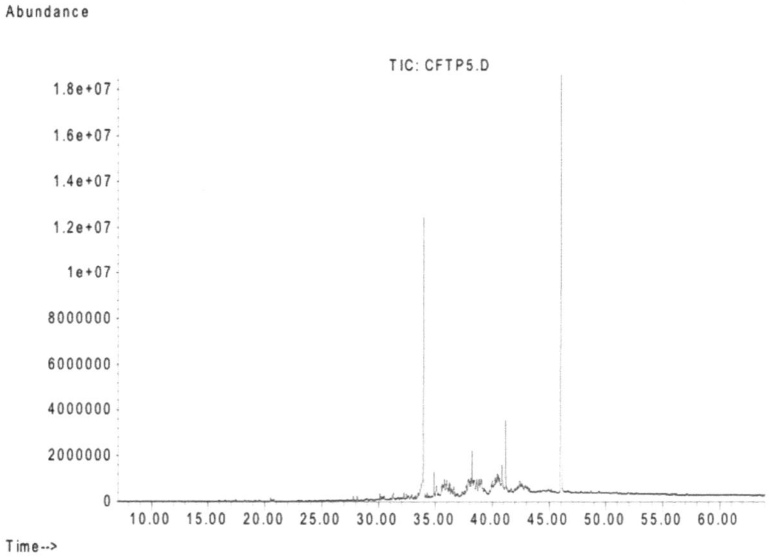


**Figure 23.6** GC-MS spectrum of the filtrate of test 5.

The pristane/phytane (Pr/Ph) ratios can provide information about the origin of HCs. It was almost equal to 1, reflecting the petrogenic origin of the HCs [517]. After filtration of the raw effluent, the spectrum of the first test showed a decrease in the peak areas of n-alkanes with disappearance of the lightest n-alkanes: n-$C_{11}$, n-$C1_{12}$ and n-$C1_{13}$. The second test showed a decrease in TPH levels compared to the first test. They decreased from 59.2 mg/l to 36.4 mg/l. This result was demonstrated by the GC-MS spectra of the two tests. The concentration of each n-alkane in the second test (demonstrated by its peak area) decreased in comparison with those of the first test. None of the compounds disappeared. For the third test, levels of TPH continued decreasing to 23 mg/l. These levels were significantly lower than the first and second tests. Spectrum analysis showed n-alkanes limited to n-$C1_{15}$, n-$C1_{16}$, n-$C1_{17}$ and n-$C1_{18}$. The disappearance of the other n-alkanes was clearly visual. The remaining n-alkanes (n-$C1_{15}$, n-$C1_{16}$, n-$C1_{17}$ and n-$C1_{18}$) showed low concentrations compared to previous test results.

TPH levels in the fourth test were about 17 mg/l, decreasing further compared to previous tests. This result was verified by the GC-MS spectra: there was a disappearance of almost all the n-alkanes except n-$C1_{16}$ and n-$C1_{17}$ which demonstrated very low or negligible concentrations. The fifth test was studied with the finest particle size fraction (between 0.20 mm and 0.10 mm). Levels of TPH decreased until 11 mg/l. The GC-MS spectrum showed concentrations, translated by their peak areas, similar to those of the fourth test with the same types of n-alkanes. Thus, the slight decrease observed in the TPH levels could be demonstrated in the GC-MS spectrum. It can be concluded that the decrease in the particle size fraction promotes more the retention of TPH (Fig. 23.7). This result is in agreement with the work of Patel et al. (2004) [518], who found that the adsorption kinetics of oil in water is much higher with a smaller particle size. Wang et al. (2011) [519] reported also the effect of particle size on the retention of PAH on the sediments of the Yellow River Delta of China. They found that the smaller particle size of sediments has a higher adsorption capacity. PCP results of the crude effluent and filtration tests are given in Table 23.3.

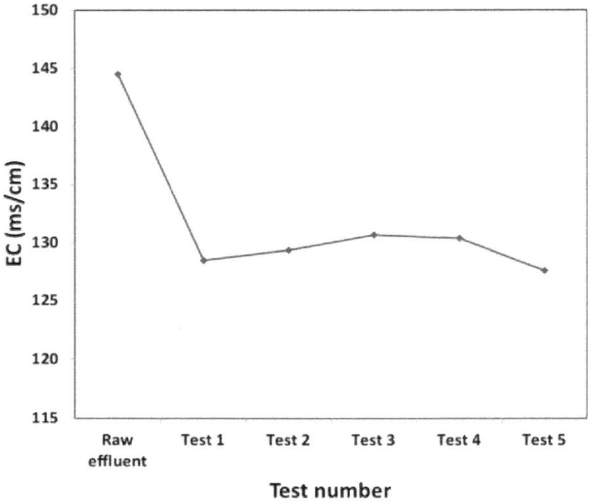


**Figure 23.7** Evolution of electrical conductivity of the different tests.

**Table 23.3** PCP of the filtration tests

| Sample | pH | EC (ms/cm) | SM (mg/l) | Turbidity |
|---|---|---|---|---|
| Raw effluent | 6.16 | 144.5 | 337.00 | 202.00 |
| Test 1 | 6.72 | 128.5 | 132.22 | 84.80 |
| Test 2 | 6.87 | 129.4 | 95.52 | 48.90 |
| Test 3 | 6.95 | 130.7 | 42.70 | 23.40 |
| Test 4 | 7.48 | 130.4 | 14.87 | 7.97 |
| Test 5 | 7.08 | 127.6 | 7.98 | 4.26 |

(1) Evolution of pH: After filtration, values of pH increased between the raw effluent and the first three tests. The fourth and fifth tests showed neutral pH under the value of 7. In general, values of pH of the raw effluent and the filtration tests were within the range of standards.

(2) Evolution of EC: The evolution of the EC showed a decrease between the raw effluent and the filtration tests. As demonstrated in Fig. 23.7, the EC decreased from 144.5 ms/cm for the raw effluent to 128.5 ms/cm for the first filtration test. After filtration, values of EC were almost stable, with a slight fluctuation around the value of 128 ms/cm.

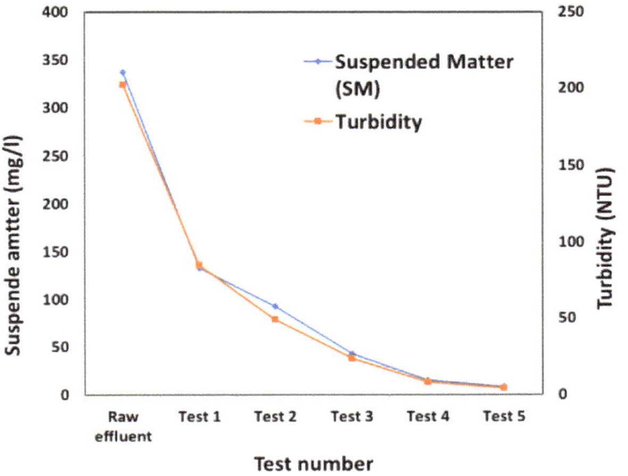


**Figure 23.8** Evolution of SM and turbidity of the different tests.

(3) Evolution of turbidity and the SM: The evolution of turbidity and the SM were in agreement with the levels of the TPH, as shown in Fig. 23.8. There was a decrease in the values of the SM and turbidity as a function of a decrease of the particle size of the sand. The fourth and fifth tests showed SM levels within the range of standards.

(4) Heavy metal retention: Analysis of heavy metals of the Sidi Litayem PW investigated the following elements: Cu, Mn, Cr, Ni, Zn, Pb and Cd (Table 23.4). Heavy metals exceeding the standards were Ni, Mn and Cr.

(5) Results of filtration tests are summarised in Table 23.5.

**Table 23.4** Levels of heavy metals in the raw effluent

| | Cu | Mn | Cr | Ni | Zn | Pb | Cd |
|---|---|---|---|---|---|---|---|
| Raw effluent (ppm) | 0.236 | 1.079 | 37.267 | 0.271 | 0.226 | 0.048 | QL* |
| NT106-02 Standards (ppm) | 0.500 | 0.500 | 0.500 | 0.200 | 5.000 | 0.100 | 0.005 |

*Quantification limit of the used setup.

**Table 23.5** Levels of heavy metals in filtration tests

| Sample | Mn | Cr | Ni |
|---|---|---|---|
| Raw effluent | 1.079 | 144.500 | 0.271 |
| Test 1 | 0.933 | 31.266 | 0.188 |
| Test 2 | 0.870 | 30.809 | 0.146 |
| Test 3 | 0.817 | 10.976 | 0.125 |
| Test 4 | 0.403 | 3.032 | 0.067 |
| Test 5 | 0.185 | 2.717 | 0.056 |

Filtration tests showed a retention capacity varying from one metal to another and depending on the initial concentration of the metal. In the case of Mn, levels reached the standards for the two last tests, with a retention percentage of 63% for test 4 and 83.5% for test 5. Test 3 resulted in a yield of 26%, exceeding the required levels. For Ni, levels reached the standards after the first test of filtration. In fact, the initial concentration slightly exceeded the required values. The adsorption rate was the maximum for test 4 and test 5, with a retention percentage of 78% and 81.5%, respectively. Cr had the largest initial concentration. Yield of retention varied from 16% to 93% between the first and fifth tests.

Therefore, as the particle size decreases, the retention of heavy metals by sand is more effective. Other studies have reported the same finding. Gunawardana et al. (2014) studied the effect of metal adsorption on particle size of built-up solids [520]. They found that the high content of quartzon a particle size more than 150 mm decreases the effect of adsorption.

## 23.4 Conclusion

PW is one of the most polluted wastewaters of the oil and gas industry. The huge organic phase and the significant presence of heavy metals are important parameters for the characterisation of oilfield PW. In this study, we studied the adsorption of the TPH and heavy metals from the PW of the Sidi Litayem oilfield through a sand filtration process, as a function of different particle sizes and in comparison to Tunisian standards. The sand filtration column showed great performance for the removal of HCs from water, with a decrease from 1026 ppm in the crude effluent to 11 ppm for the finest particle size, indicating an adsorption yield of 98.92%. Despite this important decrease, TPH levels remain slightly larger than the standards. This result was validated by the GC-MS spectra of the different tests. As we decreased the particle size, there was a decrease in the peak areas of the n-alkanes until the disappearance of certain compounds. The fourth and fifth tests indicated the absence of the majority of the compounds identified in the GC-MS spectrum of the raw effluent.

For the heavy metals, the analysis of the crude effluent showed a concentration beyond the standards for Ni, Mn and Cr. Filtration tests showed a progressive decrease in the concentrations of these metals related to a decrease in particle size. The fifth filtration test, the finest particle size, reduced the initial concentrations with a yield of 93% for Cr, 83.5% for Mn and 81.5% for Ni. Finally, levels decreased to reach Tunisian standards for Mn and Ni. Only Cr remained higher than the standards due to its initially higher concentration.

## Acknowledgements

We would like to sincerely acknowledge the laboratory facilities offered by Laboratoire de Génie de l'Environnement et Echo-Technologie GEET; Ecole Nationale d'Ingénieurs de Sfax, Université de and La compagnie Franco-Tunisienne des Pétroles CFTP, QHSE, Sfax, Tunisie.

## Chapter 24

# A Review on Characterisation and Treatment of Oilfield Produced Water

**Pratiksha Rajkumari and Rishav Chetia**
*Amazon.in, Corporate Office, Bangalore, India*
pratiksharajkumari6@gmail.com; chetia.rishav@gmail.com

Extraction of oil and gas from underground reservoirs is often accompanied by water or brine, which is referred to as oilfield produced water (OFPW). As reservoirs mature, especially if secondary or tertiary recovery methods are used, the quantity of water climbs and often exceeds the volume of the hydrocarbons (HCs) before the reservoir is exhausted. The cost of producing, handling and disposing of OFPW often defines the economic lifetime of a reservoir and the actual HC reserves; therefore, understanding and predicting the aspects, behaviour and problems induced by the OFPW flow is important. This chapter provides an introduction to OFPW, its production, its characterisation and its treatment. This chapter also reviews current technologies for the management of OFPW from oilfields and examines how electrochemical techniques may be used in these areas for future development.

*Advances in Petroleum Technology*
Edited by Subrata Borgohain Gogoi

ISBN 978-981-4877-19-0 (Hardcover), 978-1-003-04993-7 (eBook)
www.jennystanford.com

## 24.1 Introduction

Oilfield produced water (OFPW) is brought to the surface during oil extraction, and OFPW can also be called brine, saltwater or formation water (FW) [521]. The quality of OFPW varies considerably depending on the geograpHic location of the reservoir, geochemistry of the producing formation, type of hydrocarbon (HC) being produced and type of the producing well [429]. When a HC reservoir is penetrated by a well, the produced fluids may contain FW, in addition to the oil, natural gas and/or gas liquids [448]. Fresh water, brine/seawater and production chemicals are injected into a reservoir to enhance both recovery rates and the safety of operations. These surface waters and chemicals penetrate to the production zone and are recovered with oil and gas during production. Therefore, OFPW comprises of formation and injected water containing chemicals used during drilling and production operations and which represents the largest volume waste stream in the oil and gas industry, accounting for 80% of the wastes and residuals produced from natural gas production operations [522].

### 24.1.1 Production of OFPW

In the original reservoir, the pores in the mineral matrix contain natural fluids at chemical equilibrium. As reservoir rock is largely of sedimentary origin, water was present at the time of rock genesis and, therefore, is trapped in the pores of the rock [523]. Water may also move or migrate according to the hydraulic pressures induced by geological processes which also form the reservoirs [524].

In HC reservoirs, some of the water is displaced by the HC, but some water always remains. If the rock originated in a sea or ocean, then it will be saline [525]. Rocks deposited in lakes, rivers or estuaries have fresher water. Originally, the water was in chemical equilibrium with the mineral suite of the rock, but on invasion of the oil and gas, a new equilibrium with those phases was achieved [526]. Thus, there are both equilibria and chemical reaction dynamics associated with the inorganic (mineral) phases and the oil and gas phases, which are important to understand. Water is an excellent solvent; it will react to dissolve many of the

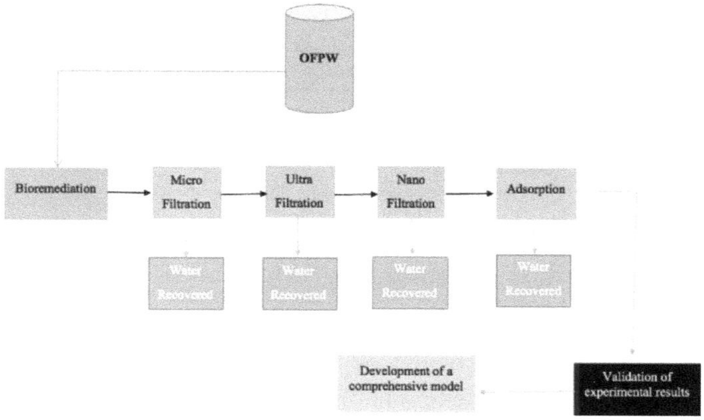


**Figure 24.1** Treatment technologies of OFPW.

phases it contacts. The OFPW-to-oil ratio (WOR) or the water-to-gas ratio (WGR) produced from a well varies widely from essentially 0 to more than 50 (98% water and 2% oil). The WGR usually is higher than the WOR. The average worldwide WOR is about 2–3. The volume of OFPW generated usually increases as oil and gas production decreases (WOR and WGR increase) with the age of the well. OFPW cannot be disposed of into the ecosystem without treatment; therefore different processes are studied in this work to find the most suitable treatment process (processes are highlighted in Fig. 24.1).

## 24.2 Composition of Produced Water

The organic and inorganic compounds normally present in OFPW are given in Table 24.1.

### 24.2.1 Dissolved and Dispersed Oil Components

Dispersed and dissolved oil components are a mixture of HCs, including benzene, toluene, ethylbenzene and xylene (BTEX), polycyclic aromatic hydrocarbons (PAHs) and phenols. Dissolved oils are the polar constituent organic compounds in OFPW, while

**Table 24.1** Organic and inorganic compounds in OFPW

| S. no. | Organic compounds | Inorganic compounds |
|---|---|---|
| 1 | n-Alkanes | **Anions** : $Cl^-$, $SO_4^{2-}$, $CO_3^{2-}$, $HCO_3^-$, $NO_3^-$, $Br^-$, $I^-$, $PO_4^{3-}$ |
| 2 | Isoprenoidspristane | **Cations** : $NH_4^+$, $Fe^{2+}$, $Cu^{2+}$, $Mn^{2+}$, $Na^+$, $K^+$, $Li^+$, $Ca^{2+}$ |
| 3 | Phytane | **NORM** : $^{238}U$, $^{235}U$, $^{232}Th$, $^{226}Ra$, $^{214}Pb$, $^{214}Bi$ |
| 4 | Branched HCs | |

small droplets of oil suspended in the aqueous phase are called dispersed oil. BTEX, phenols, aliphatic HCs, carboxylic acid and low-molecular-weight aromatic compounds are classified as dissolved oil, while less soluble PAHs and heavy alkyl phenols are present in OFPW as dispersed oil. The dissolved and dispersed oil content in OFPW is dangerous to the environment, and their concentration can be very high at some oilfields. The quantity of oil present in OFPW is governed by a number of complex but interrelated factors.

## 24.2.2 Dissolved Minerals

Dissolved inorganic compounds or minerals are usually high in concentration and are classified as cations and anions, naturally occurring radioactive materials and heavy metals. Cations and anions play a significant role in the chemistry of OFPW. $Na^+$ and $Cl^-$ are responsible for salinity, ranging from a few milligrams per litre to 300,000 mg/l. $Cl^-$, $SO_4^{2-}$, $CO_3^{2-}$, $HCO^-$, $Na^+$, $K^+$, $Ca^{2+}$, $Ba^{2+}$, $Mg^{2+}$, $Fe^{2+}$ and $Sr^{2+}$ affect conductivity and scale-forming potential. Typical OFPW contains heavy metals in varied concentrations, depending on the formation geology and the age of the oil well. Heavy metal concentrations in OFPW are usually higher than those of receiving water (for enhanced oil recovery [EOR]) and those found in seawater.

$^{226}Ra$ and $^{228}Ra$ are the most abundant naturally occurring radioactive elements present in OFPW. Radioactivity of OFPW results primarily from radium which is co-precipitated with $BaSO_4$ (scale) or other types of scales. The concentration of $Ba^{2+}$ ions in OFPW could give a strong indication of radium isotopes present in it. In some oilfields, up to 21 Bq/l of $^{228}Ra$ has been detected in produced OFPW samples.

### 24.2.3 Production Chemicals

Production chemicals can be pure compounds or compounds containing active ingredients dissolved in a solvent or a co-solvent and are used for inhibition of corrosion, hydrate formation, scale deposition, foam production, wax deposition, bacterial growth, gas dehydration and emulsion breaking in order to improve the separation of oil and water. These chemicals enter OFPW in traces and sometimes significant amounts and vary from platform to platform. Active ingredients partition themselves into all phases present depending on their relative solubilities in oil, gas or water. The fate of these chemicals is difficult to determine because some active ingredients are consumed within the process.

### 24.2.4 Produced Solids

Produced solids include clays, precipitated solids, waxes, bacteria, carbonates, sand and silt, corrosion and scale products, proppant, formation solids and other suspended solids. Their concentrations vary from one platform to another. Produced solids could cause serious problems during oil production, for example, common scales and bacteria can clog flow lines, form oily sludge and emulsion, which must be removed.

### 24.2.5 Dissolved Gases

The major dissolved gases in OFPW are $CO_2$, $O_2$ and $H_2S$. They are formed naturally by the activities of bacterial or by chemical reactions in the water.

## 24.3 Technologies

### 24.3.1 Bioremediation

Microbial biodegradation is extremely effective in treating various types of dissolved molecules, including petroleum HCs, and therefore are very useful in lowering the organic contamination levels in produced water (PW).

Biological treatments of wastewater are very well known and implemented in oily water treatment in other fields (such as oil refineries and storage farms). The most common biological method is known as activated sludge. Treating OFPW with activated sludge works because the dissolved contaminants are consumed by the microorganisms within the tank to form additional biomass (sludge). Next, water and sludge are washed out of the bioreactor and into a clarifier.

The solid waste or sludge which settles on the bottom of the clarifier is composed of both live and dead bacteria. Approximately 50%–70% of this sludge is transported back into the bioreactor in order to continually re-activate the biological process. The other half of the sludge is fed into a de-watering system for removing additional water and is then disposed of in a landfill. The effluent from the clarifier passes through a filtration unit (usually an Ulta filter) into a settling pond, and only then is it ready for discharge or reuse.

## 24.3.2 Membranes

Membranes are microporous films with specific pore ratings, which selectively separate a fluid from its components. There are four established membrane separation processes, as shown in Fig. 24.1, including microfiltration (MF), ultrafiltration (UF), reverse osmosis (RO) and nanofiltration (NF). RO separates dissolved and

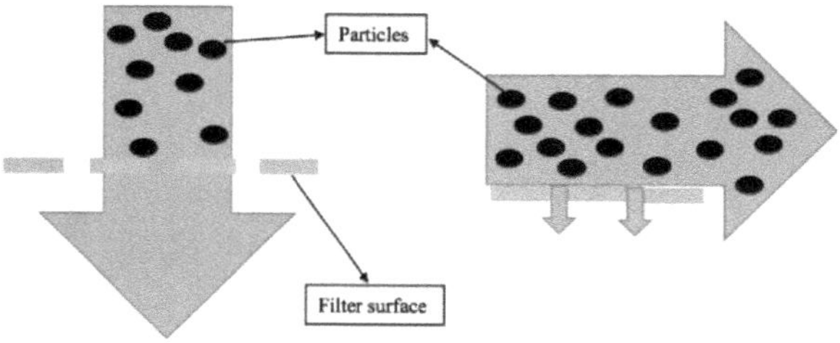


**Figure 24.2** MF and UF technologies.

ionic components, MF separates suspended particles, UF separates macromolecules and NF is selective for multivalent ions. MF and UF can be used as a stand-alone technology for treating industrial wastewater, but RO and NF are usually employed in water desalination. Membrane technology operates two types of filtration processes, crossflow filtration and dead-end filtration, which can be a pressure- or vacuum-driven system.

MF has the largest pore size (0.1–3 μm) and is typically used for the removal of suspended solids and turbidity reduction. It can operate in either crossflow or dead-end filtration. UF pore sizes are between 0.01 and 0.1 μm. They are employed in the removal of colour, odour, viruses and colloidal organic matter. UF is the most effective method for oil removal from OFPW in comparison to traditional separation methods, and it is more efficient than MF for the removal of HCs, suspended solids (SS) and dissolved constituents from OFPW. Both MF and UF operate at low transmembrane pressure (1–30 psi) and can serve as a pretreatment to desalination but cannot remove salt from water.

#### 24.3.2.1 RO and NF

RO and NF are pressure-driven membrane processes. The osmotic pressure of the feed solution is suppressed by applying hydraulic pressure which forces permeate (clean water) to diffuse through a dense, nonporous membrane. Seawater RO can remove contaminants as small as 0.0001 μm, but its major disadvantages are membrane fouling and scaling. Early studies on using RO to treat OFPW failed due to insufficient process integration and poor treatment. Nicolaisen and Lien (2009) [527], however, reported successful RO treatment of OFPW in Bakersfield, CA. The pilot system which was operated for over 1700 h in 6 months produced 20 gpm of clean water. Studies have shown the potential of brackish water RO membranes to successfully treat oil and gas PW. Experiments indicated that RO membrane technology would be excellent for OFPW treatment with appropriate pretreatment technology.

### 24.3.3 Adsorption

Adsorption is generally utilised as a polishing step in a treatment process rather than as a stand-alone technology since adsorbents can be easily overloaded with organics. It has been used to remove Mn, Fe, total organic carbon (TOC), BTEX, oil and more than 80% of heavy metals present in OFPW. There are a variety of adsorbents, such as activated carbon, organoclays, activated alumina and zeolites. The adsorption process is applicable to OFPW treatment irrespective of salinity. It requires a vessel to contain the media and pumps to implement backwashes which happen periodically to remove particulates trapped in the voids of the media. Replacement or regeneration of the media may be required depending on feed water quality and media type. The rate of media usage is one of the main operational costs of adsorption technology. Chemicals are used to regenerate media when all active sites are blocked, which often results in liquid waste disposal, and media replacement results in solid waste management.

### 24.3.4 Electrochemistry

Electrochemistry is rarely employed in OFPW treatment, as shown in Fig. 24.3, even though it has been widely used in the treatment of other wastewaters. So far, only electrodialysis (ED) and electrodialysis reversal (EDR) are established electrochemical treatment technologies of OFPW and are mainly useful when removing salts from OFPW for irrigation use. However, heavy metals, oil, produced solids and other contaminants present in OFPW can be as harmful to the soil as its salt content. Progress in electrochemistry knowledge and research suggests that electrochemistry could be the future treatment technology of OFPW. Although current treatment technologies have been used to carry out desalination, de-oiling, removal of SS and in some cases naturally occurring radioactive material (NORM) removal from OFPW, they are accompanied by many setbacks. High treatment cost, production and discharge of secondary waste, high energy requirement and use of chemicals in some cases are common problems facing these technologies.

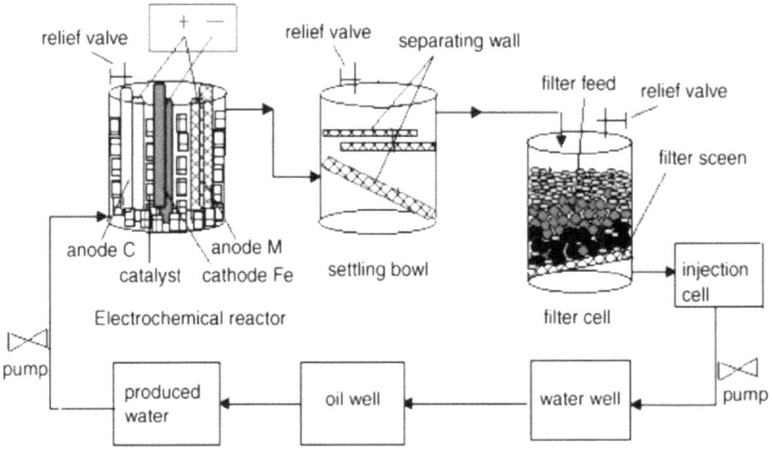


**Figure 24.3** Electrochemistry of OFPW.

Electrochemistry, on the other hand, is a relatively cheap green technology. It does not generate secondary waste nor does it involve the use of additional chemicals, and it offers improved beneficial uses of PW. It can generate and store energy, remove organics, produce clean water and recover valuable materials from OFPW with little or no negative impact on the environment. This is achievable by harmonising photoelectrochemistry (photoelectrolysis, photocatalysis and photoelectrocatalysis), water electrolysis, fuel cells, electrodeposition and other electrochemical techniques into a single electrochemical process technology, as shown in Fig. 24.3. Photoelectrolysis is a chemical process of breaking down molecules into smaller units by light. This process has played significant roles in hydrogen production and removal of organics from wastewater. This method has been investigated for the removal of organics from OFPW and used successfully for a variety of organic pollutant treatments. Ma and Wang (2006) [528] set up a catalytic electrochemical pilot-scale plant for the removal of organics from OFPW using double anodes with an active metal and graphite, and iron as the cathode and a noble metal catalyst with a big surface. They found that chemical oxygen demand (COD) and biochemical oxygen demand (BOD) were reduced by over 90% in 6 min, suspended solids by 99%, $Ca^{2+}$ content by 22%, corrosion rate by 98% and bacteria (sulphate-reducing bacteria and iron bacteria) by 99% in 3 min under 15 V/120 A.

Photoelectrolysis also offers a great promise for inexpensive production of $H_2$ and has widely been reported for the generation of $H_2$ through water splitting. Although not yet competitive on a commercial scale, photoelectrolysis has the potential to become a major $H_2$ production process. Powder semiconductor photocatalysts, nanophotocatalysts, photoanodes and several metal oxides are being investigated for improved $H_2$ production from OFPW [529]. As these technologies develop, generation of $H_2$ from OFPW would become a reality. Thus, it may be possible to reduce the energy cost of PW treatment significantly if removal of organics and generation of hydrogen from PW is efficiently carried out by photoelectrolysis. A fuel cell is another major electrochemical technology, as shown in Fig. 24.4, which is important in the future of OFPW treatment technology. A fuel cell converts chemical energy contained in, for example, $H^2$ gas, into electricity and generates water and heat as by-products. Electrodeposition is a mature technology which is widely applied in various fields of electrochemistry, particularly in material

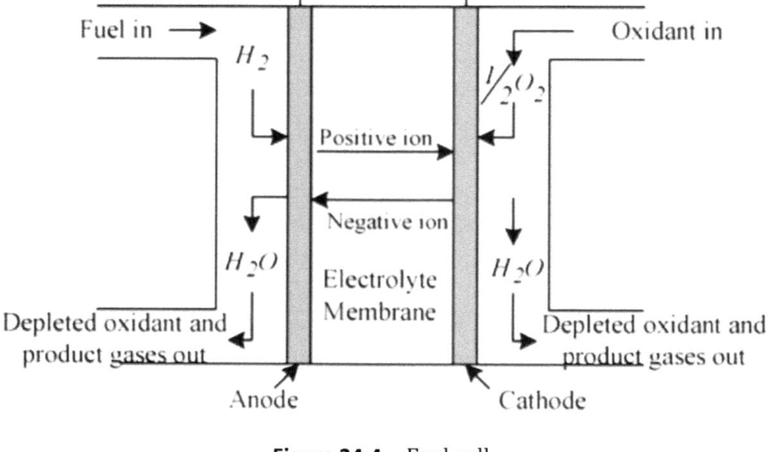


**Figure 24.4** Fuel cell.

coating, fabrication of magnetic films and metal recovery [104–106]. It is a cathodic reaction where a metal ion gains electrons to become a metal.

In the future, an electrochemical process unit for PW treatment would integrate photoelectrochemistry, water electrolysis, fuel cells and electrodeposition technologies to achieve production of clean water, storage of energy and recovery of valuable metals from OFPW.

## 24.4 Conclusion

OFPW is commonly regarded as a high-volume toxic waste but can be beneficial to humans if properly managed. The treatment of OFPW is very important due to legislation and environmental concerns. In the future, demand for the treatment of OFPW as a source of fresh water is very likely with the world population now above 7 billion and the demand for fresh water on the increase. In this report, we have reviewed major OFPW treatment technologies and their application in future management. Current thermal OFPW treatment technologies are mature but may not be relevant in future management unless significant reductions are made in energy costs. This may be achieved if low-pressure steam is available through cogeneration arrangements. Membrane technologies are some of the finest for OFPW treatment today; however, significant progress must be made to reduce membrane fouling and secondary waste generation for them to compete well in the future management of OFPW. High costs of rigorous pretreatment, fouling and regular backwashing are major setbacks of these technologies for future application. Macroporous polymer extraction (MPPE) technology is a fairly new OFPW treatment technology which may well compete in the future management of OFPW [530]. Its potential to achieve zero pollutant discharge and a significant reduction in energy consumption compared to thermal technologies is very promising, but advancement is needed to bring down its relatively high cost. It has been reported that to optimise PW treatment, two or more technologies must be combined or employed in a hybrid system, but a cost-effective technology with zero pollutant discharge will be the technology of choice for the future management of PW, and this can be potentially achieved by electrochemistry. The application of electrochemically driven treatment technologies can lead to the production of clean water, production/storage of energy and recovery of valuable materials from PW by integrating photoelectrochemistry, electrodeposition, fuel cells and other electrochemical techniques into a single electrochemical unit.

## Chapter 25

# Application of Membrane Bioreactors for the Modification of Microfiltration Membrane Surface for Enhanced Antibiofouling Capability in Wastewater Treatment

**Akshayjit Podder**

*Department of Chemistry, Assam University, Silchar, Assam, India*

chem.akshayjit.aus@gmail.com

The continuous depletion of freshwater resources has shifted the focus more towards water recovery, reuse and recycling, which requires an extension of conventional wastewater treatment technologies. Downstream external factors like stricter compliance requirements for wastewater discharge, rising treatment costs and spatial constraints necessitate renewed investigation of alternative technologies. Membrane bioreactors (MBRs) are now widely implemented in wastewater treatment plants due to their compact design and high effluent water quality. However, membrane biofouling has a significant impact on MBR performance and energy cost. Quorum sensing inhibiting (QSI) compounds have the ability to disrupt bacterial communication pathways which lead to

*Advances in Petroleum Technology*

Edited by Subrata Borgohain Gogoi



ISBN 978-981-4877-19-0 (Hardcover), 978-1-003-04993-7 (eBook)

www.jennystanford.com

biofilm formation and biofouling. In this chapter, an antibiofouling membrane was developed by incorporating vanillin (VA), a known QSI molecule, onto the surface of a commercial chlorinated polyethylene (CPE) microfiltration (MF) membrane by Williamson ether synthesis reaction. The CPE membrane was reacted with the VA in a solvent mixture containing potassium carbonate and *N*,*N*-dimethylformamide for VA incorporation. The solvent mixture without VA was also used to create a control membrane to understand the QSI antibiofouling capabilities. Raman and Fourier transform infrared spectroscopy indicated that VA was retained on the modified membrane surface under crossflow condition and sodium hypochlorite (NaOCl) cleaning. The modified membrane had significantly less biofouling compared to the unmodified and control membranes tested using activated sludge collected from wastewater treatment plants. There are many advantages in using an MBR process, the prime ones being the treated water quality, the small footprint of the plant, less sludge production and flexibility of operation. This research showed that VA-modified membranes have potential to improve the membrane antibiofouling capabilities and MBR performance by reduction in cells in the fouling layer.

## 25.1 Introduction

Membrane bioreactor (MBR) technology has been extensively employed for various industrial wastewater treatments due to its distinct advantages over conventional technologies. MBR processes, which use membrane filtration units to replace the secondary settler, offer several well-known advantages over the conventional activated sludge processes, such as excellent effluent quality, good disinfection capability, higher volumetric loading, reduced footprint and sludge production, process flexibility towards influent changes and improved nitrification. MBR technology has proven quite efficient in removing organic and inorganic contaminants as well as microorganisms from wastewaters and has gained increasing popularity due to more stringent environmental regulations and growing water reuse initiatives in recent years. MBRs are now widely used in municipal and industrial wastewater treatment plants. MBR technology integrates two processes, activated sludge and low-pressure filtration, into a single system that degrades organic matter and clarifies the effluent [531]. A microfiltration (MF) or ultrafiltration (UF) membrane is typically used in low-pressure membrane filtration. An MBR removes the need for a secondary sedimentation basin, which greatly decreases the space needed in a densely populated urban area. MBR technology offers stable effluent water quality, shorter sludge retention times and a potential 30%–50% smaller energy footprint than the conventional process [532, 533]. Also, MBR systems are subjected to membrane fouling from particulate deposition, bacteria accumulation and biofilm formation on the membrane surface. Microbial community composition and cell density in activated sludge also influence membrane fouling [534]. Fouling reduces permeate flux and increases the transmembrane pressure, which leads to elevated energy demand and operational costs for the unit [535]. Currently, MBR fouling is mitigated using physical and chemical cleaning methods to remove particulate matter and bacterial biofilm on the membrane surface [535]. Yet, even with repeated cleanings, biofouling is still inevitable due to the dense mixture of bacteria and suspended particles in activated sludge. Therefore, it is essential to find alternative methods of mitigating MBR biofouling.

Past research found that bacteria in wastewater MBRs rely on cell-to-cell signalling systems known as quorum sensing (QS) to coordinate community behaviour leading to biofilm formation. The QS pathway is a concentration-dependent system where production of sensing molecules depends on the amount of bacteria present in the local environment. Blocking these bacterial communication pathways may offer a promising approach to control membrane biofouling without exerting external forces which often lead to bacterial resistance. Several natural or synthetic compounds have been found to impede QS functions and reduce bacterial biofilm formation [536, 537]. Vanillin (VA), ($C_8H_8O_3$), as shown in Fig. 25.1, a natural favouring agent, was shown to inhibit a broad range of gram-negative bacterial QS functions relevant to MBR biofouling. VA is known to modify the signalling molecule in the cell culture to

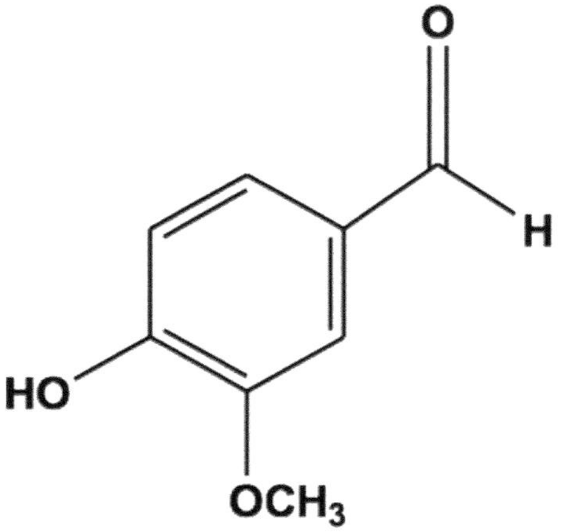


**Figure 25.1** Structure of vanillin.

prevent it from binding to the appropriate transcription regulator so that biofilm formation is not activated [535, 538]. Past studies showed that VA dosed in the bulk fluid reduced biofilm production and MBR biofouling [539, 540].

However, continuously dosing VA in the bulk fluid is not economical because a constant high dose is required to effectively block bacterial communication pathways. Modification of membrane surface with VA could be a promising approach to combat biofilm formation at membrane surfaces instead of in the bulk fluid. We recently reported that adsorption of VA onto the surface of polyamide reverse osmosis (RO) membranes [541] can enhance the membranes' antibiofouling property [542, 543]. The additional VA layer also did not impact membrane performance in respect to pure water permeate flux and salt rejection. In a high-pressure RO system subjected to high density of marine bacteria and nutrients, biofilm production was suppressed on VA-modified membranes and the water flux was improved compared to the unmodified membrane [541]. However, VA coated by physical adsorption was not stable on the membrane surface. Hence, VA was washed off after subjecting to crossflow for longer than 48 h.

### 25.1.1 Goal of the Study

The goal is to develop an in situ antibiofouling membrane by chemically linking VA to the surface of an MF membrane to improve VA's retention. Williamson ether synthesis reaction was used to modify the membrane surface to maintain VA's QS inhibition (QSI) activity. Raman and Fourier transform infrared (FTIR) spectroscopy were used to verify VA attachment and retention on the modified membrane surface. The modified membrane performance was evaluated in a membrane biofouling–monitoring system and compared with control and unmodified membranes. Confocal scanning laser microscopy (CSLM) was used to examine biofilm formation on the fouled membrane samples.

### 25.1.2 The MBR Process

MBR systems [544] essentially consist of a combination of membrane units responsible for physical separation and biological reactor systems responsible for biodegradation of the waste compounds. These systems are implemented on the basis of two main configurations: external/side-stream configuration (Fig. 25.2a) and submerged/immersed configuration (Fig. 25.2b).

The external configuration, which involves the recirculation of the mixed liquor through a membrane module which is outside the bioreactor, usually employs high crossflow velocity (CFV) along the membrane surface to provide a membrane driving force and control

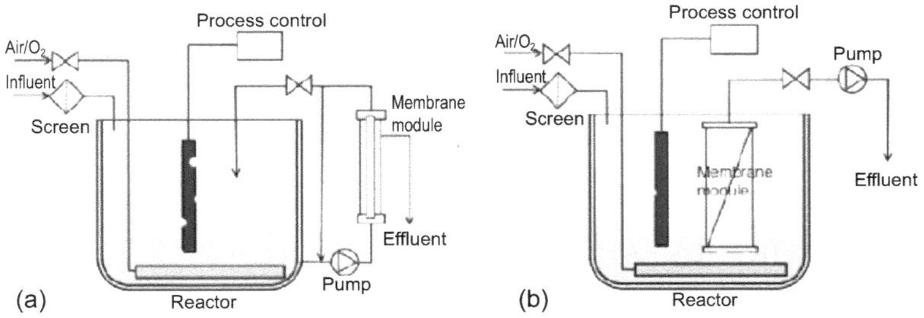


**Figure 25.2** Williamson ether reaction mechanism for vanillin attachment to a chlorinated polyethylene (CPE) membrane.

membrane fouling. As a result, this configuration provides more direct hydrodynamic control of membrane fouling and offers the advantages of easier membrane replacement and high fluxes but at the expense of frequent cleaning and high energy consumption (2–12 kWh/$m^3$ product). For submerged configuration, membrane modules are directly placed in the mixed liquor. The driving force across the membrane is achieved by pressurising the bioreactor or creating a negative pressure on the permeate side. Several distinct advantages [545] of submerged MBRs are their much lower energy consumption and less rigorous cleaning procedures.

### 25.1.3 Characteristics of Industrial Wastewaters

Characteristics of industrial wastewaters can usually be represented by the basic parameters, including chemical oxygen demand (COD), biochemical oxygen demand (BOD), suspended solids (SS), ammonium nitrogen ($NH_4$–N), heavy metals, pH, colour, turbidity and biological parameters. The characteristics of industrial wastewaters strongly depend on the type of industrial wastewaters and industrial processes. Compared with municipal wastewater, industrial wastewaters usually have high organic strength and extreme physicochemical nature (e.g., pH, temperature, salinity) and contain synthetic and natural substances which may be toxic to or inhibit biological treatment processes. Municipal wastewater is characterised by low-organic-strength (250–800 mg COD/L), whereas industrial operations often generate strong ($\geq$1000 mg COD/L) to extremely strong wastewaters. Extremely strong wastewaters, with COD concentrations which may even exceed 200 g/l, are generated in olive mills and the textile and beverage production industries.

In general, extreme pH and salinity may give rise to difficulties in the biological treatment of industrial waste streams. Inhibition of many microorganisms and deflocculation of sludge flocs were frequently encountered problems when wastewater treatment systems were operated under these conditions. In addition, industrial wastewaters may contain a large variety of potentially inhibiting or toxic compounds, such as aromatics, phenols, chlorinated or fluorinated compounds, volatile organic compounds (VOCs), heavy metals, surfactants, biocides. Some of the toxic compounds may be mostly inert to biodegradation and may require additional physico-chemical treatment, either as pre- or posttreatment, depending on their effects on the biological process (e.g., heavy metals).

## 25.2 Experimental

### 25.2.1 Materials for Membrane Modification

A commercially available and commonly used chlorinated polyethylene (CPE) MF membrane was adopted for surface modification in this study due to its proven efficiency in MBR operation. The following chemicals from Finar Chemicals were used for membrane surface modification without further purification: VA (4-hydroxy-3-methoxybenzaldehyde), potassium carbonate (PC), and *N*,*N*-dimethylformamide (DMF).

### 25.2.2 Methodology

Williamson ether synthesis reaction [546], as shown in Fig. 25.3, was adapted to bond VA to the CPE membrane surface through the following anticipated steps, as described by the Williamson ether reaction mechanism:

(1) PC deprotonates VA to produce an alkoxide ion.
(2) DMF slows down the reaction rate of the highly reactive alkoxide ion.
(3) The alkoxide ion replaces the haloalkane bond on CPE to form an ether bond between VA and CPE [547].

In preparation for this reaction, the CPE membrane was first cleaned with 80% ethanol and then rinsed with deionised (DI) water. VA-PC-DMF testing solution was prepared by adding VA (0.11 mM final concentration) and PC (0.6 mM final concentration) to DMF. The CPE membrane was soaked in the VA-PC-DMF solution for various time periods to determine the best duration to complete the reaction.

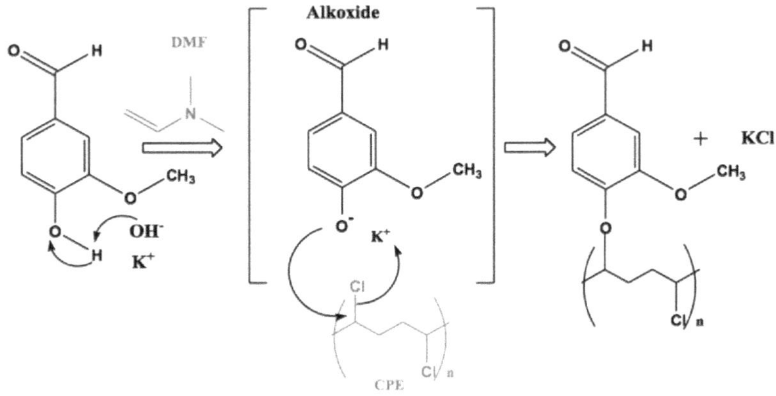


**Figure 25.3** Williamson ether reaction mechanism for vanillin attachment to a chlorinated polyethylene (CPE) membrane.

To further understand changes in the membrane surface chemical structure during VA incorporation and the new antifouling properties, a second modification solution was prepared by mixing PC with DMF without VA. This PC-DMF testing solution was used for CPE membrane modification following the same reaction procedure as described before. The PC-DMF-modified membrane was used as a VA negative control for evaluating the membrane performance. All membrane samples were rinsed by DI water thoroughly after the modification reaction to avoid chemical carry-over and dried before further analysis.

The reaction was monitored and product confirmation along with separation was done using column chromatography and subsequent silica-coated plates were checked for product formation, where E is the starting material, F is the reaction mixture spot of the starting material and intermediate and G is the reaction mixture with two spots having a significant difference in the retention factor (Rf) value by which we can ascertain the formation of an ether bond in the product stage (Fig. 25.4).

### 25.2.2.1 Spectroscopic studies

Unmodified and modified membrane samples were analysed using Raman and FTIR spectroscopy. In Raman in vivo microscopy

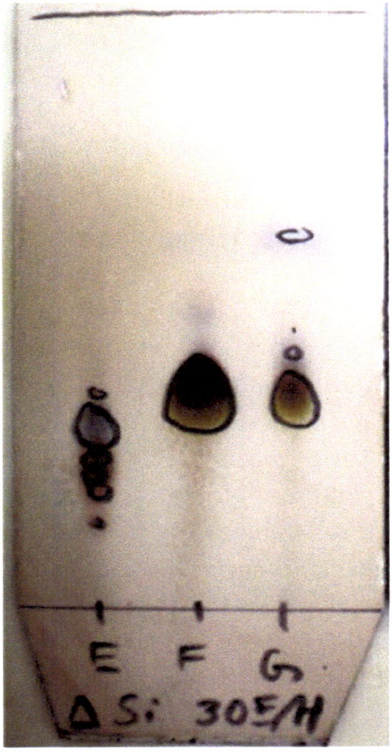


**Figure 25.4** TLC plate showing the product formation spot.

analysis, samples were placed directly under the diode laser at an excitation of 532 nm (visible light range) under 50,000 magnifications. Spectrums were collected using either 3 mW or 15 mW of laser power and 1 s exposure time from about 10 random locations along the membrane surface. The baseline was corrected for all spectrums using WIRE 3.4 software where the samples were sent. For FTIR (JASCO R-4700 type) analysis, each membrane sample was examined at room temperature between 400 and 4000 $cm^{-1}$. The signals were collected for 32 scans with a scan interval of 0.241 $cm^{-1}$ and optical resolution of 1 $cm^{-1}$. The target ether bond and other relevant peaks were determined by comparing the FTIR and Raman spectra of reference samples (i.e., unmodified CPE membrane, VA, DMF and PC in its natural state) with VA-PC-DMF- and PC-DMF-modified CPE membrane samples. Spectral peaks were also identified and confirmed by comparing with literature reports.

#### 25.2.2.2 Membrane antibiofouling potential

Biofilm formation on the surface of modified and unmodified membranes was evaluated in the membrane cassette system in recirculation mode. For comparisons, unmodified and modified (both PC-DMF-CPE and VA-PC-DMF-CPE) membranes were inserted in parallel membrane cassette holders (dimensions were 9.4 cm × 25.1 cm × 19.3 cm and the cross-sectional area of the water path defined by the spacer and o-ring was 0.01 $m^2$) and were subjected to the same crossflow driven by the same peristalic pump. To test the membrane antifouling potential, aerobic-activated sludge was collected from the City Management Sanitation District wastewater treatment facility (Chandmari, Guwahati). The concentration of mixed liquor suspended solids (MLSS) and volatile suspended solids (VSS) was 7.6 g/l and 6.1 g/l, respectively, according to the plant's daily report. Activated sludge was added to the feed tank with no further purification. The pump was set to a crossflow of 180 $l/m^2/h$ for 48 h with no permeation following several previous works, indicating crossflow can be an adequate representation of membrane surface-fouling propensity [548, 549]. After the experimental period, membrane samples were collected from the cassette for further analysis. Fouling studies were repeated using activated sludge sampled on a separate day to confirm the reproducibility of the results.

#### 25.2.2.3 Stability of VA-modified membranes

VA-modified membrane stability was assessed under crossflow conditions without high pressure and recirculation of the return line. The modified VA-PC-DMF-CPE membrane was framed in a membrane cassette holder. The membrane surface (dimensions 9.4 cm × 25.1 cm × 19.3 cm and cross-sectional area of the water path defined by the spacer and o-ring 0.01 $m^2$) was exposed to DI water with a crossflow of 360 $l/m^2/h$ for 24 h driven by a peristaltic pump. After drying, the modified membranes were analysed using

Raman spectroscopy. This test was repeated twice to confirm the reproducibility of the results. Raman spectra on membrane surfaces were compared before and after the flushing experiments to identify any loss of peaks. VA-PC-DMF-CPE- and PC-DMF-CPE-modified membranes were also subjected to cleaning by sodium hypochlorite (NaOCl) solution [550] (free chlorine, ~8 ppm; pH, ~9.6) for 30 and 60 min. Afterwards, membranes were rinsed using DI water and dried for Raman spectroscopy analysis. A handheld photometer (ChemiLine, Gujarat) was used before and after cleaning tests to determine total chlorine and free chlorine concentrations. pH values were measured using Fisher Scientific™ accumet™ AB15 Basic and BioBasic™ pH/mV/°C meters.

#### 25.2.2.4 CSLM

CSLM (Zeiss LSM 700, Carl Zeiss) was used to characterise live cells, dead cells and polysaccharides on membrane surfaces. Live cells were stained green with SYTO 9, and dead cells were stained red with propidium iodide following guidelines in the FilmTracer™ LIVE/DEAD biofilm viability kit.

## 25.3 Results and Discussion

### 25.3.1 Membrane Modification

Raman spectroscopy results confirmed there were abundant haloalkane (C–Cl) bonds on the virgin CPE membrane surface with major peaks at 637 and 698 $cm^{-1}$. In comparison, these haloalkane bonds were removed on VA-PC-DMF-CPE membranes. Additionally, several new bonds on the VA-PC-DMF-CPE Raman spectra were observed at 1180, 1291 and 1497 $cm^{-1}$ which corresponded to pure VA functional groups [551]. During the Williamson ether reaction, the $OH^-$ group of VA was deprotonated, resulting in the alkoxide ion, which then formed an ether bond with the CPE surface [460]. Consistent with this reaction, an ether bond was identified at a Raman shift of 1098 $cm^{-1}$ on the VA-PC-DMF-CPE membrane, which did not correspond to any other reactant [552].

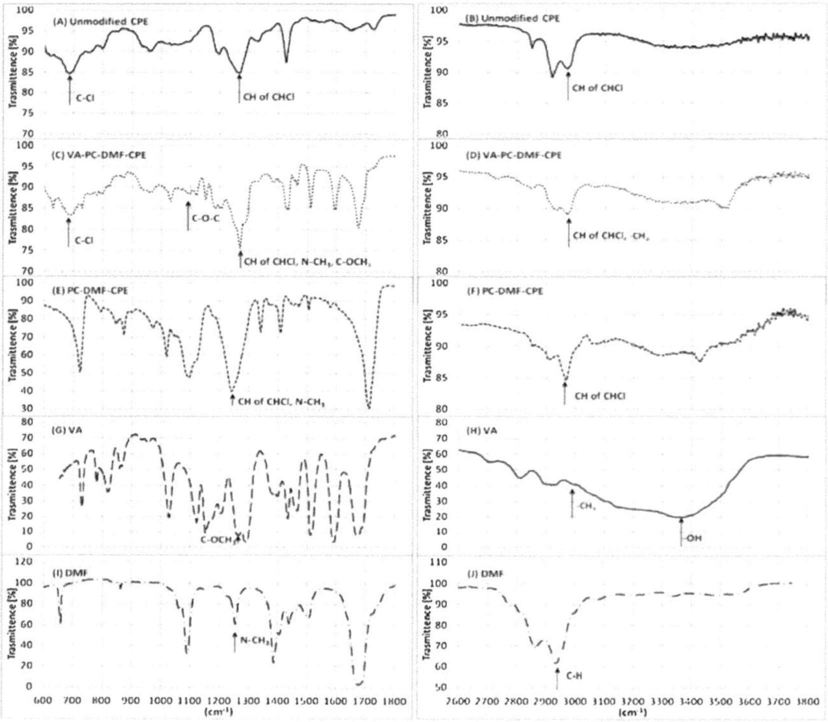


**Figure 25.5** FTIR spectra from two regions of interest, 600 to 1800 $cm^{-1}$ and 2600 to 3800 $cm^{-1}$, respectively, for the unmodified CPE membrane (A, B), VA-PC-DMF-CPE-modified membrane (C, D) and PC-DMF-CPE-modified membrane (E, G).

Under Raman spectroscopy, the loss of the hydroxyl group from VA could not be well observed, and thus FTIR was used to confirm the reaction mechanism (Fig. 25.5). FTIR spectral data showed that the $OH^-$ group of VA between 3000 to 3600 $cm^{-1}$ (Fig. 25.5H) was absent on the VA-PC-DMF-CPE membrane surface (Fig. 25.5D). FTIR spectroscopy also verified the formation of an ether bond at 1098 $cm^{-1}$ on the VA-PC-DMF-CPE membrane (Fig. 25.5C) [551]. This result confirmed the replacement of OH with a C–O–C bond. However, FTIR data revealed the C–Cl bond was reduced but still present on the VA-PC-DMF-CPE membrane surface (Fig. 25.5C). Spectral peaks associated with the CH bond of CHCl at 1265 and 2970 $cm^{-1}$ (Fig. 25.5C,D) were also present on the VA-PC-DMF-CPE membrane surface [551, 553].

To further understand the change in the membrane's chemical structure with respect to the removal of C–Cl bonds, the FTIR spectra of the VA-PC-DMF-CPE membrane (Fig. 25.2C,D) was compared with the PC-DMF-CPE membrane (Fig. 25.5E,F) and the pure chemical reactants, VA (Fig. 25.5G,H) and DMF (Fig. 25.5I,J). FTIR chlorine spectra peaks (1265 $cm^{-1}$ and 2970 $cm^{-1}$) corresponding to the CH bond of CHCl were identified on the PC-DMF-CPE membrane (Fig. 25.5E,F). However, the FTIR peak at 1265 $cm^{-1}$ overlapped with the N-$CH_3$ bond of the amide in DMF (Fig. 25.5I) and C-$OCH_3$ peak in VA (Fig. 25.5G) [551]. The 2970 $cm^{-1}$ peak overlapped with the methyl group of VA (Fig. 25.5H) and the C–H stretch of DMF (Fig. 25.5J) [551]. Therefore, it was difficult to assess the extent of VA replacing the haloalkane bonds from these peaks. It is possible chlorine was not completely replaced by VA during the reaction. This could be due to the kinetics and energy threshold needed to break the haloalkane bonds on the CPE surface. It is also important to note that PC-DMF testing solution altered the CPE membrane surface with spectral peaks representing pure DMF and new peaks, not associated with any reactant, which likely corresponded to N–H, C–N, C–H and C–C bonds [554–556] (Figs. 25.5E,I and 25.5F,J). Since the reaction was not run to completion, FTIR and Raman spectroscopy were used to determine a suitable reaction time. Overall, as the reaction between VA-PC-DMF and CPE progressed, the C–Cl bond decreased and the ether bond increased under FTIR. There were significant differences in FTIR and Raman spectral peaks associated with chlorine, ether and VA bonds from 1 to 3 min reaction times (data not shown). On the basis of the spectral data, a 3 min reaction time was selected for the following experiments.

The reference spectra for VA as a powder (Fig. 25.5G,H) and DMF as a powder (Fig. 25.5I,J) are adapted from the NIST data centre. The arrows in the spectrum indicate the peaks associated with chloride, hydroxyl, amine and ether functional groups.

## 25.3.2 Modified Membrane Stability

When VA-PC-DMF-CPE membranes were tested under rapid 24 h crossflow flushing using neutral pH DI water, no significant change in spectral peaks under Raman spectroscopy was observed on the membrane before and after experiments (data not shown). This data indicated that the modified membrane was stable under crossflow conditions and likely stable during water treatment.

In addition, stability tests using sodium hypochlorite (8 ppm free chlorine) as a cleaning solution indicated that chemical structures of the VA-PC-DMF-CPE membrane were stable. No significant change was observed on membrane surfaces on the basis of peaks revealed by Raman spectroscopy (data not shown) after 1 h of sodium hypochlorite cleaning. There was also no considerable change in pH of the cleaning solution (pH 9.6). A decrease in free chlorine after the cleaning experiment was observed, which indicated a possible reaction between the modified membrane and hypochlorite, such as oxidation of the aldehyde functional group of VA to carboxylic acid [557]. It should be mentioned that the free chlorine concentration used in this study was lower than that applied in industrial cleaning (100–1000 ppm) [550]. The target pH was in range with industry guidelines [539, 540, 544] to remove organic foulants and biofilm. Therefore, future studies should examine membrane surface stability under a large range and higher concentration of free chlorine.

## 25.3.3 Membrane Biofilm Production

The mechanism for biofilm reduction on VA-modified membranes is not entirely clear. Pure VA inhibits the QS pathway by altering the QS molecule in solution so that QS cannot bind to the transcription regulator to activate biofilm formation [558]. Figure 25.6 illustrates the coverage of live bacteria, dead bacteria and polysaccharides on the modified and unmodified CPE surfaces using the Image J volume viewer plugin [559]. The unmodified membrane had an uneven distribution of dead bacteria, with live bacteria densely covering the membrane surface (Fig. 25.6A). The PC-DMF-CPE

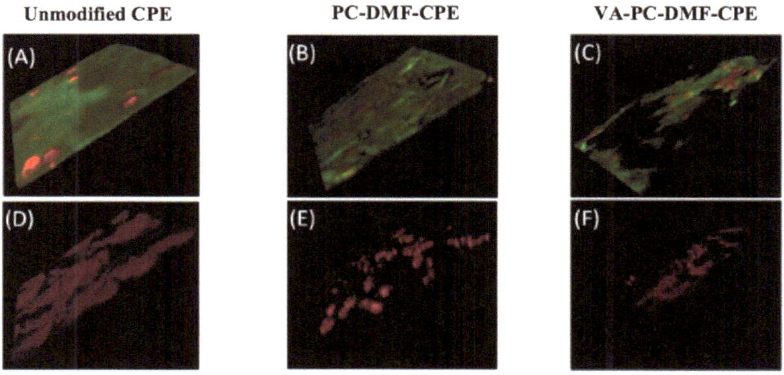


**Figure 25.6** 3D biofilm images of the distribution of live (green) and dead (red) bacteria in the top row for (A) the unmodified, (B) PC-DMF-CPE and (C) VA-PC-DMF-CPE membranes. The bottom row (D–F) illustrates the amount of polysaccharides on each membrane.

membrane experienced a reduction in live and dead bacteria on the surface compared to the unmodified membrane (Fig. 25.6B). The VA-PC-DMF-CPE membrane had live and dead bacteria sparsely covering the surface (Fig. 25.6C). The polysaccharide formation on the membrane surface followed a similar trend. The unmodified membrane had the highest production of polysaccharides, whereas the PC-DMF-CPE and VA-PC-DMF-CPE membranes had much lower amounts of polysaccharides (Figs. 25.3a, 25.6B and 25.6D–F).

The addition of VA on the membrane surface enhanced the anti-biofilm potential of the modified membrane. Raman spectroscopy also indicated that VA was still present on modified membranes after use in membrane biofouling experiments but slightly reduced. Overall, the addition of VA on the modified membrane surface led to a greater resistance towards bacterial colonisation and biofilm production from activated sludge compared to DMF-modified and DMF-unmodified CPE membranes.

In the current study with VA incorporated on the membrane surface, the VA-PC-DMF-CPE membrane had a significant reduction of polysaccharides as well as live and dead cell density. The overall biofilm biomass significantly reduced in comparison with the unmodified membrane. A similar, but to a lesser extent, reduction of biofilm biomass was also observed on the PC-DMF-CPE membrane in the absence of VA. This result suggested that VA and DMF have a synergetic effect for biofilm reduction, which may be due to DMF's antimicrobial activities.

Furthermore, one concern of membrane surface modification is the loss of water permeability. Specifically, adding organic molecules to the membrane layer may have a pore-blocking effect. However, the pore-blocking effect cannot be completely ruled out for VA binding to CPE. Since membrane biofouling is the main cause of water flux reduction mitigation of biofouling on the membrane surface, linking of VA to the CPE membrane would likely result in improved water flux during wastewater treatment operation.

## 25.4 Conclusion

This study showed that an ether bond was established between VA and the surface of a CPE membrane. VA was retained and the ether bond was stable under filtration conditions and chemical cleaning. Under fouling conditions, the VA-PC-DMF-CPE membrane inhibited live and dead bacteria and suppressed polysaccharide production compared to the unmodified CPE membrane. This research showed that CPE membranes have the potential to reduce membrane biofouling in wastewater treatment plants.

## 25.5 Future Prospects

We can examine membrane surface stability under a large range and higher concentration of free chlorine.

Anchoring VA to the membrane surface by an ether bond alters the chemical structure of VA and changes its QSI mechanism. These changes are still in progress and would be interesting to study in future work.

## Acknowledgements

I would like to acknowledge the following:

(1) SAIF, NEHU, Bose Institute (Kolkata) and CSIR-CDRI, Lucknow, India
(2) City Management Sanitation District wastewater treatment facility (Chandmari, Guwahati)
(3) All the faculty members of the Department of Chemistry, Assam University, Silchar
(4) All the technical staff members and other members of the Chemistry Department
(5) Arohan Beverages and Foods Pvt. Ltd. for membrane study in its plant and processed water supply and collaborative support
(6) Assam University, CIF, for laboratory works

# Bibliography


1. M Patric and B Donald. The effects of sediment transport on grain size distribution. *Journal of Sedimentary Petrology*, **55**(4):0457–0470, 1985.
2. K Mandal and R Dasgupta. Upper Assam Basin and its basinal depositional history. In *10th Biennial International Conference and Exposition, Society of Petroleum Geophysicists*, Kochi, Vol. 292, 2013.
3. G Saikia and P Borgohain. Petrography and sandstone chemistry of oil bearing tipam sandstone formation in parts of Upper Assam shelf. *Journal of Earth Science*, Special Volume, 95–105, 2013.
4. MP Gogoi, SB Gogoi, P Borgohain, and D Konwar. Grain size analysis of Naharkatiya and Jorajan oil fields of Upper Assam basin. In *Proceeding Report of International Conference on Advances in Petroleum, Chemical, and Energy Challenges*, pp. 251–262, 2017.
5. RL Folk. *Petrology of Sedimentary Rocks*. Hemphill Publishing Co., Austin, Texas, 1974.
6. GM Friedman. On sorting, sorting coefficients, and the lognormality of the grain-size distribution of sandstones. *Journal of Geology*, **70**(6):737–753, 1962.
7. AR Rao and MK Samanta. Structural style of the Naga overthrust belt and its implication on exploration. *Bulletin of the Oil and Natural Gas Commission*, **24**:69–109, 1987.
8. J Craig, N Hakhoo, GM Bhat, M Hafiz, MR Khan, R Misra, SK Pandita, BK Raina, J Thurow, B Thusu, et al. Petroleum systems and hydrocarbon potential of the North-West Himalaya of India and Pakistan. *Earth-Science Reviews*, **187**:109–185, 2018.
9. HS Rajkumar, I Soibam, KS Khaidem, SS Sanasam, and Ch M Khuman. Ichnological significance of upper Disang formation and lower Barail formation (late Eocene to early Oligocene) of Nagaland, northeast India, in the Indo-Myanmar ranges. *Journal of the Geological Society of India*, **93**(4):471–481, 2019.

10. LP Mathur and P Evans. Oil in India: a review prepared by the staffs of the Oil and Natural Gas Commission, the Assam Oil Co. Ltd., and Oil India Ltd., 1964.
11. BK Gogoi and RK Sarmah. Evaluation of geochemical characteristics of shale of Disang Group in a part of Assam-Arakan basin in perspective of its hydrocarbon potential. *Science Vision*, **13**(3):103–112, 2013.
12. D Bezbaruah, K Kashyap, and MP Boruah. Field geology and morphometric evidences supporting active tectonics of Amguri-Tuli-Merangkong area in the Assam-Nagaland border, India. *South East Asian Journal of Sedimentary Basin Research*, 37–46, 2016.
13. RL Folk and WC Ward. Brazos River bar [Texas]; a study in the significance of grain size parameters. *Journal of Sedimentary Research*, **27**(1):3–26, 1957.
14. GM Friedman. Distinction between dune, beach, and river sands from their textural characteristics. *Journal of Sedimentary Research*, **31**(4):514–529, 1961.
15. BK Sahu. Depositional mechanisms from the size analysis of clastic sediments. *Journal of Sedimentary Research*, **34**(1):73–83, 1964.
16. RA Berner and R Raiswell. C/s method for distinguishing freshwater from marine sedimentary rocks. *Geology*, **12**(6):365–368, 1984.
17. BP Tissot, R Pelet, and PH Ungerer. Thermal history of sedimentary basins, maturation indices, and kinetics of oil and gas generation. *AAPG Bulletin*, **71**(12):1445–1466, 1987.
18. AK Burnham. Van Krevelen diagrams. In *Encyclopedia of Petroleum Geoscience*, Springer, Cham, pp. 1–5, 2018.
19. KR Ghori and PW Haines. Paleozoic petroleum systems of the Canning Basin, Western Australia. *Search and Discovery Article*, 10120, 2007.
20. K Borgohain, P Borgohain, B Bharali, and J Baruah. Sedimentological characteristics of the Barail arenaceous unit of Makum-North Hapjan oil field, Assam. *International Journal of Earth Science and Geology*, **1**(2):66–73, 2019.
21. BD Mandaokar. Palynology and palaeoenvironment of the Bhuban Formation (early Miocene) of Ramrikawn, near Aizawl, Mizoram, India. *Palaeobotanist*, **49**:317–324, 2000.
22. RJ Hayman. Reconnaissance map of part of Lushai Hills. *Unpublished Report Burma Oil Company, RJH*, 11, 1937.
23. WA Franklin. Photogeological map of Assam and Tripura. *Assam Oil Corporation Letter*, Unpublished report, 1948.

24. JL Ganju. Geology of Mizoram. *Bulletin of the Geological, Mining and Metallurgical Society of India*, **48**:17–26, 1975.
25. DR Nandy, SD Gupta, K Sarkar, and A Ganguly. Tectonic evolution of Tripura Mizoram fold belt, Surma basin, northeast India. *Quarterly Journal of the Geological, Mining, and Metallurgical Society of India*, **35**(4):186–194, 1983.
26. S Dasgupta. Tectonic trends in Surma basin and possible genesis of the folded belt. *Records of the Geological Survey of India*, **113**(4):58–61, 1984.
27. RP Tiwari and RP Kachhara. Molluscan biostratigraphy of the tertiary sediments of the Mizoram, India. *Journal of the Palaeontological Society of India*, **48**:59–82, 2003.
28. RP Tiwari, J Malsawma, SJ Sangode, and BR Arora. Magnetostratigraphy of a part of middle Bhuban sequence (Surma Group), Aizawl, Mizoram. *Journal-Geological Society of India*, **70**(4):667, 2007.
29. RP Tiwari, C Rajkonwar, Lalchawimawii, PLJ Malsawma, VZ Ralte, SJ Patel, et al. Trace fossils from Bhuban Formation, Surma Group (Lower to Middle Miocene) of Mizoram India and their palaeoenvironmental significance. *Journal of Earth System Science*, **120**(6):1127–1143, 2011.
30. P Gazzi. Le arenarie del flysch sopracretaceo dell'appennino modenese; correlazioni coni flysch di Monghidoro. *Mineralogica et Petrographica Acta*, **12**:69–97, 1966.
31. C Karunakaran. Geology and mineral resources of the states of India. Part IV: Arunachal Pradesh, Assam, Manipur, Meghalaya, Mizoram, Nagaland and Tripura. *Miscellaneous publication (Geological Survey of India)*, **30**(4):93–101, 1974.
32. SB Gogoi. Effluent as surfactant for enhanced oil recovery. *Innovative Energy Policies*, **3**(1):46–54, 2014.
33. SB Gogoi and BM Das. Use of an effluent for enhanced oil recovery. *Indian Journal of Chemical Technology*, **19**:366–370, 2012.
34. SB Gogoi. Characterization of vesicle for enhanced oil recovery. *Indian Journal of Chemical Technology*, **17**:282–290, 2010.
35. SB Gogoi and KD Gogoi. Emulsion flow in bhogpara core sample with emphasis on enhanced oil recovery. *Indian Journal of Petroleum Geology*, **14**(2):148–168, 2005.
36. S Sarmah, SB Gogoi, J Barman, and D Dutta. Study on the effects of the rheological properties for flow behaviour of Assam crude oil with various chemicals. In D Singh and A Galaa (eds) *GeoMEast 2017:*

*Contemporary Issues in Geoenvironmental Engineering*, Sustainable Civil Infrastructures. Springer, Cham, pp. 148–168, 2017.

37. SB Gogoi. Adsorption–desorption of surfactant for enhanced oil recovery. *Transport in Porous Media*, **90**(2):589, 2011.
38. BM Das and SB Gogoi. Relating IFT with oil recovery with special reference to bhogpara porous media of Upper Assam Basin. *Journal of Petroleum Engineering & Technology*, **5**(1):1–8, 2015.
39. BM Das and SB Gogoi. Adsorption study of a natural product in to the oil producing porous media of Upper Assam Basin. *International Journal of Research in Engineering and Applied Sciences*, 2249–3905, 2015.
40. R Phukan, SB Gogoi, and P Tiwari. Enhanced oil recovery by alkaline-surfactant-alternated-gas/$CO_2$ flooding. *Journal of Petroleum Exploration and Production Technology*, **9**(1):247–260, 2019.
41. SB Gogoi and M Kakoty. A study of $CO_2$ flooding on wave velocities in the Naharkatiya oil reservoir of Upper Assam Basin. *Resource-Efficient Technologies*, **3**(1):101–112, 2017.
42. K Hazarika, R Yadav, SB Gogoi, and UK Bhui. Characterisation of crude oil for enhanced oil recovery: study with anionic surfactant. *International Journal of Ambient Energy*, **40**(6):645–656, 2019.
43. K Hazarika and SB Gogoi. Comparative study of an enhanced oil recovery process with various chemicals for Naharkatiya oil field. *International Journal of Applied Sciences and Biotechnology*, **2**(4):432–436, 2014.
44. S Sarmah, SB Gogoi, K Jagatheesan, and K Hazarika. Formulation of a combined low saline water and polymer flooding for enhanced oil recovery. *International Journal of Ambient Energy*, 1–9, 2019.
45. AB Dyes, BH Caudle, RA Erickson, et al. Oil production after breakthrough as influenced by mobility ratio. *Journal of Petroleum Technology*, **6**(04):27–32, 1954.
46. KA Elraies and IM Tan. *The Application of a New Polymeric Surfactant for Chemical EOR*. INTECH Open Access Publisher, 2012.
47. Y Wang, F Zhao, B Bai, et al. Optimized surfactant IFT and polymer viscosity for surfactant-polymer flooding in heterogeneous formations. In *SPE Improved Oil Recovery Symposium*. Society of Petroleum Engineers, 2010.
48. M Buchgraber, T Clemens, LM Castanier, A Kovscek, et al. A microvisual study of the displacement of viscous oil by polymer solutions. *SPE Reservoir Evaluation & Engineering*, **14**(03):269–280, 2011.

49. AR Kovscek. Screening criteria for CO2 storage in oil reservoirs. *Petroleum Science and Technology*, **20**(7–8):841–866, 2002.

50. B Habermann, et al. The efficiency of miscible displacement as a function of mobility ratio. Society of Petroleum Engineers, 1960.

51. J Bryan, A Kantzas, et al. Potential for alkali-surfactant flooding in heavy oil reservoirs through oil-in-water emulsification. *Journal of Canadian Petroleum Technology*, **48**(02):37–46, 2009.

52. E Claridge. Control of viscous fingering in enhanced oil recovery process: effect of heterogeneities. Society of Petroleum Engineers, 1978.

53. GA Pope and C Huh. Residual oil saturation from polymer floods: laboratory measurements and theoretical interpretation. In *SPE Symposium on Improved Oil Recovery*. Society of Petroleum Engineers, 2008.

54. W Lv, C Xie, and M Wang. Shear-thinning or shear-thickening fluid for better EOR?: a direct pore-scale study. *Journal of Petroleum Science and Engineering*, **161**:683–691, 2018.

55. MT Balhoff, GA Pope, MZ Erincik, P Qi. New method to reduce residual oil saturation by polymer flooding. *SPE Journal*, **23**(05):1–944, 2018.

56. J Chen, YB Zhu, DT Lu, H Liu, HA Wu, JC Fan, FC Wang. Molecular mechanism of viscoelastic polymer enhanced oil recovery in nanopores. *Royal Society Open Science*, **5**(6):180076, 2018.

57. V Kippe, K Sandengen, T Skauge, J Solbakken, D Zhu, B Vik, A Kedir. Viscous oil recovery by polymer injection; impact of in-situ polymer rheology on water front stabilization. In *SPE Europec Featured at 80th EAGE Conference and Exhibition*. Society of Petroleum Engineers, 2018.

58. JC Berg. *An Introduction to Interfaces & Colloids: The Bridge to Nanoscience*. World Scientific, 2010.

59. A Shukla and H Rehage. Zeta potentials and debye screening lengths of aqueous, viscoelastic surfactant solutions (cetyltrimethylammonium bromide/sodium salicylate system). *Langmuir*, **24**(16):8507–8513, 2008.

60. MA Brown, A Goel, and Z Abbas. Effect of electrolyte concentration on the stern layer thickness at a charged interface. *Angewandte Chemie International Edition*, **55**(11):3790–3794, 2016.

61. D Man-li and G Xiao-li. The viscoelasticity of polymer solution in porous media. *Chem. Oil Field*, **9**(2):145–150, 1992.

62. M Guangyan. The viscoelasticity of pam solution in displacing process and distribution of reservoir pressure. *Chem. Oil Field*, **13**(4):353–356, 1996.

63. ST Gao. Chemical flooding in America. *Petroleum Geology and Oilfield Development in Daqing*, **8**(1):45–52, 1989.

64. A Bera and A Mandal. Microemulsions: a novel approach to enhanced oil recovery: a review. *Journal of Petroleum Exploration and Production Technology*, **5**(3):255–268, 2015.

65. Z Jeirani, B Mohamed Jan, B Si Ali, CH See, and W Saphanuchart. Pre-prepared microemulsion flooding in enhanced oil recovery: a review. *Petroleum Science and Technology*, **32**(2):180–193, 2014.

66. HL Rosano, JL Cavallo, DL Chang, JH Whittam, et al. Microemulsions: a commentary on their preparation. *Journal of the Society of Cosmetic Chemists*, 39:201–209, 1988.

67. S Sarmah, SB Gogoi, F Xianfeng, and AA Baruah. Characterization and identification of the most appropriate nonionic surfactant for enhanced oil recovery. *Journal of Petroleum Exploration and Production Technology*, 1–9, 2019.

68. A Demirbas, HE Alsulami, and WS Hassanein. Utilization of surfactant flooding processes for enhanced oil recovery (EOR). *Petroleum Science and Technology*, **33**(12):1331–1339, 2015.

69. AH Alagorni, ZB Yaacob, and AH Nour. An overview of oil production stages: enhanced oil recovery techniques and nitrogen injection. *International Journal of Environmental Science and Development*, **6**(9):693–701, 2015.

70. A Muggeridge, A Cockin, K Webb, H Frampton, I Collins, T Moulds, and P Salino. Recovery rates, enhanced oil recovery and technological limits. *Philosophical Transactions of the Royal Society A: Mathematical, Physical and Engineering Sciences*, **372**(2006):20120320, 2014.

71. I Sandrea and R Sandrea. Recovery factors leave vast target for EOR technologies. *Oil & Gas Journal*, **105**(41):44–47, 2007.

72. M Kakoty and SB Gogoi. Evaluation of surfactant formulation for EOR in some depleted oil fields of Upper Assam. In *International Congress and Exhibition on Sustainable Civil Infrastructures: Innovative Infrastructure Geotechnology*, Springer, pp. 57–75, 2018.

73. SO Isehunwa and O Olanisebe. Interfacial tension of crude oil- brine systems in the Niger Delta. *International Journal of Research and Reviews in Applied Sciences*, **10**:460–465, 2012.

74. S Trabelsi, A Hutin, J-F Argillier, C Dalmazzone, B Bazin, and D Langevin. Effect of added surfactants on the dynamic interfacial tension behaviour of alkaline/diluted heavy crude oil system. *Oil & Gas Science and Technology–Revue d'IFP Energies Nouvelles*, **67**(6):963–968, 2012.

75. D Langevin, S Poteau, I Hénaut, and JF Argillier. Crude oil emulsion properties and their application to heavy oil transportation. *Oil & gas Science and Technology*, **59**(5):511–521, 2004.

76. T Al-Sahhaf, A Suttar Ahmed, and A Elkamel. Producing ultralow interfacial tension at the oil/water interface. *Petroleum Science and Technology*, **20**(7–8):773–788, 2002.

77. S Gao, H Li, H Li, et al. Laboratory investigation of combination of alkaline-surfactant-polymer for daqing EOR. *SPE Reservoir Engineering*, **10**(03):194–197, 1995.

78. FD Martin, JC Oxley, H Lim, et al. Enhanced recovery of a "j" sand crude oil with a combination of surfactant and alkaline chemicals. In *SPE Annual Technical Conference and Exhibition*. Society of Petroleum Engineers, 1985.

79. M Abbas, et al. An extension of Johnson, Bossler and Neumann JBN method for calculating relative permeabilities. In *SPE Annual Technical Conference and Exhibition*. Society of Petroleum Engineers, 2016.

80. E-R Pérez-Carrillo, J-F Zapata-Arango, and N Santos-Santos. A new method for the experimental determination of three-phase relative permeabilities. *CT&F-Ciencia, Tecnologí a y Futuro*, **3**(4):23–43, 2008.

81. I Shikhov, MN d'Eurydice, J Arns, et al. An experimental and numerical study of relative permeability estimates using spatially resolved $T_{1-z}$ NMR. *Transport in Porous Media*, **118**:225–250, 2017.

82. D Xie, J Hou, A Doda, JJ Trivedi, et al. Organic alkali for heavy oil chemical EOR improves the performance over inorganic alkali. In *SPE International Heavy Oil Conference and Exhibition*. Society of Petroleum Engineers, 2014.

83. NC Thomas, B Ghosh, WS AlAmeri, and A Kilybay. Alkali and hybrid-alkali flooding as a tertiary oil recovery mode: prospects and challenges. *International Journal of Petroleum and Petrochemical Engineering (IJPPE)*, **2**(2):22–31, 2016.

84. A Samanta, A Bera, K Ojha, and A Mandal. Comparative studies on enhanced oil recovery by alkali–surfactant and polymer flooding. *Journal of Petroleum Exploration and Production Technology*, **2**(2):67–74, 2012.

85. JJ Sheng. Investigation of alkaline–crude oil reaction. *Petroleum*, **1**(1):31–39, 2015.

86. AM Sarem, et al. Secondary and tertiary recovery of oil by MCCF process. In *SPE California Regional Meeting*. Society of Petroleum Engineers, 1974.

87. JJ Sheng. Status of surfactant EOR technology. *Petroleum*, **1**(2):97–105, 2015.

88. EC Donaldson, GV Chilingarian, and TF Yen. *Enhanced Oil Recovery, I: Fundamentals and Analyses*. Elsevier, 1985.

89. S Carlino and MJ Hudson. A thermal decomposition study on the intercalation of tris-(oxalato) ferrate (iii) trihydrate into a layered (Mg/Al) double hydroxide. *Solid State Ionics*, **110**(1–2):153–161, 1998.

90. F Prinetto, G Ghiotti, P Graffin, and D Tichit. Synthesis and characterization of sol–gel Mg/Al and Ni/Al layered double hydroxides and comparison with co-precipitated samples. *Microporous and Mesoporous Materials*, **39**(1–2):229–247, 2000.

91. S Carlino and MJ Hudson. Thermal intercalation of layered double hydroxides: capric acid into an Mg–Al LDH. *Journal of Materials Chemistry*, **5**(9):1433–1442, 1995.

92. Y Xi, W Martens, H He, and RL Frost. Thermogravimetric analysis of organoclays intercalated with the surfactant octadecyltrimethylammonium bromide. *Journal of Thermal Analysis and Calorimetry*, **81**(1):91–97, 2005.

93. ZB Ibrahim, AA Abdul Manap, P Abdul Hamid, VY Hon, PH Lim, K Wyatt, et al. Laboratory aspect of chemical EOR processes evaluation for Malaysian oilfields. In *SPE Asia Pacific Oil & Gas Conference and Exhibition*. Society of Petroleum Engineers, 2006.

94. JE Moreno, OM Gurpinar, Y Liu, A Al-Kinani, N Cakir, et al. EOR advisor system: a comprehensive approach to EOR selection. In *International Petroleum Technology Conference*, 2014.

95. JJ Taber, FD Martin, RS Seright, et al. EOR screening criteria revisited, part 1: introduction to screening criteria and enhanced recovery field projects. *SPE Reservoir Engineering*, **12**(03):189–198, 1997.

96. A Al Adasani and B Bai. Analysis of EOR projects and updated screening criteria. *Journal of Petroleum Science and Engineering*, **79**(1–2):10–24, 2011.

97. A Kamari and AH Mohammadi. Screening of enhanced oil recovery methods. *Handbook on Oil Production Research*, Nova Science Publishers, Inc., USA, 2014.
98. L Akanji and R Sandrea. A neuro-fuzzy approach to screening reservoir candidates for EOR. *Advances in Petroleum Exploration and Development*, **12**:1–14, 2016.
99. A Aladasani. Updated EOR screening criteria and modeling the impacts of water salinity changes on oil recovery. Dissertation, Missouri University of Science and Technolog, 2012.
100. SA Kalwar, KA Elraies, MK Memon, S Kumar, G Abbas, AH Mithani, et al. A new approach to asp flooding in high saline and hard carbonate reservoirs. In *International Petroleum Technology Conference*, 2014.
101. LS Romsted. *Surfactant Science and Technology: Retrospects and Prospects*. CRC Press, 2014.
102. BJ Ben Shiau, T-P Hsu, P Lohateeraparp, MR Rojas, M Budhathoki, A Raj, W Wan, S Bang, JH Harwell, et al. Recovery of oil from high salinity reservoir using chemical flooding: from laboratory to field tests. In *SPE Enhanced Oil Recovery Conference*. Society of Petroleum Engineers, 2013.
103. KA Elraies. The effect of a new in situ precipitation inhibitor on chemical EOR. *Journal of Petroleum Exploration and Production Technology*, **3**(2):133–137, 2013.
104. S Alian, AA Omar, AF Alta'ee, and I Hani. Study of asphaltene precipitation induced formation damage during $CO_2$ injection for a Malaysian light oil. *World Academy of Science, Engineering and Technology*, **78**:733–737, 2011.
105. M Nobakht, S Moghadam, and Y Gu. Effects of viscous and capillary forces on $CO_2$ enhanced oil recovery under reservoir conditions. *Energy & Fuels*, **21**(6):3469–3476, 2007.
106. HL Chang, ZQ Zhang, QM Wang, ZS Xu, ZD Guo, HQ Sun, XL Cao, Q Qiao, et al. Advances in polymer flooding and alkaline/surfactant/polymer processes as developed and applied in the people's republic of China. *Journal of Petroleum Technology*, **58**(02):84–89, 2006.
107. SA Shedid, et al. Experimental investigation of alkaline/surfactant/polymer (ASP) flooding in low permeability heterogeneous carbonate reservoirs. In *SPE North Africa Technical Conference and Exhibition*. Society of Petroleum Engineers, 2015.

108. FM Orr, JP Heller, and JJ Taber. Carbon dioxide flooding for enhanced oil recovery: promise and problems. *Journal of the American Oil Chemists Society*, **59**(10):810A–817A, 1982.

109. J Amyx, D Bass, RL Whiting, et al. *Petroleum Reservoir Engineering: Physical Properties*. McGraw-Hill Book Company, Inc., 1960.

110. M Kennedy. Core and other real rock measurements. In *Developments in Petroleum Science*, Elsevier, Vol. 62, pp. 73–88, 2015.

111. P Talukdar and SB Gogoi. A study on the role of pre-gelatinized starch (PGS) in the non damaging drilling fluid (NDDF) for the Tipam sand of Geleki oilfield of Upper Assam Basin. *International Journal of Applied Sciences and Biotechnology*, **3**(2):291–300, 2015.

112. PH Nelson and KJ Bird. *Porosity-Depth Trends and Regional Uplift Calculated from Sonic Logs, National Petroleum Reserve in Alaska*. US Department of the Interior, US Geological Survey, 2005.

113. M Ramm and K Bjørlykke. Porosity/depth trends in reservoir sandstones: assessing the quantitative effects of varying pore-pressure, temperature history and mineralogy, Norwegian shelf data. *Clay Minerals*, **29**(4):475–490, 1994.

114. S Ma and NR Morrow. Relationships between porosity and permeability for porous rocks. In *International Symposium of the Society of Core Analysts*, pp. 8–10, 1996.

115. TEW Nind. *Principles of Oil Well Production*. McGraw-Hill, New York, 1981.

116. WH Al-Bazzaz, A Gupta, et al. Reservoir characterization and data integration of complex Mauddud-Burgan carbonate reservoir. In *SPE Europec/EAGE Annual Conference and Exhibition*. Society of Petroleum Engineers, 2006.

117. MS Kamal, IA Hussein, and AS Sultan. Review on surfactant flooding: phase behavior, retention, IFT, and field applications. *Energy & Fuels*, **31**(8):7701–7720, 2017.

118. JE Welton, et al. *SEM Petrology Atlas*. American Association of Petroleum Geologist, 1984.

119. L Wilson, MJ Wilson, J Green, and I Patey. The influence of clay mineralogy on formation damage in north sea reservoir sandstones: a review with illustrative examples. *Earth-Science Reviews*, **134**:70–80, 2014.

120. P Julius, PN Ananthanarayanan, and V Srinivasan. Investigation of enhanced oil recovery (EOR) surfactants on clay mixed sandstone

reservoirs for adsorption. *Indian Journal of Science and Technology*, **8**(14):1, 2015.

121. E Shashi Menon. *Liquid Pipeline Hydraulics*. CRC Press, 2004.
122. S Jafarinejad. *Petroleum Waste Treatment and Pollution Control*. Butterworth-Heinemann, 2016.
123. T Ahmed. *Reservoir Engineering Handbook*. Gulf Professional Publishing, 2018.
124. AW Drews. *Manual on Hydrocarbon Analysis*, Vol. 3. ASTM International, 1998.
125. VC Kandwal, KM Agrawal, SP Nautiyal, and HU Khan. Paraffin deposition and viscosity temperature behaviour of Assam crude oil. *Petroleum Science and Technology*, **18**(7–8):755–769, 2000.
126. AR Rabbani, G Hassanzadeh, and E Dehyadegari. Asphaltene deposition under $CO_2$ injection: an experimental study of the bangestan reservoir of Ahwaz oilfield, SW Iran. *Petroleum Science and Technology*, **30**(1):9–15, 2012.
127. S Ghedan. Global laboratory experience of $CO_2$-EOR flooding. In *SPE/EAGE Reservoir Characterization & Simulation Conference*, 2009.
128. RK Srivastava, SS Huang, et al. Asphaltene deposition during $CO_2$ flooding: a laboratory assessment. In *SPE Production Operations Symposium*. Society of Petroleum Engineers, 1997.
129. J Bon, HK Sarma, et al. A technical evaluation of a $CO_2$ flood for EOR benefits in the Cooper Basin, South Australia. In *SPE Asia Pacific Oil and Gas Conference and Exhibition*. Society of Petroleum Engineers, 2004.
130. KJ Leontaritis, GA Mansoori, et al. Asphaltene flocculation during oil production and processing: a thermodynamic collodial model. In *SPE International Symposium on Oilfield Chemistry*. Society of Petroleum Engineers, 1987.
131. MM Ramirez-Corredores. *The Science and Technology of Unconventional Oils: Finding Refining Opportunities*. Academic Press, 2017.
132. N Aske, H Kallevik, and J Sjöblom. Determination of saturate, aromatic, resin, and asphaltenic (SARA) components in crude oils by means of infrared and near-infrared spectroscopy. *Energy & Fuels*, **15**(5):1304–1312, 2001.
133. A Samanta, K Ojha, and A Mandal. Interactions between acidic crude oil and alkali and their effects on enhanced oil recovery. *Energy & Fuels*, **25**(4):1642–1649, 2011.

134. R Wan. *Advanced Well Completion Engineering*. Gulf Professional Publishing, 2011.

135. EJ Burcik. *Properties of Petroleum Reservoir Fluids*. Wiley, 1957.

136. H Lee, SG Lee, and PS Doyle. Photopatterned oil-reservoir micromodels with tailored wetting properties. *Lab on a Chip*, **15**(14):3047–3055, 2015.

137. VA Lifton. Microfluidics: an enabling screening technology for enhanced oil recovery (EOR). *Lab on a Chip*, **16**(10):1777–1796, 2016.

138. MH Schneider and P Tabeling. Lab-on-chip methodology in the energy industry: wettability patterns and their impact on fluid displacement in oil reservoir models. *American Journal of Applied Sciences*, **8**(10):927, 2011.

139. L Zhong, A Mayer, and RJ Glass. Visualization of surfactant-enhanced nonaqueous phase liquid mobilization and solubilization in a two-dimensional micromodel. *Water Resources Research*, **37**(3):523–537, 2001.

140. C Negin, S Ali, and Q Xie. Most common surfactants employed in chemical enhanced oil recovery. *Petroleum*, **3**(2):197–211, 2017.

141. H Al-Sulaimani, S Joshi, Y Al-Wahaibi, S Al-Bahry, A Elshafie, and A Al-Bemani. Microbial biotechnology for enhancing oil recovery: current developments and future prospects. *Biotechnology, Bioinformatics and Bioengineering Journal*, **1**(2):147–158, 2011.

142. S Thomas. Enhanced oil recovery: an overview. *Oil & Gas Science and Technology-Revue de l'IFP*, **63**(1):9–19, 2008.

143. I Lazar, IG Petrisor, and TF Yen. Microbial enhanced oil recovery (MEOR). *Petroleum Science and Technology*, **25**(11):1353–1366, 2007.

144. R Sen. Biotechnology in petroleum recovery: the microbial EOR. *Progress in Energy and Combustion Science*, **34**(6):714–724, 2008.

145. DKF Santos, RD Rufino, JM Luna, VA Santos, and LA Sarubbo. Biosurfactants: multifunctional biomolecules of the 21st century. *International Journal of Molecular Sciences*, **17**(3):401, 2016.

146. IM Banat, RS Makkar, and SS Cameotra. Potential commercial applications of microbial surfactants. *Applied Microbiology and Biotechnology*, **53**(5):495–508, 2000.

147. A Chandra and S Ghosh. Smart sankey picturization for energy management systems in India. *AI & Society*, pp. 1–7, 2019.

148. JD Van Hamme, A Singh, and OP Ward. Petroleum microbiology, part 1: underlying biochemistry and physiology. *Chimica Oggi*, **24**(1):52–56, 2006.

149. AS Nerurkar, HG Suthar, and AJ Desai. Biosystem development for microbial enhanced oil recovery (MEOR). In *Microorganisms in Sustainable Agriculture and Biotechnology*, Springer, pp. 711–737, 2012.

150. DW Green, GP Willhite, et al. *Enhanced Oil Recovery*, Vol. 6. Henry L Doherty Memorial Fund of AIME, Society of Petroleum Engineers, 1998.

151. Z Shi, Y Zhang, M Liu, DAH Hanaor, and Y Gan. Dynamic contact angle hysteresis in liquid bridges. *Colloids and Surfaces A: Physicochemical and Engineering Aspects*, **555**:365–371, 2018.

152. K Guo, H Li, and Z Yu. In-situ heavy and extra-heavy oil recovery: a review. *Fuel*, **185**:886–902, 2016.

153. X Dong, H Liu, Z Chen, K Wu, N Lu, and Q Zhang. Enhanced oil recovery techniques for heavy oil and oilsands reservoirs after steam injection. *Applied Energy*, **239**:1190–1211, 2019.

154. C Cheih, et al. State-of-the-art review of fireflood field projects (includes associated papers 10901 and 10918). *Journal of Petroleum Technology*, **34**(01):19–36, 1982.

155. JJ Taber, et al. Technical screening guides for the enhanced recovery of oil. In *SPE Annual Technical Conference and Exhibition*. Society of Petroleum Engineers, 1983.

156. M Cokar, ID Gates, MS Kallos, et al. Reservoir simulation of steam fracturing in early-cycle cyclic steam stimulation. *SPE Reservoir Evaluation & Engineering*, **15**(06):676–687, 2012.

157. W Littmann. Application of surface-active agents in petroleum production. In *Hand Book of Surface and Colloid Chemistry*, CRC Press, pp. 689–699, 1997.

158. A-Y Huc. *Heavy Crude Oils: From Geology to Upgrading: An Overview*. Editions Technip, 2010.

159. G Zerkalov. Polymer flooding for enhanced oil recovery. PH240, Stanford University, 2015.

160. AA Olajire. Review of ASP EOR (alkaline surfactant polymer enhanced oil recovery) technology in the petroleum industry: prospects and challenges. *Energy*, **77**:963–982, 2014.

161. A Andrianov, R Farajzadeh, MM Nick, M Talanana, and PLJ Zitha. Immiscible foam for enhancing oil recovery: bulk and porous media experiments. *Industrial & Engineering Chemistry Research*, **51**(5):2214–2226, 2012.

162. L Yarborough, LR Smith, et al. Solvent and driving gas compositions for miscible slug displacement. *Society of Petroleum Engineers Journal*, **10**(03):298–310, 1970.

163. S Maudgalya, RM Knapp, M McInerney, et al. Microbially enhanced oil recovery technologies: a review of the past, present and future. In *Production and Operations Symposium*. Society of Petroleum Engineers, 2007.

164. LR Brown. Microbial enhanced oil recovery (MEOR). *Current Opinion in Microbiology*, **13**(3):316–320, 2010.

165. EC Donaldson, GV Chilingarian, and TF Yen. *Microbial Enhanced Oil Recovery*, Vol. 22. Newnes, 1989.

166. IM Banat. Biosurfactants production and possible uses in microbial enhanced oil recovery and oil pollution remediation: a review. *Bioresource Technology*, **51**(1):1–12, 1995.

167. CE Zobell. Bacteriological process for treatment of fluid- bearing earth formations. US Patent 2,413,278, 1946.

168. DM Updegraff. Recovery of petroleum oil. US Patent 2,807,570, 1957.

169. GM Evans and JC Furlong. Phytotechnology and photosynthesis. In *Environmental Biotechnology: Theory and Applications*, 1st ed. Wiley Publishing, New York, pp. 163–169, 2003.

170. RS Bryant, TE Burchfield, et al. Review of microbial technology for improving oil recovery. *SPE Reservoir Engineering*, **4**(02):151–154, 1989.

171. AY Zekri, RA Almehaideb, and O Chaalal. Microbes isolation from MEOR using hot water streams. SPE 26827, 1999.

172. I Lazar, S Dobrota, M Stefanescu, L Sandulescu, P Constantinescu, C Morosanu, N Botea, and O Iliescu. Ch. F-2 preliminary results of some recent MEOR field trials in Romania. In *Developments in Petroleum Science*, Vol. 31, Elsevier, pp. 365–385, 1991.

173. G Castorena-Cortés, I Zapata-Peñasco, T Roldán-Carrillo, J Reyes-Avila, M Mayol-Castillo, S Román-Vargas, and P Olguín-Lora. Evaluation of indigenous anaerobic microorganisms from Mexican carbonate reservoirs with potential MEOR application. *Journal of Petroleum Science and Engineering*, **81**:86–93, 2012.

174. E Sunde, J Beeder, RK Nilsen, T Torsvik, et al. Aerobic microbial enhanced oil recovery for offshore use. In *SPE/DOE Enhanced Oil Recovery Symposium*. Society of Petroleum Engineers, 1992.
175. H Zhu, HK Carlson, and JD Coates. Applicability of anaerobic nitrate-dependent Fe (II) oxidation to microbial enhanced oil recovery (MEOR). *Environmental Science & Technology*, **47**(15):8970–8977, 2013.
176. SJ Geetha, IM Banat, and SJ Joshi. Biosurfactants: production and potential applications in microbial enhanced oil recovery (MEOR). *Biocatalysis and Agricultural Biotechnology*, **14**:23–32, 2018.
177. DI Astuti, IA Purwasena, RE Putri, M Amaniyah, and Y Sugai. Screening and characterization of biosurfactant produced by pseudoxanthomonas sp. G3 and its applicability for enhanced oil recovery. *Journal of Petroleum Exploration and Production Technology*, 1–11, 2019.
178. C Gao. Experiences of microbial enhanced oil recovery in Chinese oil fields. *Journal of Petroleum Science and Engineering*, **166**:55–62, 2018.
179. JG Speight and NS El-Gendy. *Introduction to Petroleum Biotechnology*. Gulf Professional Publishing, 2017.
180. JM Khire. Bacterial biosurfactants, and their role in microbial enhanced oil recovery (MEOR). In *Biosurfactants*, Springer, pp. 146–157, 2010.
181. RS Bryant. Potential uses of microorganisms in petroleum recovery technology. In *Proceedings of the Oklahoma Academy of Science*, Vol. 67, pp. 97–104, 1987.
182. SJ Joshi. Isolation and characterization of biosurfactant producing microorganisms and their possible role in microbial enhanced oil recovery MEOR. Maharaja Sayajirao University of Baroda, 2008.
183. JG Leahy and RR Colwell. Microbial degradation of hydrocarbons in the environment. *Microbiology and Molecular Biology Reviews*, **54**(3):305–315, 1990.
184. S Maneerat, et al. Biosurfactants from marine microorganisms. *Songklanakarin Journal of Science and Technology*, **27**(6):1263–1272, 2005.
185. JE Zajic, H Guignard, and DF Gerson. Properties and biodegradation of a bioemulsifier from corynebacterium hydrocarboclastus. *Biotechnology and Bioengineering*, **19**(9):1303–1320, 1977.
186. G Georgiou, S-C Lin, and MM Sharma. Surface-active compounds from microorganisms. *Biotechnology*, **10**(1):60, 1992.

187. MPS Pirôllo, AP Mariano, RB Lovaglio, SGVAO Costa, V Walter, R Hausmann, and J Contiero. Biosurfactant synthesis by Pseudomonas aeruginosa LBI isolated from a hydrocarbon-contaminated site. *Journal of Applied Microbiology*, **105**(5):1484–1490, 2008.

188. EZ Ron and E Rosenberg. Biosurfactants and oil bioremediation. *Current Opinion in Biotechnology*, **13**(3):249–252, 2002.

189. E Rosenberg and EZ Ron. Surface active polymers from the genus acinetobacter. In *Biopolymers from Renewable Resources*, Springer, pp. 281–291, 1998.

190. RS Reis, AG Pereira, BC Neves, and DMG Freire. Gene regulation of rhamnolipid production in Pseudomonas aeruginosa: a review. *Bioresource Technology*, **102**(11):6377–6384, 2011.

191. G Soberón-Chávez, F Lépine, and E Déziel. Production of rhamnolipids by Pseudomonas aeruginosa. *Applied Microbiology and Biotechnology*, **68**(6):718–725, 2005.

192. J Araújo, J Rocha, MO Filho, S Matias, SO Júnior, and C Padilha. Rhamnolipids biosurfactants from Pseudomonas aeruginosa: a review. *Biosciences Biotechnology Research Asia*, **15**(4):767–781, 2018.

193. S Bergstrom, H Theorell, and H Davide. Pyolipic acid, a metabolic product of Pseudomonas- pyocyanea, active against mycobacterium-tuberculosis. *Archives of Biochemistry*, **10**(1):165–166, 1946.

194. FG Jarvis and MJ Johnson. A glyco-lipide produced by Pseudomonas aeruginosa. *Journal of the American Chemical Society*, **71**(12):4124–4126, 1949.

195. JR Edwards and JA Hayashi. Structure of a rhamnolipid from Pseudomonas aeruginosa. *Archives of Biochemistry and Biophysics*, **111**(2):415–421, 1965.

196. E Déziel, F Lépine, D Dennie, D Boismenu, OA Mamer, and R Villemur. Liquid chromatography/mass spectrometry analysis of mixtures of rhamnolipids produced by Pseudomonas aeruginosa strain 57RP grown on mannitol or naphthalene. *Biochimica et Biophysica Acta (BBA)-Molecular and Cell Biology of Lipids*, **1440**(2–3):244–252, 1999.

197. SGVAO Costa, M Nitschke, R Haddad, MN Eberlin, and J Contiero. Production of Pseudomonas aeruginosa LBI rhamnolipids following growth on Brazilian native oils. *Process Biochemistry*, **41**(2):483–488, 2006.

198. RM Maier and G Soberon-Chavez. Pseudomonas aeruginosa rhamnolipids: biosynthesis and potential applications. *Applied Microbiology and Biotechnology*, **54**(5):625–633, 2000.

199. E Déziel, F Lépine, S Milot, and R Villemur. Mass spectrometry monitoring of *Rhamnolipids* from a growing culture of *Pseudomonas aeruginosa* strain 57rp. *Biochimica et Biophysica Acta (BBA)-Molecular and Cell Biology of Lipids*, **1485**(2–3):145–152, 2000.

200. AM Abdel-Mawgoud, F Lépine, and E Déziel. Rhamnolipids: diversity of structures, microbial origins and roles. *Applied Microbiology and Biotechnology*, **86**(5):1323–1336, 2010.

201. KK Gautam and VK Tyagi. Microbial surfactants: a review. *Journal of Oleo Science*, **55**(4):155–166, 2006.

202. DA White, LC Hird, and ST Ali. Production and characterization of a trehalolipid biosurfactant produced by the novel marine bacterium Rhodococcus sp., strain PML026. *Journal of Applied Microbiology*, **115**(3):744–755, 2013.

203. B Tuleva, N Christova, R Cohen, G Stoev, and I Stoineva. Production and structural elucidation of trehalose tetraesters (biosurfactants) from a novel alkanothrophic Rhodococcus wratislaviensis strain. *Journal of Applied Microbiology*, **104**(6):1703–1710, 2008.

204. PAJ Gorin, JFT Spencer, and AP Tulloch. Hydroxy fatty acid glycosides of sophorose from torulopsis magnoliae. *Canadian Journal of Chemistry*, **39**(4):846–855, 1961.

205. AP Tulloch, JFT Spencer, and MH Deinema. A new hydroxy fatty acid sophoroside from Candida bogoriensis. *Canadian Journal of Chemistry*, **46**(3):345–348, 1968.

206. JFT Spencer, PAJ Gorin, and AP Tulloch. Torulopsis bombicola sp. n. *Antonie Van Leeuwenhoek*, **36**(1):129–133, 1970.

207. J Chen, X Song, H Zhang, and Y Qu. Production, structure elucidation and anticancer properties of sophorolipid from wickerhamiella domercqiae. *Enzyme and Microbial Technology*, **39**(3):501–506, 2006.

208. INA Van Bogaert, K Saerens, C De Muynck, D Develter, W Soetaert, and EJ Vandamme. Microbial production and application of sophorolipids. *Applied Microbiology and Biotechnology*, **76**(1):23–34, 2007.

209. P Zhao, Y Xue, X Li, J Li, Z Zhao, C Quan, W Gao, X Zu, X Bai, and S Feng. Fungi-derived lipopeptide antibiotics developed since 2000. *Peptides*, **113**:52–65, 2019.

210. K Arima, A Kakinuma, and G Tamura. Surfactin, a crystalline peptidelipid surfactant produced by Bacillus subtilis: isolation, characterization and its inhibition of fibrin clot formation. *Biochemical and Biophysical Research Communications*, **31**(3):488–494, 1968.

211. C Ullrich, B Kluge, Z Palacz, and J Vater. Cell-free biosynthesis of surfactin, a cyclic lipopeptide produced by Bacillus subtilis. *Biochemistry*, **30**(26):6503–6508, 1991.

212. A Ohno, T Ano, and M Shoda. Effect of temperature on production of lipopeptide antibiotics, iturin a and surfactin by a dual producer, Bacillus subtilis RB14, in solid-state fermentation. *Journal of Fermentation and Bioengineering*, **80**(5):517–519, 1995.

213. JD Desai and IM Banat. Microbial production of surfactants and their commercial potential. *Microbiology and Molecular Biology Reviews*, **61**(1):47–64, 1997.

214. S Vijayakumar and V Saravanan. Biosurfactants-types, sources and applications. *Research Journal of Microbiology*, **10**(5):181–92, 2015.

215. T Janek, M Lukaszewicz, and A Krasowska. Identification and characterization of biosurfactants produced by the arctic *Bacterium Pseudomonas Putida* BD2. *Colloids and Surfaces B: Biointerfaces*, **110**:379–386, 2013.

216. X Vecino, L Barbosa-Pereira, R Devesa-Rey, JM Cruz, and AB Moldes. Optimization of extraction conditions and fatty acid characterization of lactobacillus pentosus cell-bound biosurfactant/bioemulsifier. *Journal of the Science of Food and Agriculture*, **95**(2):313–320, 2015.

217. GE Jenneman, MJ McInerney, RM Knapp, JB Clark, JM Feero, DE Revus, and DE Menzie. Halotolerant, biosurfactant-producing Bacillus species potentially useful for enhanced oil recovery. *Developments in Industrial Microbiology (United States)*, **24**(CONF-8208164-), 1983.

218. JL Shennan and JD Levi. In situ microbial enhanced oil recovery. In *Biosurfactants and Biotechnology*, Routledge, pp. 163–181, 2017.

219. DO Hitzman, M Dennis, DC Hitzman, et al. Recent successes: MEOR using synergistic H2S prevention and increased oil recovery systems. In *SPE/DOE Symposium on Improved Oil Recovery*. Society of Petroleum Engineers, 2004.

220. N Youssef, MS Elshahed, and MJ McInerney. Microbial processes in oil fields: culprits, problems, and opportunities. *Advances in Applied Microbiology*, **66**:141–251, 2009.

221. N Youssef, DR Simpson, KE Duncan, MJ McInerney, M Folmsbee, T Fincher, and RM Knapp. In situ biosurfactant production by Bacillus strains injected into a limestone petroleum reservoir. *Applied and Environmental Microbiology*, **73**(4):1239–1247, 2007.

222. EJ Gudiña, JFB Pereira, LR Rodrigues, JAP Coutinho, and JA Teixeira. Isolation and study of microorganisms from oil samples for application

in microbial enhanced oil recovery. *International Biodeterioration & Biodegradation*, **68**:56–64, 2012.

223. SN Al-Bahry, YM Al-Wahaibi, AE Elshafie, AS Al-Bemani, SJ Joshi, HS Al-Makhmari, and HS Al-Sulaimani. Biosurfactant production by Bacillus subtilis B20 using date molasses and its possible application in enhanced oil recovery. *International Biodeterioration & Biodegradation*, **81**:141–146, 2013.

224. H She, D Kong, Y Li, Z Hu, and H Guo. Recent advance of microbial enhanced oil recovery (MEOR) in China. *Geofluids*, **2019**: 1871392, 2019.

225. C-Y Ke, G-M Lu, Y-B Li, W-J Sun, Q-Z Zhang, and X-L Zhang. A pilot study on large-scale microbial enhanced oil recovery (MEOR) in baolige oilfield. *International Biodeterioration & Biodegradation*, **127**:247–253, 2018.

226. X Wang, X Li, L Yu, Y Li, L Huang, W Lin, and D Li. Distinctive microbial communities imply the main mechanism in a MEOR trial in high pour-point reservoir. *Journal of Petroleum Science and Engineering*, **175**:97–107, 2019.

227. LJ Chai, F Zhang, YH She, I Banat, and DJ Hou. Impact of a microbial-enhanced oil recovery field trial on microbial communities in a low-temperature heavy oil reservoir. *Nature Environment and Pollution Technology*, **14**(3):455–462, 2015.

228. J Li, S Xue, C He, H Qi, F Chen, and Y Ma. Effect of exogenous inoculants on enhancing oil recovery and indigenous bacterial community dynamics in long-term field pilot of low permeability reservoir. *World Journal of Microbiology and Biotechnology*, **34**(4):53, 2018.

229. M Henkel, M Geissler, F Weggenmann, and R Hausmann. Production of microbial biosurfactants: status quo of rhamnolipid and surfactin towards large-scale production. *Biotechnology Journal*, **12**(7):1600561, 2017.

230. SNRL Silva, CBB Farias, RD Rufino, JM Luna, and LA Sarubbo. Glycerol as substrate for the production of biosurfactant by Pseudomonas aeruginosa UCP0992. *Colloids and Surfaces B: Biointerfaces*, **79**(1):174–183, 2010.

231. J Chen, XJ Wang, JD Hu, and S Tao. Effect of surfactants on biodegradation of PAHs by white-rot fungi. *Huan jing ke xue= Huanjing kexue*, **27**(1):154–159, 2006.

232. DG Cooper and DA Paddock. Production of a biosurfactant from torulopsis bombicola. *Applied and Environmental Microbiology*, **47**(1):173–176, 1984.

233. N Kosaric. Biosurfactants in industry. *Pure and Applied Chemistry*, **64**(11):1731–1737, 1992.

234. S Mukherjee, P Das, and R Sen. Towards commercial production of microbial surfactants. *Trends in Biotechnology*, **24**(11):509–515, 2006.

235. IM Banat, SK Satpute, SS Cameotra, R Patil, and NV Nyayanit. Cost effective technologies and renewable substrates for biosurfactants' production. *Frontiers in Microbiology*, **5**:697, 2014.

236. RS Makkar, SS Cameotra, and IM Banat. Advances in utilization of renewable substrates for biosurfactant production. *AMB Express*, **1**(1):5, 2011.

237. O Pornsunthorntawee, P Wongpanit, and R Rujiravanit. Rhamnolipid biosurfactants: production and their potential in environmental biotechnology. In *Biosurfactants*, Springer, pp. 211–221, 2010.

238. NS Shaligram and RS Singhal. Surfactin: a review on biosynthesis, fermentation, purification and applications. *Food Technology and Biotechnology*, **48**(2):119–134, 2010.

239. F Sineriz, RK Hommel, and HP Kleber. Production of biosurfactants. In *Encyclopedia of Life Support Systems*, Eolls Publishers, Oxford, 2001.

240. A Wittgens, F Kovacic, MM Müller, M Gerlitzki, B Santiago-Schübel, D Hofmann, T Tiso, LM Blank, M Henkel, R Hausmann, et al. Novel insights into biosynthesis and uptake of rhamnolipids and their precursors. *Applied Microbiology and Biotechnology*, **101**(7):2865–2878, 2017.

241. RS Makkar and SS Cameotra. Biosurfactant production by microorganisms on unconventional carbon sources. *Journal of Surfactants and Detergents*, **2**(2):237–241, 1999.

242. S George and K Jayachandran. Production and characterization of rhamnolipid biosurfactant from waste frying coconut oil using a novel Pseudomonas aeruginosa D. *Journal of Applied Microbiology*, **114**(2):373–383, 2013.

243. FASL Reis, EFC Sérvulo, and FP De França. Lipopeptide surfactant production by Bacillus subtilis grown on low-cost raw materials. In *Proceedings of the Twenty-Fifth Symposium on Biotechnology for Fuels and Chemicals Held May 4–7, 2003*, Breckenridge, CO, Springer, pp. 899–912, 2004.

244. IM Ramírez, K Tsaousi, M Rudden, R Marchant, EJ Alameda, MG Roman, and IM Banat. Rhamnolipid and surfactin production from olive oil mill waste as sole carbon source. *Bioresource Technology*, **198**:231–236, 2015.

245. MA Pele, DR Ribeaux, ER Vieira, AF Souza, MAC Luna, DM Rodríguez, RFS Andrade, DS Alviano, CS Alviano, E Barreto-Bergter, et al. Conversion of renewable substrates for biosurfactant production by Rhizopus arrhizus UCP 1607 and enhancing the removal of diesel oil from marine soil. *Electronic Journal of Biotechnology*, **38**:40–48, 2019.

246. EF Santos, MFS Teixeira, A Converti, ALF Porto, and LA Sarubbo. Production of a new lipoprotein biosurfactant by Streptomyces sp. DPUA1566 isolated from lichens collected in the Brazilian Amazon using agroindustry wastes. *Biocatalysis and Agricultural Biotechnology*, **17**:142–150, 2019.

247. V Walter, C Syldatk, R Hausmann, and MC Flickinger. Microbial production of rhamnolipid biosurfactants. In *Encyclopedia of Industrial Biotechnology*. Wiley-VCH, Weinheim, Germany, 2010.

248. DA Davis, HC Lynch, and J Varley. The application of foaming for the recovery of surfactin from b. subtilis ATCC 21332 cultures. *Enzyme and Microbial Technology*, **28**(4–5):346–354, 2001.

249. S Alonso and PJ Martin. Impact of foaming on surfactin production by Bacillus subtilis: implications on the development of integrated in situ foam fractionation removal systems. *Biochemical Engineering Journal*, **110**:125–133, 2016.

250. DA Davis, HC Lynch, and J Varley. The production of surfactin in batch culture by Bacillus subtilis ATCC 21332 is strongly influenced by the conditions of nitrogen metabolism. *Enzyme and Microbial Technology*, **25**(3–5):322–329, 1999.

251. JB Winterburn and PJ Martin. Foam mitigation and exploitation in biosurfactant production. *Biotechnology Letters*, **34**(2):187–195, 2012.

252. C-Y Chen, SC Baker, and RC Darton. Batch production of biosurfactant with foam fractionation. *Journal of Chemical Technology & Biotechnology: International Research in Process, Environmental & Clean Technology*, **81**(12):1923–1931, 2006.

253. M-S Yeh, Y-H Wei, and J-S Chang. Bioreactor design for enhanced carrier-assisted surfactin production with Bacillus subtilis. *Process Biochemistry*, **41**(8):1799–1805, 2006.

254. DKF Santos, HM Meira, RD Rufino, JM Luna, and LA Sarubbo. Biosurfactant production from Candida lipolytica in bioreactor and

evaluation of its toxicity for application as a bioremediation agent. *Process Biochemistry*, **54**:20–27, 2017.

255. J Beuker, A Steier, A Wittgens, F Rosenau, M Henkel, and R Hausmann. Integrated foam fractionation for heterologous rhamnolipid production with recombinant Pseudomonas putida in a bioreactor. *AMB Express*, **6**(1):11, 2016.
256. KV Deepika, S Kalam, PR Sridhar, AR Podile, and PV Bramhachari. Optimization of rhamnolipid biosurfactant production by mangrove sediment bacterium Pseudomonas aeruginosa KVD-HR42 using response surface methodology. *Biocatalysis and Agricultural Biotechnology*, **5**:38–47, 2016.
257. S Sharma, P Datta, B Kumar, P Tiwari, and LM Pandey. Production of novel rhamnolipids via biodegradation of waste cooking oil using Pseudomonas aeruginosa MTCC7815. *Biodegradation*, **30**(4):301–312, 2019.
258. K Zhu and CO Rock. RhlA converts $\beta$-hydroxyacyl-acyl carrier protein intermediates in fatty acid synthesis to the $\beta$-hydroxydecanoyl-$\beta$-hydroxydecanoate component of rhamnolipids in Pseudomonas aeruginosa. *Journal of Bacteriology*, **190**(9):3147–3154, 2008.
259. NJ Hadia, C Ottenheim, S Li, NQ Hua, LP Stubbs, and HC Lau. Experimental investigation of biosurfactant mixtures of surfactin produced by Bacillus subtilis for EOR application. *Fuel*, **251**:789–799, 2019.
260. S Ha, H Kim, H Chun, I Hwang, J-H Lee, J-C Kim, I Kim, and H Park. Effect of oxygen supply on surfactin production and sporulation in submerged culture of Bacillus subtilis y9. *Applied Sciences*, **8**(9):1660, 2018.
261. A Wittgens, B Santiago-Schuebel, M Henkel, T Tiso, LM Blank, R Hausmann, D Hofmann, S Wilhelm, K-E Jaeger, and F Rosenau. Heterologous production of long-chain *Rhamnolipids* from burkholderia glumae in *Pseudomonas Putida*—a step forward to tailor-made *Rhamnolipids*. *Applied Microbiology and Biotechnology*, **102**(3):1229–1239, 2018.
262. S Steller, A Sokoll, C Wilde, F Bernhard, P Franke, and J Vater. Initiation of surfactin biosynthesis and the role of SRFD-thioesterase protein. *Biochemistry*, **43**(35):11331–11343, 2004.
263. LFD Tavares, PM Silva, M Junqueira, DCO Mariano, FCS Nogueira, GB Domont, DMG Freire, and BC Neves. Characterization of rhamnolipids produced by wild-type and engineered burkholderia kururiensis. *Applied Bicrobiology and Biotechnology*, **97**(5):1909–1921, 2013.

264. Q Wang, X Fang, B Bai, X Liang, PJ Shuler, WA Goddard III, and Y Tang. Engineering bacteria for production of rhamnolipid as an agent for enhanced oil recovery. *Biotechnology and Bioengineering*, **98**(4):842–853, 2007.

265. K Bailey, SK Kon, F Dickens, JRP O'Brien, EJ King, C Rimington, BCJG Knight, and AR Todd. Biochemical. In *Engineering Fundamentals*, McGraw-Hill, New York, p. 984, 1986.

266. C-Y Ke, G-M Lu, Y-L Wei, W-J Sun, J-F Hui, X-Y Zheng, Q-Z Zhang, and X-L Zhang. Biodegradation of crude oil by chelatococcus daeguensis HB-4 and its potential for microbial enhanced oil recovery (MEOR) in heavy oil reservoirs. *Bioresource Technology*, **287**:121442, 2019.

267. MR Couto, EJ Gudiña, D Ferreira, JA Teixeira, and LR Rodrigues. The biopolymer produced by *Rhizobium viscosum* CECT 908 is a promising agent for application in microbial enhanced oil recovery. *New Biotechnology*, **49**:144–150, 2019.

268. M Bhattacharya, S Guchhait, D Biswas, and R Singh. Evaluation of a microbial consortium for crude oil spill bioremediation and its potential uses in enhanced oil recovery. *Biocatalysis and Agricultural Biotechnology*, **18**:101034, 2019.

269. C-Y Ke, W-J Sun, Y-B Li, G-M Lu, Q-Z Zhang, and X-L Zhang. Microbial enhanced oil recovery in baolige oilfield using an indigenous facultative anaerobic strain *Luteimonas Huabeiensis* sp. nov. *Journal of Petroleum Science and Engineering*, **167**:160–167, 2018.

270. H Gao, H Lai, J Zhang, and Q Xue. Effects of bacterial cell density and alternating microbial- and enzymolysis-enhanced oil recovery on oil displacement efficiency. *Chemical Engineering Journal*, **327**:28–38, 2017.

271. P Sivasankar and GS Kumar. Influence of pH on dynamics of microbial enhanced oil recovery processes using biosurfactant producing *Pseudomonas Putida*: mathematical modelling and numerical simulation. *Bioresource Technology*, **224**:498–508, 2017.

272. MD Dasetal. Application of biosurfactant produced by an adaptive strain of c. tropicalis MTCC230 in microbial enhanced oil recovery (MEOR) and removal of motor oil from contaminated sand and water. *Journal of Petroleum Science and Engineering*, **170**:40–48, 2018.

273. G Dhanarajan, V Rangarajan, C Bandi, A Dixit, S Das, K Ale, and R Sen. Biosurfactant-biopolymer driven microbial enhanced oil recovery (MEOR) and its optimization by an ANN-GA hybrid technique. *Journal of Biotechnology*, **256**:46–56, 2017.

274. I Vasconcelos, L Girbal, and P Soucaille. Regulation of carbon and electron flow in *Clostridium acetobutylicum* grown in chemostat culture at neutral pH on mixtures of glucose and glycerol. *Journal of Bacteriology*, **176**(5):1443–1450, 1994.

275. G-C Yang, L Zhou, SM Mbadinga, J You, H-Z Yang, J-F Liu, S-Z Yang, J-D Gu, and B-Z Mu. Activation of $CO_2$-reducing methanogens in oil reservoir after addition of nutrient. *Journal of Bioscience and Bioengineering*, **122**(6):740–747, 2016.

276. J Wang, L Yu, and L-X Huang. Study on enhancing oil recovery by indigenous microorganisms in a hypersalinity reservoir. *Petroleum Science and Technology*, **29**(14):1504–1511, 2011.

277. PL Fernandes, EM Rodrigues, FR Paiva, BAL Ayupe, MJ McInerney, and MR Totola. Biosurfactant, solvents and polymer production by Bacillus subtilis RI4914 and their application for enhanced oil recovery. *Fuel*, **180**:551–557, 2016.

278. JC Mata-Sandoval, J Karns, and A Torrents. High-performance liquid chromatography method for the characterization of rhamnolipid mixtures produced by *Pseudomonas aeruginosa* UG2 on corn oil. *Journal of Chromatography A*, **864**(2):211–220, 1999.

279. M Benincasa, A Abalos, I Oliveira, and A Manresa. Chemical structure, surface properties and biological activities of the biosurfactant produced by *Peudomonas aeruginosa* IBI from soapstock. *Antonie Van Leeuwenhoek*, **85**(1):1–8, 2004.

280. SA Monteiro, GL Sassaki, LM de Souza, JA Meira, JM de Araujo, DA Mitchell, LP Ramos, and N Krieger. Molecular and structural characterization of the biosurfactant produced by pseudomonas aeruginosa DAUPE 614. *Chemistry and Physics of Lipids*, **147**(1):1–13, 2007.

281. J-F Liu, G Wu, S-Z Yang, and B-Z Mu. Structural characterization of rhamnolipid produced by *Pseudonomas aeruginosa* strain FIN2 isolated from oil reservoir water. *World Journal of Microbiology and Biotechnology*, **30**(5):1473–1484, 2014.

282. K Patowary, R Patowary, MC Kalita, and S Deka. Characterization of biosurfactant produced during degradation of hydrocarbons using crude oil as sole source of carbon. *Frontiers in Microbiology*, **8**:279, 2017.

283. T Ndlovu, M Rautenbach, JA Vosloo, S Khan, and W Khan. Characterisation and antimicrobial activity of biosurfactant extracts produced by

bacillus amyloliquefaciens and *Pseudonomas aeruginosa* isolated from a wastewater treatment plant. *AMB Express*, **7**(1):108, 2017.

284. C Chen, N Sun, D Li, S Long, X Tang, G Xiao, and L Wang. Optimization and characterization of biosurfactant production from kitchen waste oil using *Pseudonomas aeruginosa*. *Environmental Science and Pollution Research*, **25**(15):14934–14943, 2018.

285. S Häussler, M Nimtz, T Domke, V Wray, and I Steinmetz. Purification and characterization of a cytotoxic exo lipid of *Burkholderia pseudomallei*. *Infection and Immunity*, **66**(4):1588–1593, 1998.

286. J Du, A Zhang, X Zhang, X Si, and J Cao. Comparative analysis of rhamnolipid congener synthesis in neotype *Pseudonomas aeruginosa* ATCC 10145 and two marine isolates. *Bioresource Technology*, **286**:121380, 2019.

287. VK Gaur, A Bajaj, RK Regar, M Kamthan, RR Jha, JK Srivastava, and N Manickam. *Rhamnolipid* from a *Lysinibacillus sphaericus* strain IITR51 and its potential application for dissolution of hydrophobic pesticides. *Bioresource Technology*, **272**:19–25, 2019.

288. D Dubeau, E Déziel, DE Woods, and F Lépine. Burkholderia thailandensis harbors two identical rhl gene clusters responsible for the biosynthesis of rhamnolipids. *BMC Microbiology*, **9**(1):263, 2009.

289. S Haloi and T Medhi. Optimization and characterization of a glycolipid produced by *Achromobacter* sp. to use in petroleum industries. *Journal of Basic Microbiology*, **59**(3):238–248, 2019.

290. P Datta, P Tiwari, and LM Pandey. Isolation and characterization of biosurfactant producing and oil degrading Bacillus subtilis MG495086 from formation water of Assam oil reservoir and its suitability for enhanced oil recovery. *Bioresource Technology*, **270**:439–448, 2018.

291. SR Routhu, RN Chary, AB Shaik, S Prabhakar, CG Kumar, and A Kamal. Induction of apoptosis in lung carcinoma cells by antiproliferative cyclic lipopeptides from marine algicolous isolate Bacillus atrophaeus strain AKLSR1. *Process Biochemistry*, **79**:142–154, 2019.

292. D Hentati, A Chebbi, F Hadrich, I Frikha, F Rabanal, S Sayadi, A Manresa, and M Chamkha. Production, characterization and biotechnological potential of lipopeptide biosurfactants from a novel marine *Bacillus stratospshericus* strain FLU5. *Ecotoxicology and Environmental Safety*, **167**:441–449, 2019.

293. SS Balan, CG Kumar, and S Jayalakshmi. *Aneurinifactin*, a new lipopeptide biosurfactant produced by a marine *Aneurinibacillus*

*aneurinilyticus* SBP-11 isolated from Gulf of Mannar: purification, characterization and its biological evaluation. *Microbiological Research*, **194**:1–9, 2017.

294. FA Bezza and EMN Chirwa. Pyrene biodegradation enhancement potential of lipopeptide biosurfactant produced by *Paenibacillus dendritiformis* CN5 strain. *Journal of Hazardous Materials*, **321**:218–227, 2017.

295. M Ohadi, G Dehghannoudeh, H Forootanfar, M Shakibaie, and M Rajaee. Investigation of the structural, physicochemical properties, and aggregation behavior of lipopeptide biosurfactant produced by Acinetobacter junii B6. *International Journal of Biological Macromolecules*, **112**:712–719, 2018.

296. M Morikawa, H Daido, T Takao, S Murata, Y Shimonishi, and T Imanaka. A new lipopeptide biosurfactant produced by *Arthrobacter* sp. strain MIS38. *Journal of Bacteriology*, **175**(20):6459–6466, 1993.

297. P Das, S Mukherjee, and R Sen. Antimicrobial potential of a lipopeptide biosurfactant derived from a marine *Bacillus circulans*. *Journal of Applied Microbiology*, **104**(6):1675–1684, 2008.

298. R Shi, Q Cui, S Han, H Dong, and Y Zhang. Biosurfactant production under diverse conditions by two kinds of biosurfactant-producing bacteria for microbial enhanced oil recovery. *Journal of Petroleum Science and Engineering*, **157**:124–130, 2017.

299. F Zhao, P Li, C Guo, R-J Shi, and Y Zhang. Bioaugmentation of oil reservoir indigenous *Pseudonomas aeruginosa* to enhance oil recovery through in-situ biosurfactant production without air injection. *Bioresource Technology*, **251**:295–302, 2018.

300. P Sarafzadeh, A Niazi, V Oboodi, M Ravanbakhsh, AZ Hezave, SS Ayatollahi, and S Raeissi. Investigating the efficiency of MEOR processes using *Enterobacter cloacae* and *Bacillus stearothermophilus* SUCPM#14 (biosurfactant-producing strains) in carbonated reservoirs. *Journal of Petroleum Science and Engineering*, **113**:46–53, 2014.

301. AN Marghmaleki, S Kord, A Hashemi, and H Motamedi. Experimental investigation of efficiency of MEOR process in a carbonate oil reservoir using alcaligenes faecalis: impact of interfacial tension reduction and wettability alteration mechanisms. *Fuel*, **232**:27–35, 2018.

302. P Sarafzadeh, AZ Hezave, M Ravanbakhsh, A Niazi, and S Ayatollahi. Enterobacter cloacae as biosurfactant producing bacterium: differentiating its effects on interfacial tension and wettability alteration

mechanisms for oil recovery during MEOR process. *Colloids and Surfaces B: Biointerfaces*, **105**:223–229, 2013.

303. F Zhao, J Zhang, R Shi, S Han, F Ma, and Y Zhang. Production of biosurfactant by a Pseudomonas aeruginosa isolate and its applicability to in situ microbial enhanced oil recovery under anoxic conditions. *RSC Advances*, **5**(45):36044–36050, 2015.

304. QF Cui, SS Sun, YJ Luo, L Yu, and ZZ Zhang. Comparison of in-situ and ex-situ microbial enhanced oil recovery by strain Pseudomonas aeruginosa WJ-1 in laboratory sand-pack columns. *Petroleum Science and Technology*, **35**(21):2044–2050, 2017.

305. AR Najafi, R Roostaazad, M Soleimani, D Arabian, MT Moazed, MR Rahimpour, and S Mazinani. Comparison and modification of models in production of biosurfactant for Paenibacillus alvei and Bacillus mycoides and its effect on MEOR efficiency. *Journal of Petroleum Science and Engineering*, **128**:177–183, 2015.

306. HS El-Sheshtawy, I Aiad, ME Osman, AA Abo-ELnasr, and AS Kobisy. Production of biosurfactants by *Bacillus licheniformis* and *Candida albicans* for application in microbial enhanced oil recovery. *Egyptian Journal of Petroleum*, **25**(3):293–298, 2016.

307. KV Pathak and H Keharia. Application of extracellular lipopeptide biosurfactant produced by endophytic Bacillus subtilis K1 isolated from aerial roots of banyan (Ficus benghalensis) in microbially enhanced oil recovery (MEOR). *3 Biotech*, **4**(1):41–48, 2014.

308. H Amani, MM Müller, C Syldatk, and R Hausmann. Production of microbial rhamnolipid by Pseudomonas aeruginosa MM1011 for ex situ enhanced oil recovery. *Applied Biochemistry and Biotechnology*, **170**(5):1080–1093, 2013.

309. SJ Varjani and VN Upasani. Core flood study for enhanced oil recovery through ex-situ bioaugmentation with thermo- and halo-tolerant rhamnolipid produced by *Pseudomonas aeruginosa* NCIM 5514. *Bioresource Technology*, **220**:175–182, 2016.

310. T Hema, GS Kiran, A Sajayyan, A Ravendran, GG Raj, and J Selvin. Response surface optimization of a glycolipid biosurfactant produced by a sponge associated marine bacterium Planococcus sp. MMD26. *Biocatalysis and Agricultural Biotechnology*, **18**:101071, 2019.

311. FE Mouafi, MM Abo Elsoud, and ME Moharam. Optimization of biosurfactant production by *Bacillus brevis* using response surface methodology. *Biotechnology Reports*, **9**:31–37, 2016.

312. A Nickzad, C Guertin, and E Déziel. Culturemedium optimization for production of rhamnolipids by Burkholderia glumae. *Colloids and Interfaces*, **2**(4):49, 2018.

313. MJ Veshareh, EG Azad, T Deihimi, A Niazi, and S Ayatollahi. Isolation and screening of *Bacillus subtilis* MJ01 for MEOR application: biosurfactant characterization, production optimization and wetting effect on carbonate surfaces. *Journal of Petroleum Exploration and Production Technology*, **9**(1):233–245, 2019.

314. S Shekhar, A Sundaramanickam, K Saranya, M Meena, S Kumaresan, and T Balasubramanian. Production and characterization of biosurfactant by marine bacterium *Pseudomonas stutzeri* (SSASM1). *International Journal of Environmental Science and Technology*, **16**(8):4697–4706, 2019.

315. Asia Fadhile Almansoory, Hassimi Abu Hasan, Mushrifah Idris, Siti Rozaimah Sheikh Abdullah, Nurina Anuar, et al. Biosurfactant production by the hydrocarbon-degrading bacteria (HNDH) serratia marcescens: Optimization using central composite design (ccd). Journal of industrial and engineering chemistry, **47**:272–280, 2017.

316. P Fd Almeida, RS Moreira, RCC Almeida, AK Guimaraes, AS Carvalho, C Quintella, MCA Esperidia, and CA Taft. Selection and application of microorganisms to improve oil recovery. *Engineering in Life Sciences*, **4**(4):319–325, 2004.

317. SA Bailey, TM Kenney, DR Schneider, et al. Microbial enhanced oil recovery: diverse successful applications of biotechnology in the oil field. In *SPE Asia Pacific Improved Oil Recovery Conference*. Society of Petroleum Engineers, 2001.

318. FG Brown, et al. Microbes: the practical and environmental safe solution to production problems, enhanced production, and enhanced oil recovery. In *Permian Basin Oil and Gas Recovery Conference*. Society of Petroleum Engineers, 1992.

319. I Lazar. Microbial systems for enhancement of oil recovery used in Romanian oil fields. *Mineral Procesing and Extractive Metallurgy Review*, **19**(1):379–393, 1998.

320. BM Das. Microbial enhanced oil recovery: a comprehensive approach. *Everyman's Science*, **XLVII**:234–237, 2012.

321. OA Omoniyi and FA Abdulmalik. Review of microbial enhanced oil recovery: current development and future prospects. *International Journal of Scientific & Engineering Research*, **6**(1):1378–1389, 2015.

322. CE ZoBell. Bacterial release of oil from sedimentary materials. *Oil and Gas Journal*, **46**:62–65, 1926.

323. JD Van Hamme, A Singh, and OP Ward. Recent advances in petroleum microbiology. *Microbiology and Molecular Biology Reviews*, **67**(4):503–549, 2003.

324. H Suthar, K Hingurao, A Desai, and A Nerurkar. Evaluation of bioemulsifier mediated microbial enhanced oil recovery using sand pack column. *Journal of Microbiological Methods*, **75**(2):225–230, 2008.

325. A Hiorth, K Kaster, A Lohne, OK Siqveland, H Berland, NH Giske, and A Stavland. Microbial enhanced oil recovery-mechanism. In *International Symposium of the Society of Core Analysts*, Calgary, Canada, pp. 10–13, 2007.

326. A Cui. Experimental study of microbial enhanced oil recovery and its impact on residual oil in sandstones. PhD thesis, 2016.

327. H Rashedi, F Yazdian, and S Naghizadeh. Microbial enhanced oil recovery. In *Introduction to Enhanced Oil Recovery (EOR) Processes and Bioremediation of Oil-Contaminated Sites*. IntechOpen, 2012.

328. SD Kapusta, AC Ooms, FGA Van den Berg, JW Buijs, and A Smith. Processing opportunity crudes: a new strategy for crude selection. Petrotech, Paper R075, New Delhi, 2001.

329. D Stratiev, R Dinkov, N Nikolaev, and K Stanulov. Evaluation of impact of crude oil quality on refinery profit. *Erdoel Erdgas Kohle*, **126**(1):17–21, 2010.

330. V Kumar, A Sharma, IR Chowdhury, S Ganguly, and DN Saraf. A crude distillation unit model suitable for online applications. *Fuel Processing Technology*, **73**(1):1–21, 2001.

331. D Stratiev, A Nedelchev, R Dinkov, and A Batchvarov. It's possible to derive TBP from partial distillation data. *Oil & Gas Journal*, **109**(17):114–122, 2011.

332. A Nedelchev, D Stratiev, A Ivanov, and G Stoilov. Boiling point distribution of crude oils based on TBP and ASTM D-86 distillation data. *Petroleum & Coal*, **53**(4):275–290, 2011.

333. MR Riazi. *Characterization and Properties of Petroleum Fractions*, Vol. 50. ASTM International, West Conshohocken, 2005.

334. WC Edmister and DH Pollock. Phase relations for petroleum fractions. *Chemical Engineering Progress*, **44**(12):905–926, 1948.

335. TE Daubert. Petroleum fraction distillation interconversions. *Hydrocarbon Processing (United States)*, **73**(9), 1994.

336. ASTM Committee D-2 on Petroleum Products, Lubricants, K Boldt, and BR Hall. Significance of tests for petroleum products. American Society for Testing and Materials, 1977.

337. QM Abid Ali Al-Assady. Characterization of petroleum fractions. *Iraqi Journal of Mechanical and Material Engineering*, **9**(2):223–238, 2009.

338. MS Graboski and TE Daubert. A modified soave equation of state for phase equilibrium calculations. 1. Hydrocarbon systems. *Industrial & Engineering Chemistry Process Design and Development*, **17**(4):443–448, 1978.

339. MS Al-Yasiri and WT Al-Sallami. How the drilling fluids can be made more efficient by using nanomaterials. *American Journal of Nano Research and Applications*, **3**(3):41–45, 2015.

340. AR Ismail. Improve performance of water-based drilling fluids. In *Sriwijaya International Seminar on Energy-Environmental Science and Technology*, Vol. 1, pp. 43–47, 2014.

341. L Quintero, AE Cardenas, and DE Clark. Nanofluids and methods of use for drilling and completion fluids. US Patent 8,822,386, 2014.

342. DJ Oakley, K Morton, A Eunson, A Gilmour, D Pritchard, A Valentine, et al. Innovative drilling fluid design and rigorous pre-well planning enable success in an extreme HTHP well. In *IADC/SPE Asia Pacific Drilling Technology*. Society of Petroleum Engineers, 2000.

343. M Amanullah, AM Al-Tahini, et al. Nano-technology: its significance in smart fluid development for oil and gas field application. In *SPE Saudi Arabia Section Technical Symposium*. Society of Petroleum Engineers, 2009.

344. P Talukdar and SB Gogoi. Optimization of the composition of KCl and KCl+NaCl as a clay/shale stabilizing component in the NDDF for Upper Assam Basin. *International Journal of Research in Engineering and Applied Sciences*, **5**(12):16–38, 2015.

345. AS Dhiman, et al. Rheological properties & corrosion characteristics of drilling mud additives. Dalhousie University, Halifax, Nova Scotia, p. 22, 2012.

346. HCH Darley and GR Gray. *Composition and Properties of Drilling and Completion Fluids*. Gulf Professional Publishing, 1988.

347. OV Onu, O Ikponmwosa, et al. Antimony Sulphide (Stibnite) and Potash as local substitute to barite and lignosulfonate as drilling mud

weighting additives. In *SPE Nigeria Annual International Conference and Exhibition*. Society of Petroleum Engineers, 2014.

348. M Khodja, M Khodja-Saber, JP Canselier, N Cohaut, and F Bergaya. Drilling fluid technology: performances and environmental considerations. In *Products and Services; from R&D to Final Solutions*. IntechOpen, 2010.
349. H Behnamanhar, et al. Environmentally friendly water-based drilling fluid for drilling of water-sensitive formations. *Journal of Petroleum and Gas Exploration Research*, **4**(4):60–71, 2014.
350. SO Blkoor and KA Fattah. The influence of XC-polymer on drilling fluid filter cake properties and formation damage. *Journal of Petroleum & Environmental Biotechnology*, **4**(5), 2013.
351. KA Fattah and A Lashin. Investigation of mud density and weighting materials effect on drilling fluid filter cake properties and formation damage. *Journal of African Earth Sciences*, **117**:345–357, 2016.
352. U Osokogwu, JA Ajeinka, and AN Okon. Evaluating the effects of additives on drilling fluid characteristics. *International Journal of Engineering Sciences & Research Technology (IJESRT)*, **3**(6):676–687, 2014.
353. RY Surampalli, et al. Journal of hazardous, toxic, and radioactive waste. American. Society of Civil Engineers.
354. GM Adewole, TM Adewale, and E Ufuoma. Environmental aspect of oil and water-based drilling muds and cuttings from Dibi and Ewan offshore wells in the Niger Delta, Nigeria. *African Journal of Environmental Science and Technology*, **4**(5), 2010.
355. AE Elemam, RAM Hussien, AA Ibrahim, and SA Mohamed. An improving rheological properties and filter loss of water-base drilling fluids using carboxymethyle cellulose (CMC). *Online International Interdisciplinary Research Journal*, **4**:55–65, 2014.
356. B Safi, S Zarouri, R Chabane-Chaouache, M Saidi, and A Benmounah. Physico-chemical and rheological characterization of water-based mud in the presence of polymers. *Journal of Petroleum Exploration and Production Technology*, **6**(2):185–190, 2016.
357. AS Kumar, V Mahto, and VP Sharma. Behaviour of organic polymers on the rheological properties of Indian bentonite-water based drilling fluid system and its effect on formation damage. *Indian Journal of Chemical Technology*, **10**(5):531–538, 2003.

358. G Jiang, T Liu, F Ning, Y Tu, L Zhang, Y Yu, and L Kuang. Polyethylene glycol drilling fluid for drilling in marine gas hydrates-bearing sediments: an experimental study. *Energies*, **4**(1):140–150, 2011.

359. D Mech, G Pandey, and JS Sangwai. Effect of molecular weight of polyethylene glycol on the equilibrium dissociation pressures of methane hydrate system. *Journal of Chemical & Engineering Data*, **60**(6):1878–1885, 2015.

360. O Omole, JO Adeleye, O Falode, S Malomo, and OA Oyedeji. Investigation into the rheological and filtration properties of drilling mud formulated with clays from Northern Nigeria. *Journal of Petroleum and Gas Engineering*, **4**(1):1–7, 2013.

361. Y Tamsilian and A Ramazani SA. Study of hydrogel modified drilling fluid: preparation, rheological properties and swelling behaviors of chitosan N-isopropylacrylamide hydrogel. *Brazilian Journal of Petroleum and Gas*, **6**(2), 2012.

362. G Cheraghian, Q Wu, M Mostofi, M-C Li, M Afrand, and JS Sangwai. Effect of a novel clay/silica nanocomposite on water-based drilling fluids: improvements in rheological and filtration properties. *Colloids and Surfaces A: Physicochemical and Engineering Aspects*, **555**:339–350, 2018.

363. K Van Dyke. *Drilling Fluids, Mud Pumps, and Conditioning Equipment*. University of Texas at Austin Petroleum, 1998.

364. M Omotioma, PCN Ejikeme, and GO Mbah. Comparative analysis of the effects of cashew and mango extracts on the rheological properties of water based mud. *Journal of Engineering Research and Applications*, **4**:56–61, 2014.

365. A Moslemizadeh and SR Shadizadeh. A natural dye in water-based drilling fluids: swelling inhibitive characteristic and side effects. *Petroleum*, **3**(3):355–366, 2017.

366. I Ismail and AK Idris. The prospect of utilising local starches as fluid loss control agents in the petroleum industry. In *Proceedings of Regional Symposium on Chemical Engineering*. UTM, Johor, Malaysia, 1997.

367. OF Joel, A Taiwo and AA Kazeem. Investigation of local polymer (cassava starches) as a substitute for imported sample in viscosity and fluid loss control of water based drilling mud. *ARPN Journal of Engineering and Applied Sciences*, **6**(12):43–48, 2006.

368. ME Hossain and M Wajheeuddin. The use of grass as an environmentally friendly additive in water-based drilling fluids. *Petroleum Science*, **13**(2):292–303, 2016.

369. AB Oriji and OF Joel. Suitability of beneficiated local starch under elevated temperature as a fluid loss control additive used in petroleum industry. *Scientia Africana*, **11**(1), 2012.

370. D Power and M Zamora. Drilling fluid yield stress: measurement techniques for improved understanding of critical drilling fluid parameters. In *AADE Technical Conference*, Houston, pp. 1–3, 2003.

371. M Kane, M Djabourov, J-L Volle, J-P Lechaire, and G Frebourg. Morphology of paraffin crystals in waxy crude oils cooled in quiescent conditions and under flow. *Fuel*, **82**(2):127–135, 2003.

372. A Farina and A Fasano. Flow characteristics of waxy crude oils in laboratory experimental loops. *Mathematical and Computer Modelling*, **25**(5):75–86, 1997.

373. C Bai and J Zhang. Effect of carbon number distribution of wax on the yield stress of waxy oil gels. *Industrial & Engineering Chemistry Research*, **52**(7):2732–2739, 2013.

374. K Oh, M Jemmett, and M Deo. Yield behavior of gelled waxy oil: effect of stress application in creep ranges. *Industrial & Engineering Chemistry Research*, **48**(19):8950–8953, 2009.

375. AS Kasumu, S Arumugam, and AK Mehrotra. Effect of cooling rate on the wax precipitation temperature of "waxy" mixtures. *Fuel*, **103**:1144–1147, 2013.

376. M Lashkarbolooki, A Seyfaee, F Esmaeilzadeh, and D Mowla. Experimental investigation of wax deposition in Kermanshah crude oil through a monitored flow loop apparatus. *Energy & Fuels*, **24**(2):1234–1241, 2010.

377. A Kamari, A Khaksar-Manshad, F Gharagheizi, AH Mohammadi, and S Ashoori. Robust model for the determination of wax deposition in oil systems. *Industrial & Engineering Chemistry Research*, **52**(44):15664–15672, 2013.

378. Z Huang, Y Lu, R Hoffmann, L Amundsen, and HS Fogler. The effect of operating temperatures on wax deposition. *Energy & Fuels*, **25**(11):5180–5188, 2011.

379. JL Creek, HJ Lund, JP Brill, and M Volk. Wax deposition in single phase flow. *Fluid Phase Equilibria*, **158**:801–811, 1999.

380. HS Lee. Computational and rheological study of wax deposition and gelation in subsea pipelines. PhD thesis, University of Michigan, Ann Arbor, Mich, USA, 2008.

381. R Dalirsefat and F Feyzi. A thermodynamic model for wax deposition phenomena. *Fuel*, **86**(10–11):1402–1408, 2007.

382. H Zhu, X Yang, J Li, and N Li. Simulation analysis of thermal influential factors on crude oil temperature when double pipelines are laid in one ditch. *Advances in Engineering Software*, **65**:23–31, 2013.

383. NF Carnahan, et al. Wax properties and behavior in offshore production and transportation systems. In *Offshore Technology Conference*, 2007.

384. R Bagherinia, M Assareh, and F Feyzi. An improved thermodynamic model for Wax precipitation using a UNIQUAC+ PC-SAFT approach. *Fluid Phase Equilibria*, **425**:21–30, 2016.

385. L Guo, J Zhang, G Sun, and Y Bao. Thixotropy and its estimation of water-in-waxy crude emulsion gels. *Journal of Petroleum Science and Engineering*, **131**:86–95, 2015.

386. A Salehi-Shabestari, A Ahmadpour, M Raisee, and K Sadeghy. Flow and displacement of waxy crude oils in a homogenous porous medium: a numerical study. *Journal of Non-Newtonian Fluid Mechanics*, **235**:47–63, 2016.

387. G Vinay, A Wachs, and J-F Agassant. Numerical simulation of non-isothermal viscoplastic waxy crude oil flows. *Journal of Non-Newtonian Fluid Mechanics*, **128**(2–3):144–162, 2005.

388. W Rukthong, W Weerapakkaroon, U Wongsiriwan, P Piumsomboon, and B Chalermsinsuwan. Integration of computational fluid dynamics simulation and statistical factorial experimental design of thick-wall crude oil pipeline with heat loss. *Advances in Engineering Software*, **86**:49–54, 2015.

389. A Tavakoli and M Baktash. Numerical approach for temperature development of horizontal pipe flow with thermal leakage to ambient. *International Journal of Modern Engineering Research*, **2**:3784–3794, 2012.

390. Z Guozhong and L Gang. Study on the wax deposition of waxy crude in pipelines and its application. *Journal of Petroleum Science and Engineering*, **70**(1–2):1–9, 2010.

391. M Theyab.Wax deposition process: mechanisms, affecting factors and mitigation methods. *Open Access Journal of Science*, **2**:112–118, 2018.

392. R Amorin. Application of minimum curvature method to well-path calculations. PhD thesis, 2009.

393. H Rabia. *Oil Well Drilling Engineering: Principle and Practice*. Graham and Trotman Inc., Gaihersburg, USA, 1985.

394. WA Zaremba, et al. Directional survey by the circular arc method. *Society of Petroleum Engineers Journal*, **13**(01):5–11, 1973.

395. MG Kramer. Angular displacement measuring apparatus with strobe means. US Patent 3,999,064, 1976.

396. JU Brackbill, DB Kothe, and C Zemach. A continuum method for modeling surface tension. *Journal of Computational Physics*, **100**(2):335–354, 1992.

397. R Breon and K Katano. Microsoft Office 97 executable content security risks and countermeasures. *National Security Agency*, pp. 1–44, 1999.

398. P Slavinić and M Cvetković. Volume calculation of subsurface structures and traps in hydrocarbon exploration—a comparison between numerical integration and cell based models. *Open Geosciences*, **8**(1):14–21, 2016.

399. S Li and W Hu. Application of petrel software in reserve estimation in Xinli-Xinbei oilfield, Jilin Province, China. In *International Conference on Materials Engineering and Information Technology Applications (MEITA 2015)*, Atlantis Press, 2015.

400. S Bhattacharya and S Verma. Application of volumetric seismic attributes for complex fault network characterization on the North Slope, Alaska. *Journal of Natural Gas Science and Engineering*, **65**:56–67, 2019.

401. A Nouri, H Vaziri, H Belhaj, R Islam, et al. Sand production prediction: a new set of criteria for modeling based on large-scale transient experiments and numerical investigation. In *SPE Annual Technical Conference and Exhibition*. Society of Petroleum Engineers, 2004.

402. SO Osisanya, et al. Practical guidelines for predicting sand production. In *Nigeria Annual International Conference and Exhibition*. Society of Petroleum Engineers, 2010.

403. AM Abanum, D Appah, et al. Laboratory studies of chemicals for sand consolidation (SCON) in the Niger Delta fields. In *SPE Nigeria Annual International Conference and Exhibition*. Society of Petroleum Engineers, 2013.

404. I Foo, R Bhasker, AF-F Foo-Karna, et al. Successful application of aqueous-based formation consolidation treatment introduced to the

North Sea. In *SPE/ICoTA Coiled Tubing & Well Intervention Conference & Exhibition*. Society of Petroleum Engineers, 2013.

405. PF Sullivan, E Nelson, J-C Gonzalez, S Danican, and G Tustin. Emulsion system for sand consolidation. US Patent 7,975,764, 2011.
406. JA Hugh and J Ramos. Guidelines to sand control, report. *Halliburton Services from Halliburton Co*, 1995.
407. OO Amund and N Nwokoye. Hydrocarbon degradation potentials of yeast isolates from a polluted lagoon. University of Lagos Library, 1993.
408. B Lal and S Khanna. Degradation of crude oil by Acinetobacter calcoaceticus and Alcaligenes odorans. *Journal of Applied Bacteriology*, **81**(4):355–362, 1996.
409. DM Jones, AG Douglas, RJ Parkes, J Taylor, W Giger, and C Schaffner. The recognition of biodegraded petroleum-derived aromatic hydrocarbons in recent marine sediments. *Marine Pollution Bulletin*, **14**(3):103–108, 1983.
410. SA Adebusoye, MO Ilori, OO Amund, OD Teniola, and SO Olatope. Microbial degradation of petroleum hydrocarbons in a polluted tropical stream. *World Journal of Microbiology and Biotechnology*, **23**(8):1149–1159, 2007.
411. N Das and P Chandran. Microbial degradation of petroleum hydrocarbon contaminants: an overview. *Biotechnology Research International*, **2011**:941810, 2011.
412. (a) A Kvenvolden and CK Cooper. Natural seepage of crude oil into the marine environment. *Geo-Marine Letters*, **23**(3–4):140–146, 2003. (b) V Kothari, M Panchal, and N Srivastava. Microbial degradation of hydrocarbons. *National Science Digital Library*, **3**, 2014.
413. M Atlas and R Bartha. Microbial evolution and biodiversity. In M Atlas, and R Bartha, *Microbial Ecology*. Menlo Park, Benjamim/Cummings Science, pp. 27–57, 1998.
414. PK Jain and V Bajpai. Biotechnology of bioremediation: a review. *International Journal of Environmental Science*, **3**(1):535–549, 2012.
415. J Liu, Y Chen, R Xu, and Y Jia. Screening and evaluation of biosurfactant-producing strains isolated from oilfield wastewater. *Indian Journal of Microbiology*, **53**(2):168–174, 2013.
416. R Thavasi, VRM Subramanyam Nambaru, S Jayalakshmi, T Balasubramanian, and IM Banat. Biosurfactant production by Pseudomonas aeruginosa from renewable resources. *Indian Journal of Microbiology*, **51**(1):30–36, 2011.

417. H Razafindralambo, M Paquot, A Baniel, Y Popineau, C Hbid, P Jacques, and P Thonart. Foaming properties of surfactin, a lipopeptide biosurfactant from Bacillus subtilis. *Journal of the American Oil Chemists' Society*, **73**(1):149–151, 1996.

418. P Agarwal and DK Sharma. Studies on the production of biosurfactant for the microbial enhanced oil recovery by using bacteria isolated from oil contaminated wet soil. *Petroleum Science and Technology*, **27**(16):1880–1893, 2009.

419. AP Karlapudi, TC Venkateswarulu, J Tammineedi, L Kanumuri, BK Ravuru, V Dirisala, and VP Kodali. Role of biosurfactants in bioremediation of oil pollution: a review. *Petroleum*, **4**(3):241–249, 2018.

420. GL Zhang, YT Wu, XP Qian, and Q Meng. Biodegradation of crude oil by Pseudomonas aeruginosa in the presence of rhamnolipids. *Journal of Zhejiang University Science B*, **6**(8):725–730, 2005.

421. S Iqbal, ZM Khalid, and KA Malik. Enhanced biodegradation and emulsification of crude oil and hyperproduction of biosurfactants by a gamma ray-induced mutant of Pseudomonas aeruginosa. *Letters in Applied Microbiology*, **21**(3):176–179, 1995.

422. E Abatenh, B Gizaw, Z Tsegaye, and M Wassie. Application of microorganisms in bioremediation: review. *Journal of Environmental Microbiology*, **1**(1), 2017.

423. Y Du, L Guan, H Liang, et al. Advances of produced water management. In *Canadian International Petroleum Conference*. Petroleum Society of Canada, 2005.

424. ME Ersahin, H Ozgun, R Kaya, BK Mutlu, C Kinaci, and I Koyuncu. Treatment of produced water originated from oil and gas production wells: a pilot study and cost analysis. *Environmental Science and Pollution Research*, **25**(7):6398–6406, 2018.

425. D Konwar, SB Gogoi, J Barman, and MP Gogoi. Correlation analysis of formation water of Upper Assam Basin for a greener ecosystem. *International Journal of Engineering Technology Science and Research (IJETSR)*, **4**(8):324–334, 2017.

426. P Jain, M Sharma, P Dureja, PM Sarma, and B Lal. Bioelectrochemical approaches for removal of sulfate, hydrocarbon and salinity from produced water. *Chemosphere*, **166**:96–108, 2017.

427. K Guerra, K Dahm, and S Dundorf. Oil and gas produced water management and beneficial use in the Western United States. US Department of the Interior, Bureau of Reclamation, 2011.

428. H Patel, D Borgohain, and S Jena. Hydro geochemistry of Upper Assam oil field. *Journal of Petroleum Technology*, **2**:11–14, 2003.

429. SB Gogoi, RK Sen, A Rajbongshi, and K Hazarika. Characterization of oil field produced waters of Upper Assam Basin, India. *International Journal of New Technologies in Science and Engineering*, **2**(1):1–7, 2015.

430. H Ozgun, ME Ersahin, S Erdem, B Atay, S Sayili, E Eren, P Hoshan, D Atay, M Altinbas, C Kinaci, et al. Comparative evaluation for characterization of produced water generated from oil, gas, and oil–gas production fields. *Clean–Soil, Air, Water*, **41**(12):1175–1182, 2013.

431. DO Kpeglo, J Mantero, EO Darko, G Emi-Reynolds, A Faanu, G Manjón, I Vioque, EHK Akaho, and R Garcia-Tenorio. Radiochemical characterization of produced water from two production offshore oilfields in Ghana. *Journal of Environmental Radioactivity*, **152**:35–45, 2016.

432. J Barman, SB Gogoi, and S Sarmah. Characterization of oil field formation water and its effect on tap water dilution for eco-friendly environment. *International Journal of Engineering Technology Science and Research*, **4**(8):309–323, 2017.

433. NG Pizzi. *Water Treatment Operator Handbook*. American Water Works Association, 2011.

434. M Kurdi, MS Ferdows, and A Maghsoudi. Sensitivity of corrosion and scaling indices based on ions; case study Iran. *Water Quality, Exposure and Health*, **7**(3):363–372, 2015.

435. 3RPWM Jacobs, ROH Grant, J Kwant, JM Marquenie, and E Mentzer. The composition of produced water from shell operated oil and gas production in the North Sea. In *Produced Water*, Springer, pp. 13–21, 1992.

436. E Hoseinzadeh, A Yusefzadeh, N Rahimi, and H Khorsandi. Evaluation of corrosion and scaling potential of a water treatment plant. *Archives of Hygiene Sciences*, **2**(2):41–47, 2013.

437. M Maeng, I Hyun, S Choi, and S Dockko. Effects of rainfall characteristics on corrosion indices in Korean river basins. *Desalination and Water Treatment*, **54**(4–5):1233–1241, 2015.

438. PR Puckorius and JM Brooke. A new practical index for calcium carbonate scale prediction in cooling tower systems. *Corrosion*, **47**(4):280–284, 1991.

439. VA Prisyazhniuk. Prognosticating scale-forming properties of water. *Applied Thermal Engineering*, **27**(8–9):1637–1641, 2007.

440. J McDonald. Alkalinity and ph relationships. *CSTN*, May 2006.

441. A Fakhru'l-Razi, P Alireza, CH Luqman, RAB Dayang, SM Sayed, and ZA Zurina. Review of technologies for oil and gas produced water. *Journal of Hazardous Materials*, **170**:530–551, 2009.

442. J Choi, BG Choi, and S Hong. Effects of NF treated water on corrosion of pipe distribution system and its implications to blending with conventionally treated water. *Desalination*, **360**:138–145, 2015.

443. HM Müller-Steinhagen and CA Branch. Comparison of indices for the scaling and corrosion tendency of water. *The Canadian Journal of Chemical Engineering*, **66**(6):1005–1007, 1988.

444. SI Korfali. Total and speciated metals and metal scales in domestic water: impact of metallic water distribution system. *Journal of Environmental Science and Engineering A*, **1**(6A):732, 2012.

445. T Asano, FL Burton, HL Leverenz, R Tsuchihashi, and G Tchobanoglous. *Wastewater Engineering: Treatment and Reuse*, Metcalf and Eddy, 2002.

446. FE DeMartini. Corrosion and the Langelier calcium carbonate saturation index. *Journal (American Water Works Association)*, **30**(1):85–111, 1938.

447. RT Duraisamy, AH Beni, and A Henni. State of the art treatment of produced water. In*Water Treatment*, INTECH Open Access, pp. 199–222, 2013.

448. JA Veil, MG Puder, D Elcock, and RJ Redweik Jr. A white paper describing produced water from production of crude oil, natural gas, and coal bed methane. Technical report, Argonne National Lab., IL (US), 2004.

449. S Deng, G Yu, Z Chen, D Wu, F Xia, and N Jiang. Characterization of suspended solids in produced water in daqing oilfield. *Colloids and Surfaces A: Physicochemical and Engineering Aspects*, **332**(1):63–69, 2009.

450. DB Burnett. Potential for beneficial use of oil and gas produced water. Global Petroleum Institute, Texas Water Resources Institute, pp. 1–11, 2004.

451. PB Tsang and CJ Martin. Economic evaluation of treating oilfield produced water for potable use. In *SPE International Thermal Operations and Heavy Oil Symposium and Western Regional Meeting*. Society of Petroleum Engineers, 2004.

452. TJ Gogoi and SB Gogoi. Analyses and treatment of oil field formation water of Upper Assam Basin (India). In *Advances in Petroleum Engineering and Petroleum Geochemistry*, Springer, pp. 117–120, 2019.

453. J Mo, Q Yang, N Zhang, W Zhang, Y Zheng, and Z Zhang. A review on agro-industrial waste (AIW) derived adsorbents for water

and wastewater treatment. *Journal of Environmental Management*, **227**:395–405, 2018.

454. A Fakhru'l-Razi, A Pendashteh, LC Abdullah, DRA Biak, SS Madaeni, and ZZ Abidin. Review of technologies for oil and gas produced water treatment. *Journal of Hazardous Materials*, **170**(2–3):530–551, 2009.
455. B Stone. Management of produced water. Oxford Publications, Washington, pp. 37–41, 2003.
456. G Voordouw. Production-related petroleum microbiology: progress and prospects. *Current Opinion in Biotechnology*, **22**(3):401–405, 2011.
457. JD Arthur, BG Langhus, and C Patel. Technical summary of oil & gas produced water treatment technologies. All Consulting, LLC, Tulsa, OK, 2005.
458. AA Al-Haleem, HH Abdulah, and E Abdul-Jalil Saeed. Components and treatments of oilfield produced water. *Al-Khwarizmi Engineering Journal*, **6**(1):24–30, 2010.
459. F Bayati, J Shayegan, and A Noorjahan. Treatment of oilfield produced water by dissolved air precipitation/solvent sublation. *Journal of Petroleum Science and Engineering*, **80**(1):26–31, 2011.
460. R Singh. Water and membrane treatment. In *Hybrid Membrane Systems for Water Purification*, Elsevier Science, 2005.
461. Z Khatib, P Verbeek, et al. Water to value-produced water management for sustainable field development of mature and green fields. *Journal of Petroleum Technology*, **55**(01):26–28, 2003.
462. SB Gogoi and AS Tamuli. Recent advances in treating oil field produced water: a review. *Journal of Petroleum Engineering & Technology*, **4**(2):1–6, 2019.
463. Q Li, C Kang, and C Zhang. Waste water produced from an oilfield and continuous treatment with an oil-degrading bacterium. *Process Biochemistry*, **40**(2):873–877, 2005.
464. L Metcalf. *Wastewater Engineering: Treatment and Reuse*. Metcalf & Eddy Inc, 2003.
465. C Gakwisiri, N Raut, A Al-Saadi, S Al-Aisri, and A Al-Ajmi. A critical review of removal of zinc from wastewater. In *Proceedings of the World Congress on Engineering*, Vol. 1, pp. 627–630, 2012.
466. M Khastar-Boroujeni, K Esmaili, H Samadi-Boroujeni, and A Ziaei. Wastewater effect on the deposition of cohesive sediment. *Journal of Environmental Engineering*, **144**(1):04017083, 2017.

467. J Cotruvo. Hardness in drinking water, background document for development of who guidelines for drinking water quality. World Health Organization, Geneva, Switzerland, 2011.

468. S Jafarinejad and SC Jiang. Current technologies and future directions for treating petroleum refineries and petrochemical plants (PRPP) wastewaters. *Journal of Environmental Chemical Engineering*, **7**:103326, 2019.

469. OF Joel, CA Amajuoyi, and CU Nwokoye. Characterization of formation water constituents and the effect of fresh water dilution from land rig location of the Niger Delta, Nigeria. *Journal of Applied Sciences and Environmental Management*, **14**(2), 2010.

470. MF Davil, AH Mahvi, M Norouzi, S Mazloomi, A Amarluie, A Tardast, Y Karamitabar, et al. Survey of corrosion and scaling potential produced water from Ilam water treatment plant. *World Applied Sciences Journal*, **7**(11):11–24, 2009.

471. G Crini and E Lichtfouse. Advantages and disadvantages of techniques used for wastewater treatment. *Environmental Chemistry Letters*, **17**(1):145–155, 2019.

472. L Li and R Lee. Purification of produced water by ceramic membranes: material screening, process design and economics. *Separation Science and Technology*, **44**(15):3455–3484, 2009.

473. J Savari, N Jaafarzadeh, AH Hassani, and GS Khoramabadi. Heavy metals leakage and corrosion potential in Ahvaz drinking water distribution network. *Water Wastewater Journal*, **18**(64):16–24, 2008.

474. RE Loewenthal, I Morrison, and MC Wentzel. Control of corrosion and aggression in drinking water systems. *Water Science and Technology*, **49**(2):9–18, 2004.

475. VS Kale. Consequence of temperature, pH, turbidity and dissolved oxygen water quality parameters. *International Advanced Research Journal in Science, Engineering and Technology*, **3**(8):186–190, 2016.

476. GT Tellez, N Nirmalakhandan, and JL Gardea-Torresdey. Performance evaluation of an activated sludge system for removing petroleum hydrocarbons from oilfield produced water. *Advances in Environmental Research*, **6**(4):455–470, 2002.

477. YK Siong, J Idris, and MM Atabaki. Performance of activated carbon in water filters. *Water Resources*, 1–19, 2013.

478. ZF Cui, S Chang, and AG Fane. The use of gas bubbling to enhance membrane processes. *Journal of Membrane Science*, **221**(1–2):1–35, 2003.

479. S Jafarinejad. Recent developments in the application of sequencing batch reactor (SBR) technology for the petroleum industry wastewater treatment. *Chemistry International*, **3**(3):342–350, 2017.

480. S Jafarinejad. Simulation for the performance and economic evaluation of conventional activated sludge process replacing by sequencing batch reactor technology in a petroleum refinery wastewater treatment plant. *ChemEngineering*, **3**(2):45, 2019.

481. M Ebrahimi, KS Ashaghi, L Engel, D Willershausen, P Mund, P Bolduan, and P Czermak. Characterization and application of different ceramic membranes for the oil-field produced water treatment. *Desalination*, **245**(1–3):533–540, 2009.

482. MK Goyal and RY Surampalli. Impact of climate change on water resources in India. *Journal of Environmental Engineering*, **144**(7):04018054, 2018.

483. SU Delin, W Jianlong, L Kaiwen, and ZHOU Ding. Kinetic performance of oil-field produced water treatment by biological aerated filter. *Chinese Journal of Chemical Engineering*, **15**(4):591–594, 2007.

484. HI Abdel-Shafy and MSM Mansour. A review on polycyclic aromatic hydrocarbons: source, environmental impact, effect on human health and remediation. *Egyptian Journal of Petroleum*, **25**(1):107–123, 2016.

485. S Mondal and SR Wickramasinghe. Produced water treatment by nanofiltration and reverse osmosis membranes. *Journal of Membrane Science*, **322**(1):162–170, 2008.

486. JE Boysen. Evaluation of the natural freeze-thaw process for the desalinization of groundwater from the North Dakota aquifer to provide water for grand forks, North Dakota. US Department of the Interior, Bureau of Reclamation, 1999.

487. KS Ashaghi, M Ebrahimi, and P Czermak. Ceramic ultra-and nanofiltration membranes for oilfield produced water treatment: a mini review. *Open Environmental Sciences*, **1**(1), 2007.

488. S Jafarinejad. Activated sludge combined with powdered activated carbon (PACT process) for the petroleum industry wastewater treatment: a review. *Chemistry International*, **3**(4):368–377, 2017.

489. E Akhondi, F Zamani, KH Tng, G Leslie, WB Krantz, AG Fane, and JW Chew. The performance and fouling control of submerged hollow fiber (HF) systems: a review. *Applied Sciences*, **7**(8):765, 2017.

490. GJ Tapan, GB Subrata, and S Shilpi. A green approach for oil field produced waters of Upper Assam Basin. In *International Congress and*

*Exhibition, Sustainable Civil Infrastructures: Innovative Infrastructure Geotechnology*, Springer, pp. 151–171, 2018.

491. K Rafferty. Scaling in geothermal heat pump systems. *Geo-Heat Center Bulletin*, Klamath Falls, OR, 1999.
492. P Ramya, AJ Babu, E Tirupathi Reddy, and L Venkateswara Rao. A study on the estimation of hardness in ground water samples by edta tritrimetric method. *International Journal of Recent Scientific Research*, **6**(6):4505–4507, 2015.
493. PB Tchounwou, CG Yedjou, AK Patlolla, and DJ Sutton. Heavy metal toxicity and the environment. In *Molecular, Clinical and Environmental Toxicology*, Springer, pp. 133–164, 2012.
494. M Kumar and A Puri. A review of permissible limits of drinking water. *Indian Journal of Occupational and Environmental Medicine*, **16**(1):40, 2012.
495. JC Campos, RMH Borges, AM de Oliveira Filho, R Nobrega, and GL Sant'Anna Jr. Oilfield wastewater treatment by combined microfiltration and biological processes. *Water Research*, **36**(1):95–104, 2002.
496. MF Hashad, S Sharma, LF Nies, and JE Alleman. Study of salt wash water toxicity on wastewater treatment. Prepared in cooperation with the Indiana Department of Transportation and Federal Highway Administration, 2006.
497. G Crini, E Lichtfouse, LD Wilson, and N Morin-Crini. Conventional and non-conventional adsorbents for wastewater treatment. *Environmental Chemistry Letters*, **17**(1):195–213, 2019.
498. R Lee, R Seright, M Hightower, A Sattler, M Cather, B McPherson, L Wrotenbery, D Martin, and M Whitworth. Strategies for produced water handling in New México. In *Ground Water Protection Council Produced Water Conference*, pp. 16–17, 2002.
499. R Kori, S Parashar, DD Basu, JS Kamyotra, and S Kumari. Guide manual: water and wastewater analysis. Central Pollution Control Board, Ministry of Environment and Forest, India, pp. 97–109.
500. S Vigneswaran, DY Kwon, HH Ngo, and JY Hu. Improvement of microfiltration performance in water treatment: is critical flux a viable solution? *Water Science and Technology*, **41**(10–11):309–315, 2000.
501. H Hashemi, M Hoseini, and AA Ebrahimi. Flat sheet membrane sequencing batch bioreactor for the removal of coliforms and heavy metals from stabilized composting leachate. *Journal of Environmental Engineering*, **144**(4):04018015, 2018.

502. Y Ye, P Le Clech, V Chen, AG Fane, and B Jefferson. Fouling mechanisms of alginate solutions as model extracellular polymeric substances. *Desalination*, **175**(1):7–20, 2005.

503. P Ekins, R Vanner, and J Firebrace. Zero emissions of oil in water from offshore oil and gas installations: economic and environmental implications. *Journal of Cleaner Production*, **15**(13–14):1302–1315, 2007.

504. R Wasylishen. Recycling flowback and produced water in tight-oil development. *Desalination & Water Reuse*, **22**(2):26–30, 2013.

505. T Hayes, BF Severin, PE Senior Process Engineer, and MI Okemos. Barnett and appalachian shale water management and reuse technologies. *Contract*, **8122**:05, 2012.

506. G Chase and P Kulkarni. Mixed hydrophilic/hydrophobic fiber media for liquid-liquid coalescence. US Patent 8,409,448, 2013.

507. KM Mousa and SA Arafat. Experimental study on treatment of produced water. *Journal of Chemical Engineering & Process Technology*, **6**(6):1, 2015.

508. CO Igwe, AAL Saadi, and SE Ngene. Optimal options for treatment of produced water in offshore petroleum platforms. *Journal of Pollution Effects & Control*, **1**(2):1–5, 2013.

509. S Rassenfoss, et al. From flowback to fracturing: water recycling grows in the marcellus shale. *Journal of Petroleum Technology*, **63**(07):48–51, 2011.

510. JM Dickhout, J Moreno, PM Biesheuvel, L Boels, RGH Lammertink, and WM de Vos. Produced water treatment by membranes: a review from a colloidal perspective. *Journal of Colloid and Interface Science*, **487**:523–534, 2017.

511. V Rajaković-Ognjanović, G Aleksić, and Lj Rajaković. Governing factors for motor oil removal from water with different sorption materials. *Journal of Hazardous Materials*, **154**(1–3):558–563, 2008.

512. D Mowla, G Karimi, and K Salehi. Modeling of the adsorption breakthrough behaviors of oil from salty waters in a fixed bed of commercial organoclay/sand mixture. *Chemical Engineering Journal*, **218**:116–125, 2013.

513. H Moazed and T Viraraghavan. Use of organo-clay/anthracite mixture in the separation of oil from oily waters. *Energy Sources*, **27**(1–2):101–112, 2005.

514. K Okiel, M El-Sayed, and MY El-Kady. Treatment of oil–water emulsions by adsorption onto activated carbon, bentonite and deposited carbon. *Egyptian Journal of Petroleum*, **20**(2):9–15, 2011.

515. M Achak, L Mandi, and N Ouazzani. Removal of organic pollutants and nutrients from olive mill wastewater by a sand filter. *Journal of Environmental Management*, **90**(8):2771–2779, 2009.

516. K Hanna. Adsorption of aromatic carboxylate compounds on the surface of synthesized iron oxide-coated sands. *Applied Geochemistry*, **22**(9):2045–2053, 2007.

517. D Broman, A Colmsjö, B Ganning, C Näf, Y Zebühr, and C Östman. 'Fingerprinting' petroleum hydrocarbons in bottom sediment, plankton, and sediment trap collected seston. *Marine Pollution Bulletin*, **18**(7):380–388, 1987.

518. C Patel, MA Barrufet, A Bonilla Petriciolet, et al. Effective resource management of produced water in oil & gas operations. In *Canadian International Petroleum Conference*. Petroleum Society of Canada, 2004.

519. J Wang, C Wang, Q Huang, F Ding, and X He. Adsorption of pahs on the sediments from the yellow river delta as a function of particle size and salinity. *Soil and Sediment Contamination: An International Journal*, **24**(2):103–115, 2015.

520. C Gunawardana, P Egodawatta, and A Goonetilleke. Role of particle size and composition in metal adsorption by solids deposited on urban road surfaces. *Environmental Pollution*, **184**:44–53, 2014.

521. J Pichtel. Oil and gas production wastewater: soil contami-nation and pollution prevention. *Applied and Environmental Soil Science*, **2016**:2707989, 2016.

522. MT Stephenson. A survey of produced water studies. In *Produced water*, Springer, pp. 1–11, 1992.

523. D Tiab and EC Donaldson. *Petrophysics: Theory and Practice of Measuring Reservoir Rock and Fluid Transport Properties*. Gulf Professional Publishing, 2015.

524. JM Hunt. Generation and migration of petroleum from abnormally pressured fluid compartments (1). *AAPG Bulletin*, **74**(1):1–12, 1990.

525. VE Khain and ID Polyakova. Oil and gas potential of deep- and ultradeep-water zones of continental margins. *Lithology and Mineral Resources*, **39**(6):530–540, 2004.

526. JR Fanchi, DJ Blumer, and C Phillips. PEH: properties of produced water. In *Petroleum Engineering Handbook*, Society of Petroleum Engineers, pp. 465–497, 2007.

527. JE Drewes, TY Cath, P Xu, J Graydon, J Veil, and S Snyder. An integrated framework for treatment and management of produced water. RPSEA Project, pp. 07122–12, 2009.

528. H Ma and B Wang. Electrochemical pilot-scale plant for oil field produced wastewater by M/C/Fe electrodes for injection. *Journal of Hazardous Materials*, **132**(2–3):237–243, 2006.

529. SS Mao, S Shen, and L Guo. Nanomaterials for renewable hydrogen production, storage and utilization. *Progress in Natural Science: Materials International*, **22**(6):522–534, 2012.

530. HM Pars, DTh Meijer, et al. Removal of dissolved hydrocarbons from production water by macro porous polymer extraction (MPPE). In *SPE International Conference on Health, Safety, and Environment in Oil and Gas Exploration and Production*. Society of Petroleum Engineers, 1998.

531. S Judd. *The MBR Book: Principles and Applications of Membrane Bioreactors for Water and Wastewater Treatment*. Elsevier, 2010.

532. C Visvanathan, RB Aim, and K Parameshwaran. Membrane separation bioreactors for wastewater treatment. *Critical Reviews in Environmental Science and Technology*, **30**(1):1–48, 2000.

533. HY Ng and SW Hermanowicz. Membrane bioreactor operation at short solids retention times: performance and biomass characteristics. *Water Research*, **39**(6):981–992, 2005.

534. A Sepehri and M-H Sarrafzadeh. Effect of nitrifiers community on fouling mitigation and nitrification efficiency in a membrane bioreactor. *Chemical Engineering and Processing-Process Intensification*, **128**:10–18, 2018.

535. I-S Chang, P Le Clech, B Jefferson, and S Judd. Membrane fouling in membrane bioreactors for wastewater treatment. *Journal of Environmental Engineering*, **128**(11):1018–1029, 2002.

536. L Katebian, E Gomez, L Skillman, D Li, G Ho, and SC Jiang. Inhibiting quorum sensing pathways to mitigate seawater desalination RO membrane biofouling. *Desalination*, **393**:135–143, 2016.

537. S-R Kim, H-S Oh, S-J Jo, K-M Yeon, C-H Lee, D-J Lim, C-H Lee, and J-K Lee. Biofouling control with bead-entrapped quorum quenching bacteria in membrane bioreactors: physical and biological effects. *Environmental Science & Technology*, **47**(2):836–842, 2013.

538. P Le-Clech, V Chen, and TAG Fane. Fouling in membrane bioreactors used in wastewater treatment. *Journal of Membrane Science*, **284**(1–2):17–53, 2006.

539. BQ Liao, DM Bagley, HE Kraemer, GG Leppard, and SN Liss. A review of biofouling and its control in membrane separation bioreactors. *Water Environment Research*, **76**(5):425–436, 2004.

540. N Cicek. A review of membrane bioreactors and their potential application in the treatment of agricultural wastewater. *Canadian Biosystems Engineering*, **45**:6–37, 2003.

541. L Katebian, MR Hoffmann, and SC Jiang. Incorporation of quorum sensing inhibitors onto reverse osmosis membranes for biofouling prevention in seawater desalination. *Environmental Engineering Science*, **35**(4):261–269, 2018.

542. J-H Kim, D-C Choi, K-M Yeon, S-R Kim, and C-H Lee. Enzyme-immobilized nanofiltration membrane to mitigate biofouling based on quorum quenching. *Environmental Science & Technology*, **45**(4):1601–1607, 2011.

543. KR Zodrow, ME Tousley, and M Elimelech. Mitigating biofouling on thin-film composite polyamide membranes using a controlled-release platform. *Journal of Membrane Science*, **453**:84–91, 2014.

544. K Yamamoto, M Hiasa, T Mahmood, and T Matsuo. Direct solid-liquid separation using hollow fiber membrane in an activated sludge aeration tank. In *Water Pollution Research and Control Brighton*, Elsevier, pp. 43–54, 1988.

545. A Nagano, E Arikawa, and H Kobayashi. The treatment of liquor wastewater containing high-strength suspended solids by membrane bioreactor system. *Water Science and Technology*, **26**(3–4):887–895, 1992.

546. T Masuda, A Jitoe, J Isobe, N Nakatani, and S Yonemori. Anti-oxidative and anti-inflammatory curcumin-related phenolics from rhizomes of curcuma domestica. *Phytochemistry*, **32**(6):1557–1560, 1993.

547. JJ Li. *Name Reactions: A Collection of Detailed Mechanisms and Synthetic Applications*, 5th ed. Springer Science & Business Media, 2014.

548. JS Vrouwenvelder, JAM Van Paassen, LP Wessels, AF Van Dam, and SM Bakker. The membrane fouling simulator: a practical tool for fouling prediction and control. *Journal of Membrane Science*, **281**(1–2):316–324, 2006.

549. ER Cornelissen, JS Vrouwenvelder, SGJ Heijman, XD Viallefont, D Van Der Kooij, and LP Wessels. Periodic air/water cleaning for control of biofouling in spiral wound membrane elements. *Journal of Membrane Science*, **287**(1):94–101, 2007.

550. A Lok, P Bérubé, and R Andrews. The effect of concentration factor on membrane fouling. *Membranes*, **7**(3):50, 2017.

551. V Balachandran and K Parimala. Vanillin and isovanillin: comparative vibrational spectroscopic studies, conformational stability and nlo properties by density functional theory calculations. *Spectrochimica Acta Part A: Molecular and Biomolecular Spectroscopy*, **95**:354–368, 2012.

552. S Masatoki, H Matsuura, and K Fukuhara. Raman spectroscopic study of the molecular conformation of $\alpha$-octyl-$\omega$-methoxyoligo (oxyethylene)s in the solid state. *Journal of Raman Spectroscopy*, **25**(7–8):641–645, 1994.

553. G Socrates. *Infrared and Raman Characteristic Group Frequencies: Tables and Charts*. John Wiley & Sons, 2004.

554. MME Jacob and AK Arof. Ftir studies of DMF plasticized polyvinyledene fluoride based polymer electrolytes. *Electrochimica Acta*, **45**(10):1701–1706, 2000.

555. VV Chalapathi and KV Ramiah. Normal vibrations of N,N-dimethylformamide and N,N-dimethylacetamide. In *Proceedings of the Indian Academy of Sciences-Section A*, Vol. 68, Springer, pp. 109–122, 1968.

556. A Sharma, S Kaur, CG Mahajan, SK Tripathi, and GSS Saini. Fourier transform infrared spectral study of N,N′-dimethylformamide-water-rhodamine 6G mixture. *Molecular Physics*, **105**(1):117–123, 2007.

557. DT Englis and M Manchester. Oxidation of vanillin to vanillic acid. *Analytical Chemistry*, **21**(5):591–593, 1949.

558. K Ponnusamy, S Kappachery, M Thekeettle, JH Song, and JH Kweon. Anti-biofouling property of vanillin on aeromonas hydrophila initial biofilm on various membrane surfaces. *World Journal of Microbiology and Biotechnology*, **29**(9):1695–1703, 2013.

559. B Schmid, J Schindelin, A Cardona, M Longair, and M Heisenberg. A high-level 3D visualization API for Java and ImageJ. *BMC Bioinformatics*, **11**(1):274, 2010.

560. N Lourith and M Kanlayavattanakul. Natural surfactants used in cosmetics: glycolipids. *International Journal of Cosmetic Science*, **31**(4):255–261, 2009.

561. NGK Karanth, PG Deo, and NK Veenanadig. Microbial production of biosurfactants and their importance. *Current Science*, 116–126, 1999.

562. RS Makkar and Swaranjit Singh Cameotra. Production of biosurfactant at mesophilic and thermophilic conditions by a strain of bacillus subtilis. *Journal of Industrial Microbiology and Biotechnology*, **20**(1):48–52, 1998.

563. R Makkar and S Cameotra. An update on the use of unconventional substrates for biosurfactant production and their new applications. *Applied microbiology and biotechnology*, **58**(4):428–434, 2002.

564. Swaranjit Singh Cameotra and RS Makkar. Synthesis of biosurfactants in extreme conditions. *Applied microbiology and Biotechnology*, **50**(5):520–529, 1998.

565. Krishnaswamy Muthusamy, Subbuchettiar Gopalakrishnan, Thiengungal Kochupappy Ravi, and Panchaksharam Sivachidambaram. Biosurfactants: properties, commercial production and application. *Current Science (00113891)*, **94**(6), 2008.

566. A Mahmound, Y Aziza, A Abdeltif, and M Rachida. Biosurfactant production by bacillus strain injected in the petroleum reservoirs. *Journal of Industrial Microbiology & Biotechnology*, **35**(2):1303–1306, 2008.

567. MO Ilori, CJ Amobi, and AC Odocha. Factors affecting biosurfactant production by oil degrading aeromonas spp. isolated from a tropical environment. *Chemosphere*, **61**(7):985–992, 2005.

568. A Tabatabaei, A Nouhi, SV Alsadat, and AM Mazaheri. Isolation of biosurfactant producing bacteria from oil reservoirs. *Iranian Journal of Environmental Health Science and Engineering (IJEHSE)*, 2005.

# Index


AAS, *see* atomic absorption spectrophotometer
acid number, 122, 126, 129, 138, 142–43
activated carbon, 348, 357, 372, 392
activated sludge processes, conventional, 398
additives, 205, 207, 213, 219, 221–23, 239, 241–42
  common WBDF, 241
  conventional, 205, 220
  natural, 239
  organic, 95
  regular, 206–7, 220
  safe, 224
adsorption, 79, 95, 109, 144, 346, 348, 359, 372, 379, 382–83, 392, 400
  metal, 382
  physical, 348, 400
alkaline-surfactant-alternated-gas (ASAG), 123–24, 126
alkalinity, 320, 322, 324–25, 328–29, 331–33
alkali-surfactant-polymer (ASP), 77–78, 81, 83, 85–86, 88, 125
alpha olefin sulphonate (AOS), 103–4, 106–7, 111, 115–16, 118
American Petroleum Institute (API), 121, 126, 138
American Standard for Testing of Materials (ASTM), 38, 129, 193–202, 263, 301, 324–25
AOS, *see* alpha olefin sulphonate
API, *see* American Petroleum Institute
API gravity, 122, 124, 137–39, 144, 240
aqueous phase, 79, 88–89, 95, 106, 108, 114, 118, 149, 177, 388
Ardatovskiy formation, 291
areal displacement, 81
aromatic hydrocarbons, 160, 347
  biodegraded petroleum-derived, 310
  polycyclic, 344, 371, 387
  soluble, 320
aromatics, 142, 310, 351, 402
ASAG, *see* alkaline-surfactant-alternated-gas
ASAG flooding, 123, 125–26, 144
ASP, *see* alkali-surfactant-polymer
ASP flooding, 78, 81–84, 87, 90
ASP floods, 125
asphaltene, 105, 122, 126, 129, 138, 140–41, 145, 351
ASP slug, 78, 80–81, 85, 87, 89–90
Assam Shelf, 2–3
ASTM, *see* American Standard for Testing of Materials
atomic absorption spectrophotometer (AAS), 326
average angle method, 275–76, 278–82, 284, 286

backwash, 357, 392
backwashing, 352, 357, 395
bacteria, 158, 164–65, 173, 178, 182, 186, 188–90, 312, 333, 350, 389, 394, 399, 411
 aerobic, 167
 anaerobic, 314
 dead, 390, 410–12
 facultative, 166
 facultative BS-producing, 160
 live, 410
 sulphate-reducing, 166, 394
Barail and Disang Groups, 29–30, 47
Barail Group, 29–30, 33, 38, 43–47, 51, 53, 59, 62, 75
Barails, 5–6, 11, 13, 30–34, 36, 38, 40, 42, 44, 46, 51–52
Barail sandstone, 39–40, 49–50, 52, 54, 56, 58, 60, 62, 64, 66, 68, 70, 72, 74
base mud, 206, 209–10, 214, 220, 224–25, 244–47, 249–50
bentonite, 206, 208, 210, 212, 214, 216, 224–25, 242–44, 246–49, 372
Bhogpara oilfields, 200
biochemical oxygen demand (BOD), 342, 344, 351, 393, 402
biocompatibility, 169
biodegradability, 157, 177, 313
biodegradation, 176, 310–12, 315, 317–18, 351, 401, 403
 contaminant, 318
 rapid, 351
bioemulsification, 176
bioequivalence, 157
biofilm, 174–75, 353, 398–401, 406, 410–12
 bacterial, 399
biofouling, 353, 398–99, 412
biogases, 158, 164, 167, 169, 182
biological treatment, 350–51, 402
biopolymers, 158, 164, 167–69, 182, 185
bioproducts, best, 186, 189
bioreactors, 179, 182, 390, 401–2
 batch mode, 179
 integrated, 179
 stirred tank, 179
bioremediation, 309, 311–12, 315, 389
biosolvents, 158, 164, 167–69
biosurfactant (BS), 157–61, 164–65, 167–79, 181–82, 185, 312–13, 315–16, 318
BL, *see* black liquor
black liquor (BL), 78, 103–7, 110–11, 116, 118
BOD, *see* biochemical oxygen demand
boiling point, 194–95, 243, 245
borehole, 239–41
Bragg angle, 136
brine, 31, 77, 80, 82, 84–85, 87, 107–10, 113–14, 117–18, 135, 163, 301, 368, 385–86
 laboratory-prepared, 103, 108, 113–14
 optimum salinity, 101
 synthetic soft, 107
brine flooding, 78, 81, 83, 87, 109
BS, *see* biosurfactant
 lipopeptide, 172
 microbial, 160
 microbial-origin, 169, 177
bulk volume (BV), 128, 131
BV, *see* bulk volume
by-products, 158, 164–66, 169, 182, 191, 319, 321–22, 394
 bacterial, 167
 disinfection, 358

$CaCO_3$, 320, 322–24, 326, 328–30, 333–35, 349, 368
carbonaceous shale, 30, 33–34

cationic exchange capacity (CEC), 208, 217, 219
CEC, *see* cationic exchange capacity
CEOR, *see* chemical enhanced oil recovery
CFV, *see* crossflow velocity
Champhai district, 51, 224, 227
Champhai–Zokhawthar Road, 49, 51, 53, 56–58, 60, 75
chemical enhanced oil recovery (CEOR), 77–84, 86–88, 90, 104–5, 110, 118, 125, 163
chemical oxygen demand (COD), 342, 349, 393, 402
chemical slugs, 77, 99, 101
Chittagong–Mizoram–Tripura fold thrust belt, 49
chlorinated polyethylene (CPE), 398, 401, 403–5, 407, 409, 412
CI, *see* correlation index
clay, 2, 7, 50, 52, 121–22, 125–27, 136–37, 144, 217, 220, 222–23, 235, 302, 305, 307
CMC, *see* critical micelle concentration
CMC mud, 225, 227–28, 235–36
CMC-PEG4000 mud, 225, 227
COD, *see* chemical oxygen demand
confocal scanning laser microscopy (CSLM), 401, 407
copolymers, 348
core analysis, 121, 126, 128, 130, 144
core flooding, 80–83, 94, 97, 101, 104–6, 109, 117–18, 126, 144, 148, 174–75
core floods, 77, 118, 175
core plugs, 79, 81, 122, 126–28, 130–35, 144
core, 1, 4, 11–27, 55–56, 79–81, 83–85, 101, 106–7, 109, 117–18, 130–31, 136–37, 174–75, 186–90, 301–7
  bare, 118
  dry, 118
  porous, 144
  whole/conventional, 130
correlation index (CI), 193, 326
corrosion, 143, 166, 240, 242, 319–22, 328, 333–36, 339, 342, 344, 389, 394
  metallic, 334
CPE, *see* chlorinated polyethylene
critical micelle concentration (CMC), 89, 94–98, 103, 105, 107, 111–12, 148–49, 151–53, 156, 159, 171–73, 177, 222–25, 227–36
crossflow, 357, 391, 398, 400, 406, 410
crossflow velocity (CFV), 401
crude oil, 94, 99–100, 104–8, 121–22, 125–29, 138–45, 158, 160–61, 176–77, 185–86, 193–96, 198, 200–202, 259–72, 310
CSLM, *see* confocal scanning laser microscopy

DAF, *see* dissolved air flotation
Darcy's equation, 303
Darcy's law, 109
Darcy's velocity, 162
Daubert's method, 196–97
DBPs, *see* disinfection by-products
de-emulsification, 185, 348
de-emulsifiers, 195, 348–49
  oil-soluble, 195
degradation, 90, 125, 164–65, 312, 317–18, 344
  alkylaromatic, 310
  thermal, 214
denser-than-water non-aqueous-phase liquids (DNAPL), 149
depleted reservoirs, 158–60, 162–63, 167, 169, 176, 182

depletion, 78, 127
continuous, 397
primary, 141
deposition, 8, 40, 49, 59, 141, 208, 261
particulate, 399
wax, 260–62, 389
depositional environment, 28–30, 38–41, 51, 57, 59, 134, 137, 345
desalination, 357–58, 391–92
Digboi oilfield, 2, 32
Dihing Group, 33
directional drilling, 241, 275–78, 280, 286
Disang sandstone, 39–40
Disang sediment, 30
Disang shale, 42, 46–47
disinfection by-products (DBPs), 358
disposal pit (DP), 362–64, 366–67, 369
disposal wells, 355–56
dissolution, 50, 68–69, 72, 362
dissolved air flotation (DAF), 372
distilled water (DW), 96, 98–99, 101, 108, 113, 144, 147, 151, 208, 224–25, 242, 324–26
DMF, *see* *N*,*N*-dimethylformamide
DNAPL, *see* denser-than-water non-aqueous-phase liquids
DP, *see* disposal pit
drilling, 2, 123, 206, 213–15, 220–24, 237, 239–41, 244, 275–76, 282, 286, 289, 355, 368, 386
horizontal, 276
salt dome, 277
drilling fluids, 205–8, 212–15, 217, 219–20, 222–23, 228–29, 232, 235–36, 239–40, 242–44, 246–47, 256–57, 368
bentonite-based, 208
conventional, 206–7
gaseous, 240
ideal, 223
liquid, 240
smart, 207, 220
sustainable, 242
drilling mud, 207, 220–22, 225–26, 229, 239–43, 245, 247, 249, 254, 256–57
improvised, 221, 223
nanoenhanced, 206–7
stable, 214, 220
water-based, 224, 239
DW, *see* distilled water

EC, *see* electrical conductivity
edge water encroachment, 305, 307
EDLs, *see* electrical double layer
Edmister's method, 193–94, 196, 198
EDR, *see* electrodialysis reversal
EFV, *see* equilibrium flash vaporisation
electrical conductivity (EC), 323, 331–32, 339, 374–75, 380
electrical double layer (EDL), 87, 89
electrochemistry, 392–94, 396
electrodeposition, 393–94, 396
electrodialysis reversal (EDR), 392
emulsification, 174–75, 177, 185, 351
enhanced oil recovery (EOR), 78–79, 93–94, 96, 102–8, 116–18, 121–23, 137–38, 147–50, 154, 156–58, 160–65, 182, 185–86, 191, 315
EOR, *see* enhanced oil recovery
EPS, *see* extracellular polymeric substance
equilibrium flash vaporisation (EFV), 196, 200–202

ex situ MEOR, 158, 160, 167, 175, 182, 187
extracellular polymeric substance (EPS), 353
extraction, 160, 164, 243, 245, 374, 385

farmland (FL), 362, 366, 368–69
FBP, *see* final boiling point
FESEM, *see* field emission scanning electron microscopy
FID, *see* flame ionisation detector
field emission scanning electron microscopy (FESEM), 121–22, 126, 136–37, 144
filtrate loss, 206, 208, 214–19, 222–23, 226, 232–33, 235, 243, 245–47, 249–51, 256, 368
filtration, 206, 214–15, 219–22, 232, 235–36, 242, 346, 353, 368, 373, 375, 379–83, 390–91, 412
  dead-end, 391
  low-pressure, 399
  low-pressure membrane, 399
final boiling point (FBP), 195
FL, *see* farmland
flame ionisation detector (FID), 44
flat sheet membrane (FSM), 352
flooding, 77–78, 81, 84, 87, 90, 109, 123, 125–26, 160, 162, 174–75, 302
  alkaline, 122–23, 142, 145
  chemical, 110, 125
  microemulsion, 95–96, 102
  miscible, 162, 164
  stream, 162
floods, 125
  chemical, 125
  prechemical, 109
flotation, 347, 372
fluid flow, 148–49, 369
fluid loss, 223–24, 236, 242
flux, 353–54, 357
foam, 103, 105, 108–9, 115–16, 123
  surfactant-containing, 179
formation water (FW), 86, 103–8, 113–14, 118, 121–22, 125–27, 129, 143–45, 188–89, 320, 368, 371, 386
fouling, 353–54, 357, 359, 395, 398–99, 406, 412
  colloidal, 353
  irreversible, 354
  organic, 353
  reversible, 353
Fourier transform infrared (FTIR), 129, 142–43, 398, 401, 404–5, 408–9
fracturing, 50, 70, 88, 299, 355
FSM, *see* flat sheet membrane
FTIR, *see* Fourier transform infrared
FW, *see* formation water

gas, 32–33, 123, 133–35, 139, 143–44, 158–59, 164–65, 206–7, 240–41, 277, 288–91, 319–22, 341–44, 355–57, 385–87
gas chromatography–mass spectrometry (GC-MS), 374–79, 383
gas fields, 94, 299, 361
gas–oil contact (GOC), 289
gas permeability, 122, 128–29, 132–35, 144
GC-MS, *see* gas chromatography–mass spectrometry
gel strength (GS), 206–7, 212–14, 219, 229, 231–32, 241–44, 246, 250, 256
geochemistry, 321, 386
Geoscience software, 288, 296, 298

Girujan clay, 31, 33–34
Global Mapper software, 35
GOC, *see* gas–oil contact
grains, 7, 30, 50, 59, 68, 73, 75
  constituent, 50
  detrital, 68
  fine, 303
  framework, 50, 63, 68, 72
  unconsolidated, 300
grain size, 1–4, 7–9, 28, 38, 40, 59, 133, 188
grain size analyses, 1, 4, 10–27, 30, 55
greywacke, 31, 75
GS, *see* gel strength

hardness, 125, 320, 322, 324, 331–33, 339
HC degradation, 309, 317
HCPV, *see* hydrocarbon pore volume
HCs, *see* hydrocarbons
  aliphatic, 320, 388
  free, 44
  heavy-molecular-weight, 167
  high-molecular-weight, 140, 165
  low-molecular-weight, 167
  recoverable, 289
  toxic, 309
  volatile, 344
heavy metals, 320, 349, 371–74, 381–83, 388, 392, 402–3
HFM, *see* hollow fibre membrane
high-molecular-weight BSs, 169
high-pressure high-temperature (HPHT), 205–7
hollow fibre membrane (HFM), 352
HPHT, *see* high-pressure high-temperature
hydrocarbon pore volume (HCPV), 297
hydrocarbons (HCs), 2, 29, 32–33, 44–47, 49–51, 288–89, 291, 297, 309–10, 312–14, 317–20, 342–43, 348, 351, 361–62, 372, 374–75, 379, 385–87
  complex, 309
  high-molecular-weight petroleum, 320
  large, 188
  prolific, 49
  residual, 361
hydrocyclones, 347
hydrogen index, 33
hydrolysis, 86, 89

IBP, *see* initial boiling point
IFT reduction, 79, 98, 107, 113, 164, 168–69, 172, 174–75, 177
IFT, *see* interfacial tension
  base, 108, 113
  high, 167
  low, 105, 110
  oil/water, 113
  ultra-low, 107
illite, 136–37, 305
immiscible phases, 348, 362
inclination, 276, 278–79, 282, 284
industrial wastewater, 402
initial boiling point (IBP), 195
interfacial tension (IFT), 78–79, 81, 89–90, 93–99, 103–8, 110–14, 118, 147–49, 151–53, 155–157, 159, 162–63, 169, 171, 173–74, 177, 185, 188, 312, 344, 362
  high, 149
iron NPs, 205–6, 208–17, 219–20

JBN, *see* Johnson, Bossler and Neumann
JN, *see* Jorajan

Johnson, Bossler and Neumann (JBN), 109
Jorajan (JN), 4, 6, 10, 199–200

kaolinite, 136–37, 305
kink bed, 35
Klinkenberg effect, 133
Klinkenberg principle, 129, 134
kurtosis, 1, 7, 9, 39–41, 59

Langelier saturation index (LSI), 320, 322, 328–30, 334–35
Laplace equation, 163
Leontaritis criterion, 141
lipopeptides, 168–69, 172–73, 175, 316
liquefied petroleum gas (LPG), 164
liquid permeability, 128–29, 132–33, 135
lithocolumn, 55, 57–58
lithofacies, 49
lithology, 30, 33, 52, 130
litho unit, 33, 38, 43, 45
LOBS, *see* lube oil base stock
low-molecular-weight BSs, 169
low-pressure low-temperature (LPLT), 206, 224, 243, 245
LPG, *see* liquefied petroleum gas
LPLT, *see* low-pressure low-temperature
LSI, *see* Langelier saturation index
lube oil base stock (LOBS), 194

macroporous polymer extraction (MPPE), 395
mango leaf extract (MLE), 240–41, 249–57
marine conditions, 40, 59
marine ecosystem, 331
marine environment, 30, 35, 42, 47, 50, 62, 75
marine sediments, 31, 310
mass spectrometry, 374
matrix, 28, 63–65, 72–73, 137, 207, 301, 304
MBR, *see* membrane bioreactor
MBR biofouling, 399–400
MBR fouling, 399
MBT, *see* methylene blue test
MCCF, *see* mobility-controlled caustic flood
MCT, *see* mud cake thickness
measurement while drilling (MWD), 278, 286
medium-gravity oil reservoirs, 125
membrane antibiofouling, 398, 406
membrane antifouling, 406
membrane biofouling, 397, 401, 411–12
membrane bioreactor (MBR), 397–99, 401–3
membrane fouling, 353, 357–58, 391, 395, 399, 402
membranes, 341, 343, 352–54, 357–59, 390–91, 395, 398–407, 409–12
  ceramic, 359
  composite, 352
  dense, 352
  flat sheet, 352
  fouled, 354
  hollow fibre, 352
  hydrophobic, 359
  ideal, 352
  nonporous, 391
  polymeric, 359
  porous, 352
  unmodified, 400–401, 406, 410–12
membrane separation, 342, 352, 372, 390
MEOR, *see* microbial enhanced oil recovery
mesokurtic, 5, 9

methylene blue test (MBT), 206, 217, 243
microbial degradation, 309–10, 312, 314, 316–18, 351
microbial enhanced oil recovery (MEOR), 157–58, 160, 162, 164–70, 172, 174, 176–78, 180, 182, 185–91
microemulsions, 93–97, 99, 101–2, 109
microfiltration, 390, 398–99
  low-pressure, 341
microfluidics, 148, 154
microorganisms, 165, 167, 169–70, 182, 185–87, 189, 310–13, 315–17, 322, 390, 398, 402
  anaerobic, 312
  biosurfactant-producing, 309
  indigenous, 165, 318
  indigenous reservoir, 187, 189
Microsoft Excel, 263, 275–76, 280–86
mineralogy, 121, 126, 132, 136
minerals, 136–37, 189, 335, 339, 351, 368, 386, 388
  constituent, 53
  detrital, 63
  dissolved, 321, 388
  dominant, 136
  heavy, 63
  micaceous, 63
  radioactive, 320
  secondary, 50
  trace, 186, 189
Miocene, 2, 31, 33, 52
Mio-Pliocene, 4, 31
mixed liquor suspended solids (MLSS), 406
MLE, *see* mango leaf extract
MLSS, *see* mixed liquor suspended solids
mobility, 82, 87, 95, 162–64, 176, 358
mobility-controlled caustic flood (MCCF), 112
mobility ratio (MR), 77–78, 81–88, 90, 123, 163, 169
  unfavourable, 163
Mohr's method, 367
MPPE, *see* macroporous polymer extraction
MR, *see* mobility ratio
mud, 73, 208–9, 213–14, 221, 223, 225–29, 232, 235–36, 240–41, 244–45, 247, 254, 278
  bentonite-based, 209, 214
  cone-shaped, 225
  conventional, 215
  efficient, 214
  nanobased, 209
  non-aqueous, 240
  oil-based, 206, 240
  reactive, 217
  stable, 214
  synthetic-based, 206
mud balance, 208, 225, 242, 244, 246
mud cake, 226, 232, 245
  tough, 215, 217, 219, 223, 235–36
mud cake thickness (MCT), 206, 214–16, 218–19, 226, 234–35, 243, 245–46, 249–51, 256
MWD, *see* measurement while drilling

Naharkatiya (NH), 4, 6, 12–13, 78, 90, 107
nanobased mud (NBM), 209, 214, 217, 220
nanoemulsions, 149
nanofiltration (NF), 341, 352, 357, 390–91
nanomud, 212–13
nanoparticles (NPs), 149, 205–10, 212, 214, 217–20

nanophotocatalysts, 394
natural gases, 161, 164, 239, 241, 341, 386
naturally occurring radioactive minerals (NORM), 319–20, 388, 392
NBM, *see* nanobased mud
neem leaf extract (NLE), 240, 249–57
net-to-gross (NTG), 288, 290, 296–97
Newtonian flow behaviour, 260, 272
Newtonian fluid behaviour, 262
Newtonian profile, 272
NF, *see* nanofiltration
NH, *see* Naharkatiya
NLE, *see* neem leaf extract
*N*,*N*-dimethylformamide (DMF), 403–5, 409, 411–12
non-Newtonian fluid characteristics, 262
non-Newtonian viscosity behaviour, 259
NORM, *see* naturally occurring radioactive minerals
NPs, *see* nanoparticles
NTG, *see* net-to-gross

OBM, *see* oil-based mud
OFAT, *see* one factor at a time
OFPW, *see* oilfield produced water
  carbonate, 329
  high-quality, 356
  saline, 333
  treated, 343, 352, 355, 357–58
OFPW-to-oil ratio (WOR), 342, 355, 387
O&G, *see* oil and grease
oil, 80–82, 93–95, 109–10, 115–16, 159–63, 185–88, 288–91, 310–11, 318–21, 347–48, 351, 355–57, 361–63, 386–89, 391–92
  cumulative, 177
  dense, 347
  dispersed, 343–45, 347, 371, 388
  displaces, 355
  dissolved, 320, 345, 387–88
  emulsified, 372
  entrapped, 174
  existing, 167
  extra, 175
  free, 372
  heavy, 158, 161, 164, 277
  kitchen waste, 178
  left, 185–86, 189–90
  liberate, 188
  light paraffin, 109
  lower, 357
  maximum, 163
  medium-gravity, 138
  producing, 342
  productive, 355
  recoverable, 290, 297
  releasing, 188
  silicon, 179
  soybean, 182
  soybean waste frying, 178
  stock tank, 290
  total, 300
  unburned, 162
  waste frying coconut, 178
  waxy, 262
  witnessed significant, 2
oil and gas, 32–33, 94, 143, 159, 206, 240, 277, 288, 310, 319, 342, 355, 371, 385–86
oil and grease (O&G) 319–20, 325–26, 331, 333, 357
oil and water, 95, 159, 163, 167, 169, 174, 188, 344, 347, 362, 389
oil-based mud (OBM), 206–7, 240

oil-dispersed-in-water (o/w), 95, 345
oil droplets, 149, 388
  discrete, 343
oilfield produced water (OFPW), 319–26, 328–31, 333, 335–39, 341–59, 371–72, 382, 385–95
oilfields, 2–4, 126–27, 186, 299–300, 302, 304, 306, 308, 318–21, 323, 339–42, 360–62, 369, 385–86, 388–89
  ageing, 300, 321
  crude, 78
oil flood, 110, 117
oil industry, 123, 138, 164–65, 167, 191, 202, 205, 220, 320, 356
oil recovery, 88, 94, 96, 100–102, 105, 121, 158–60, 162, 164, 166, 169, 174–76, 182, 185–89, 191
  crude, 94, 158
  enhanced, 157, 185–86
  residual, 105, 117, 166
  secondary, 78, 94
  tertiary, 94, 161, 182
oil reservoirs, 122, 126, 144, 148, 165, 312
  carbonate, 190
  depleted, 93
  heavy, 162
  high pour-point, 176
oil spills, 311, 315
oil viscosity, 125, 139, 162–63, 165
  crude, 99, 139, 260
  dead, 122, 129
o/w, *see* oil-dispersed-in-water
oil–water contact (OWC), 289, 294, 296
oil/water interface, 105, 108
oil–water interfacial tension, 123, 348
oil wells, 2, 343
  mature, 191
oleic phases, 85, 98, 149, 153
one factor at a time (OFAT), 178–79
OOIP, *see* original oil in place
optimum surfactant combination, 154
organic compounds, 194, 312, 315, 357–58, 372, 374, 387–88
  dissolved, 344, 351
organic matter, 30, 33, 46–47, 350, 391, 399
  mature-source, 46
original oil in place (OOIP), 84, 87, 149, 161, 176, 185–86
overburden pressure, 101, 132
OWC, *see* oil–water contact
oxidation, 349–52, 410

PAHs, *see* polycyclic aromatic hydrocarbons
particle size analyser (PSA), 206, 208–9, 373
particle sizes, 40, 94, 97, 99–102, 149, 209, 211, 242, 345–46, 352, 372–73, 375, 379, 381–83
Pashiyskiy formation, 291
PC, *see* potassium carbonate
PCPs, *see* physicochemical parameters
PEG, *see* polyethylene glycol
PEG4000, 224–25, 227–29, 232, 235–36
PEG mud, 229–31, 233–34
permeability, 117, 121–22, 126, 130, 132–35, 144, 157, 164, 167, 188, 191, 303, 307
  absolute, 109
  air, 106, 109
  effective, 110, 163
  high retained, 301
  horizontal, 130, 276
  increased, 303

initial, 354
low, 241
poor, 167
vertical, 276
permeate flux, 353, 357, 359, 399
petroleum, 158, 161, 164, 182, 187–88, 193–95, 200, 202, 287–89, 309–10, 315, 318–19, 323–25, 351, 354–55
petroleum and natural gas (PNG), 161
petroleum HCs, 310, 389
phase behaviour, 94, 97, 99, 101, 144
phospholipids, 169, 173, 316
photocatalysis, 393
photocatalyst, 349
photoelectrocatalysis, 393
photoelectrochemistry, 393, 395–96
photoelectrolysis, 393–94
photoelectron catalysis, 349
physicochemical parameters (PCPs), 374, 379–80
physicochemical properties, 126, 138, 159
PI, *see* production index
plant extracts, 239, 242, 245, 249–50, 254, 256–57
plastic viscosity, 206, 210–14, 219, 226, 229–30, 232, 244, 246–47, 249–57
platykurtic, 5–6, 8–9, 30
plugs, 82, 128
PNG, *see* petroleum and natural gas
pollutants, 313, 321, 348, 350–51, 359
common HC, 309
inorganic, 341–42, 372
organic, 309–10, 312, 372
organic toxic, 312
petroleum HC, 310
pollution, 164, 315, 321, 354, 356, 372
environmental, 313
secondary, 158, 164
polycyclic aromatic hydrocarbons (PAHs), 313, 344, 371, 379, 387–88
polyethylene glycol (PEG), 179, 221–25, 230–34, 236
polymer flooding, 84, 87–88, 90
polymers, 77–78, 80, 83, 87–90, 149, 158, 163, 169, 205, 207, 222–23, 228, 232, 235, 240
alkali or/and, 79
water-soluble, 188
polysaccharides, 169, 187–88, 353, 407, 410–11
pores, 148–49, 159, 353, 386
pore sizes, 352, 391
pore spaces, 32, 71–73, 128, 130, 223
pore volume (PV), 77, 80, 87, 118, 128, 206, 212–13, 290, 297
porosity, 49–50, 53, 56, 62–63, 68–75, 106, 109, 121–22, 126, 128, 130–34, 144, 288, 290, 296–97
average, 50–51, 62, 75, 291
effective, 130
fracture, 68
high, 132, 352
interconnected, 2
potassium carbonate (PC), 403–12
Precambrian rocks, 4
produced water (PW), 163, 309, 318–21, 331, 340, 357, 360–63, 368–69, 371–75, 382, 385–87, 390, 393–96
production index (PI), 45–47
propan-2-ol, 94, 96–102
PSA, *see* particle size analyser
pseudomatrix, 68–69, 73–75
PSI, *see* Puckorius scaling index

Puckorius scaling index (PSI), 320, 322, 336–37
PV, *see* pore volume
PW, *see* produced water

QS, *see* quorum sensing
QSI, *see* quorum sensing inhibiting
quorum sensing (QS), 397, 399, 401, 410
Quorum sensing inhibiting (QSI), 397–98, 401, 412

R/A, *see* resin-to-asphaltene
Raman shift, 407
Raman spectra, 405, 407
Raman spectroscopy, 407–11
raw effluent, 375, 379–83
recovery, 77–79, 81, 83–84, 87, 110, 137, 139–40, 148–49, 162–64, 167, 169, 182, 339–40, 360, 395–96
  effective, 159
  maximum, 78, 87, 90
  metal, 395
  poor, 163
  primary, 94, 161
  secondary, 83, 94, 104, 141
  significant, 162
  tertiary, 94
reduction, 75, 78–79, 108, 111, 113, 163, 165, 168, 206, 351, 353, 355, 357, 395, 411–12
regulations, 352, 354, 368
  environmental, 354, 398
  re-injection, 352
relative permeability, 82, 110, 123, 130, 163
removal, 342, 351, 357, 372, 383, 391–92, 409
  complete, 312
reservoir conditions, 82, 126, 158–60, 164, 186–87, 189, 191
  anaerobic, 166
  extreme, 191
reservoir pressure, 127, 161, 165, 169, 188, 320, 355
reservoir rocks, 1–2, 4, 6, 8, 10, 12, 28–29, 75, 104–5, 121–24, 126, 130, 136–38, 144, 368
  carbonate, 368
  moderate-quality, 50
  sandstone, 125
reservoirs, 78–79, 121–23, 141–43, 160–63, 165–69, 176, 187–89, 276–77, 287–89, 291, 293–94, 296–97, 299–300, 319–20, 385–86
reservoir temperature, 90, 104, 111–13, 115, 117–18, 125, 138–39, 141, 291
residual oil, 78, 94, 105, 109–10, 112, 117–18, 123, 162
resins, 121, 126, 129, 137–38, 140–41, 300–301, 348, 351
  epoxy, 301
  formaldehyde, 301
  furan, 301
  polyester, 301
  urethane, 301
resin-to-asphaltene (R/A), 122, 137–38, 141, 145
resistance, 235, 259, 272, 301, 411
  bacterial, 399
  corrosion, 322
  mechanical, 352
response surface methodology (RSM), 179–80
retention, 63, 112, 372, 379, 382, 401, 404
reverse osmosis (RO), 341, 352, 357–58, 390–91, 400
  polyamide, 400

rhamnolipids, 170–71, 173–75, 177, 181–82, 315
rheological properties, 206, 208, 210, 212–13, 219–20, 222, 226, 229, 232, 235, 240–41, 246, 256–57, 263
Riazi–Daubert method, 196–97
RO, *see* reverse osmosis
rock-eval pyrolysis, 33, 42, 46
rock fragments, 50, 63, 65, 68, 304
  sedimentary, 65
rock plugs, 132, 159, 162–63
rocks, 29–30, 32–33, 44–45, 47, 50–51, 55, 59, 62–64, 75, 122–23, 130, 136–37, 185–86, 345, 386
  low-grade metamorphic, 73
  native, 223
  porous, 122, 133, 320
  sedimentary, 345
rock–water interaction, 345
RO membranes, 357, 391
rRNA sequencing, 311–12
RS, *see* Rudrasagar
RSI, *see* Ryznar stability index
RSM, *see* response surface methodology
Rudrasagar (RS), 6, 17–18
Ryznar index, 336
Ryznar stability index (RSI), 320, 322, 328, 330, 335–36

salinity, 97, 99, 105, 108, 113–15, 143–45, 176–77, 180, 186, 188–89, 191, 320–21, 323–24, 331–33, 402
  high, 339
  optimal, 112
sand, 3, 5–6, 9, 299–302, 304–7, 344, 372–73, 375, 381–82, 389
  coarse-grained, 7
  dry, 302
  fine, 300
  fine-grained, 1, 7
sand control method, 300
sand filtration, 371, 374, 382
sand pack, 174–75, 301–3, 307
sandstone, 30–31, 33, 35, 38, 47, 49–50, 52–54, 57–58, 63, 125, 127, 134, 137, 304, 307
  gritty, 33
  massive, 34
  wacke-type, 50
SBM, *see* synthetic-based mud
Schuppen Belt, 3, 29, 32
SFB, *see* synthetic formation brine
synthetic-based mud (SBM), 206–7
synthetic formation brine (SFB), 143–44

Taguchi method, 179–80
TBP, *see* true boiling point
TBP distillation, 194, 196–97, 202
TDS, *see* total dissolved solids
tectonics, 31, 51
TGA, *see* thermogravimetric analysis
thermogravimetric analysis (TGA), 82, 90, 106, 109
thixotropy, 262
Tipam Group, 32–33
Tipam oilfield, 300–301
Tipam reservoir sandstone, 305
Tipam sand, 127
Tipam sandstone, 31, 33–34
TMP, *see* transmembrane pressure
TOC, *see* total organic carbon
total dissolved solids (TDS), 114, 129, 143, 320, 322–24, 328, 330–32, 339, 357, 362, 367
total organic carbon (TOC), 30, 42–45, 47, 358, 392
total petroleum hydrocarbons (TPH), 371–76, 378–83

toxic compounds, 319–20, 349, 358, 402–3
undetected, 358
unknown, 358
toxicity, 159, 164, 223, 346, 349, 351, 358
combined chronic, 357
low, 157, 167, 169, 313
potential chronic, 357
TPH, *see* total petroleum hydrocarbons
transmembrane pressure (TMP), 353, 391, 399
trapped oil, 105, 158, 161–62, 164–65, 182, 186
treatment, 110, 177, 341–54, 356–60, 371–72, 385–88, 390, 392, 394–97, 406
additional physicochemical, 403
biological, 350–52, 402
chemical consolidation, 300, 307
improper, 369
industrial waste, 350
poor, 391
preliminary, 319
primary, 372
sand filter, 346
true boiling point (TBP), 193–202
Tuli–Mokokchung Road, 29, 33
Tunisian standards, 371–72, 374, 382–83
turbidity, 59, 62, 114, 320, 323–24, 331–32, 339, 374–75, 380–81, 391, 402

UF, *see* ultrafiltration
ultrafiltration (UF), 341, 352, 357, 390–91, 399
ultralow interfacial tensions, 93, 95, 98, 102, 105, 108, 111–12
ultraviolet (UV), 349
undersaturated water, 328, 330, 334–35
Upper Assam, 51, 77, 79, 94, 103, 107, 110, 121, 126, 131, 133, 139, 144, 262, 299
Upper Assam Basin, 1–4, 6, 8, 10, 12, 14, 28–29, 77–80, 90–91, 93, 102–3, 156, 299–302, 319, 339
Upper Cretaceous, 31
Upper Jurassic, 31
Upper Paleocene, 4
Urals Basin, 290–91
UV, *see* ultraviolet
UV-visible spectrophotometer, 206, 208–9

VA, *see* vanillin
VA-modified membranes, 398, 400, 406, 410
vanillin (VA), 398–401, 403–12
viscosity, 89–90, 93–94, 99, 132, 139, 162–64, 169, 188, 222–23, 225–26, 228, 241–43, 251–53, 259–66, 269–72
apparent, 82, 123, 206, 226, 228, 235, 244, 246–47, 262
dead-oil, 144
dynamic, 138–40, 263
effective, 226, 228, 235, 244
high, 140, 167
kinematic, 97
Visual Basic, 276, 280–81, 285
VOCs, *see* volatile organic compounds
volatile organic compounds (VOCs), 348, 402
volatile suspended solids (VSS), 406
VSS, *see* volatile suspended solids

WAG, *see* water-alternating gas

wastewater, 390, 392–93, 398
industrial, 391
municipal, 402
polluted, 382
strong, 402
wastewater treatment plants, 397–98, 412
industrial, 399
water
condensed, 320
conductive, 322
dense, 347
emulsified, 195
excess, 93
free, 195
fresh, 223, 356, 386, 395
ground, 223, 321
high, 321
hot, 158, 162
ice, 245
irrigation, 358
meteoric, 345
oilfield, 339, 343, 360
saline, 160, 163
saturated salt, 223
supersaturated, 335
surface, 386
treated, 351, 356
warm, 331
water-alternating gas (WAG), 123
water analyser, 129, 243, 245–46, 323
water backwashing, 353
water-based drilling fluid (WBDF), 221, 223, 239, 241
water-based drilling mud, 224, 239, 242, 244–45
water-based mud (WBM), 206, 217, 221–24, 226, 228, 230, 232, 234, 236, 240–42
water bodies, 321, 331, 354, 356
water desalination, 391
water-dispersed-in-oil (w/o), 95
water electrolysis, 393, 395
water flooding, 140, 174, 188, 320
field-scale, 110
water floods, 109, 117–18
water phase, 89, 95, 105, 348
excess, 95
water quality, 321–22, 333, 353, 356, 358, 362, 392
filtered, 346
high effluent, 397
stable effluent, 399
treated, 398
water resources, 322, 342
water saturation, 290, 296, 305, 307
connate, 110
increased, 304
irreducible, 80, 109
water-to-gas ratio (WGR), 387
water treatment, 363, 410
oily, 390
wax content, 122, 126, 129, 138, 140–41, 144
wax crystallisation, 260
wax crystals, 259
waxes, 259–62, 389
WBDF, *see* water-based drilling fluid
WBM, *see* water-based mud
well, 176, 222, 259, 262, 277, 288, 299, 355, 362–63, 367
freshwater, 362
marginal, 190
relief, 277
vertical injection, 162
wellbore, 88, 141, 190, 222, 236, 239, 275–77, 279–81, 286, 299–300, 344
stable, 223
wettability, 79, 81, 90, 94, 97, 104, 130, 157, 162, 167, 174–75, 177
WGR, water-to-gas ratio

WHO, *see* World Health Organization
Williamson ether reaction, 398, 401, 403–4, 407
Winsor classification, 94–95, 97
Winsor type I, 95
Winsor type II, 95
Winsor type III, 93, 95, 97, 99
Winsor type IV, 95
w/o, *see* water-dispersed-in-oil
WOR, *see* OFPW-to-oil ratio
World Health Organization (WHO), 322, 330–31, 333, 339

X-ray diffraction (XRD), 121–22, 126, 136, 144, 302, 306, 373
XRD, *see* X-ray diffraction

Yellow River Delta of China, 379
yield point (YP), 206, 212–14, 219, 226, 229–32, 244, 246, 250, 254–57
yield stress, 241, 259, 262
YP, *see* yield point

zeolites, 348, 372, 392
Zepuhu formation, 31
zone, 127, 276, 289–90
  higher-pressure, 247
  higher-to lower-permeability, 123
  high-permeable, 182
  low-permeable, 182
  nonpay, 320
  oil-producing, 127
  permeable, 159, 214
  permeable non-HC-bearing, 355
  sand-prone, 301
  transitional, 33